POTATOES POSTHARVEST

POTATOES POSTHARVEST

by

Bob Pringle

Chris Bishop

Rob Clayton

www.cabi.org

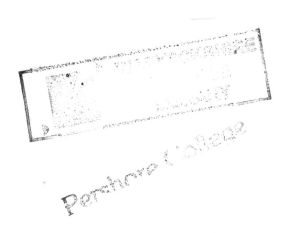

CABI is a trading name of CAB International

CABI Head Office
Nosworthy Way
Wallingford
Oxfordshire OX10 8DE
UK

CABI North American Office
875 Massachusetts Avenue
7th Floor
Cambridge, MA 02139
USA

Tel: +44 (0)1491 832111
Fax: +44 (0)1491 833508
E-mail: cabi@cabi.org
Website: www.cabi.org

Tel: +1 617 395 4056
Fax: +1 617 354 6875
E-mail: cabi-nao@cabi.org

A catalogue record for this book is available from the British Library, London, UK.

Library of Congress Cataloging-in-Publication Data

Pringle, Robert, 1944-
  Potatoes postharvest / by Robert Pringle, Chris Bishop, and Rob Clayton.
     p. cm.
  Includes bibliographical references and index.
  ISBN 978-0-85199-502-1 (alk. paper)

  1. Potatoes–Storage. 2. Potatoes–Handling. I. Bishop, Chris, 1952- II. Clayton, Rob, 1968- III. Title.

  SB211.P8P835 2009
  635'.216–dc22

                                                                              2008021193

ISBN: 978 0 85199 502 1

Typeset by SPi, Pondicherry, India.
Printed and bound in the UK by MPG Books Group.

Contents

Chapter 3: Store Climate   57

Chapter 4: Disease Control in Store   104

Chapter 5: Store Design and Structure   132

Chapter 6: Store Ventilation   164

Preface

Potatoes Postharvest is written for potato store managers, packhouse staff, academics and students wishing to know how potatoes are managed postharvest and what science underlies the practice. The text is based on the personal experience of the authors, their own research, applied research by others and laboratory work carried out to confirm findings in the field and in store. The book concentrates on the essential principles of storage, grading and dispatch of potatoes.

The focus is the potato tuber and all of the influences that can affect its final quality when sold. Background information such as the tuber's physical development, its metabolic processes, its susceptibility to damage and disease are provided where this aids understanding as to why stored crops develop the problems they do.

Potatoes are increasingly being grown all over the world, with the largest increases in the developing countries. The book is therefore written for an international audience, and includes the Dutch system of high-rate intermittent ventilation and the Scandinavian and North American system of low-rate continuous ventilation with humidified air. Refrigerated systems of storage in bags, and low-cost naturally ventilated traditional on-farm systems as used in India, Kenya and other warm areas, are also included.

While many aspects discussed are similar regardless of climate, such as the physical development of the crop, packhouse practice and quality control, other aspects are quite different. Different climates affect the diseases likely to be present on harvested crops. Storage systems in temperate continental zones are designed primarily to keep crops from freezing and to minimize desiccation when ventilating with cold dry air. In contrast, stores in maritime areas are designed primarily to ensure rapid drying of wet crops entering store, prevention of early sprout growth and the elimination of condensation on stored tubers due to the leakage of warm humid air into store. In tropical areas storage in the ground and in low-cost structures is common.

Formulas have been included to allow the reader to calculate the moisture content of air, ventilation duct sizes, fan backpressure, etc., but these have been kept to a minimum. Boxes are used to illustrate problems that occur in practice to reinforce the topics being discussed.

A certain amount of duplication is inevitable in such a publication as influences such as tuber respiration, crop temperature, ventilation, disease development, etc., all interact. Such duplication is kept to a minimum by putting a main description in one chapter and then referring to this section (e.g. Ch3.5) in the other chapters whenever this same subject matter is mentioned.

For those who want a résumé of the main points discussed in the book, they should consult the summaries at the end of each chapter.

Chapter 1 describes the developing tuber, explaining aspects such as skin set, sprouting and the points in a tuber that are vulnerable to ingress by disease. Chapter 2 describes systems of harvesting, transport to store or packhouse, cleaning, sizing and loading into store. Chapter 3 defines optimum conditions to minimize tuber dehydration and prevent sprouting over the period in store and suggests various strategies to achieve these conditions. Chapter 4 focuses on the disease contamination on tubers entering store and suggests how store climate can be manipulated to minimize their development and spread.

Chapter 5 looks at alternative building structures, their sealing and insulation. Chapter 6 covers how air is distributed through the building to dry and cool the crop and maintain a uniform crop temperature. Chapter 7 describes the use of refrigeration, when cool outside air is unavailable. Chapter 8 describes the instrumentation for both monitoring and control of store climate, and suggests how logged data can be analysed to identify why a problem occurred in store.

Chapter 9 describes store management from loading to dispatch. Chapter 10 discusses the particular requirements of seed, how it is stored, graded and dispatched, and the various ways of achieving single or multiple sprouts on seed tubers prior to planting. Chapter 11 describes grading and packhouse design and operation for pre-pack potatoes.

Quality control and traceability, vital elements in meeting farm assurance schemes and for providing feedback on problems, are described in Chapter 12. Lastly, a methodology for evaluating the cost of storage, to ascertain whether or not it is financially viable, is described in Chapter 13. These costings, combined with the quality of tubers coming out of store, will determine whether it is better to store crop near the point of production or to harvest and transport crop from different climatic areas to give year-round supply.

Bob Pringle
Chris Bishop
Rob Clayton

Acknowledgements

The authors would like to thank all those who have contributed to this book. To Dick Taylor who undertook the huge job of editing the entire manuscript for student comprehension. To Bill Leslie of Farm Electronics Ltd, Mike McLaughlin of Proctors Ltd and Rod McGovern of SAC-Aberdeen for their valuable comments on technical aspects of storage and provision of photographs. To Kees Wijngaarden of Tolsma Techniek Emmeloord BV, The Netherlands who kindly provided detailed plans of stores, photos and technical information. To Alf Johansson, Consultant, Sweden for his help and information on low-rate continuous ventilation of potatoes with humidified air. To Roger Balls for his plans of packhouses and comment on their operation. To Alistair Redpath for information on seed assessments and the ageing of seed to achieve bold or small daughter tubers. To Eric Anderson of Scottish Agronomy for his photographs. To Herbert's and Haith Tickehill for their photographs and technical information. To Nick Winmill, Greenvale for his help and comments on packhouse quality assurance. To Colin Johnston, Structural Engineer, Aberdeenshire, for detailed drawings of store floors. To Dave Ross and Fraser Milne for information on vision grading and bloom detection. To Hazel Carnegie for her drawings of potato plants and tubers. To Mary Jo Frazier, University of Idaho, and Todd Forbush for their comments on US potato storage. To Andrew Norrie, John Taylor, Tony Bambridge and Duncan Dixon for access to their farms and for their photographs. To Gavin Lishman of Martin Lishman Ltd, Frank Pirie ex of FJ Pirie, Andrew Bell of SG Baker, Bill Tennant of Linde Forklift trucks, Graeme Stroud and Steve Gerrish of BPC for photos and technical information. To Alex Hilton for providing information on wash up reports and Jay Whooton of Anderson Midlands for financial information. To John Vessey, Consultant, previously with United Biscuits, for his long-term support over a great many years.

On a more general note, we would like to thank our colleagues and ex-colleagues in SAC, BPC and Writtle College for their long-term support and contribution to potato production, in particular Stuart Wale, Phil Burgess, Claire

Hardy, John MacDonald, Kevin Potter, Roger Griffin-Walker, Eric Anderson, Alex Hilton, Dave Ross, Fraser Milne, Steve Gerrish, Adrian Cunnington, Adrian Briddon and Jeff Peters, and their excellent support team at Sutton Bridge. If there are others we have missed we apologize. There is real benefit in joint working with other disciplines and learning from each other.

Finally the authors would like to apologize to their long-suffering wives Ingrid Pringle, Stephanie Bishop and Patty Clayton, and their respective extended families, for not giving them the time they deserved.

Glossary

Term	Definition
Air blending	As air mixing
Air friction	Drag on air jet as it passes through still air or flows through ducts or potatoes
Air mixing	Mixing/blending of recirculated and ambient air, controlled by regulation of the duct temperature
Airspace ventilation	Air movement that flows around boxes in store, but is not forced through the actual potatoes themselves
Air-on/air-off temperature difference	Temperature difference between air flowing on to, and air leaving, a fridge evaporator cooling coil
Ambient air	Air external to the building structure
Ambient humidity	Relative humidity of the outside air close to the point of intake
Ambient temperature	Temperature of outside air close to the point of intake
Ambient-air cooling	Cooling of store and crop by ventilating with ambient air when it is cooler than the crop
Anaerobic conditions	Where no air is present
Auto-recirculation	Recirculation started by automatic timer, defined by a duration and an interval, until the next period
Backpressure	Pressure experienced by a fan due to the friction of air flowing through wire mesh, louvres, ducts and potatoes
BASIS Training	Training in the use of agrochemicals, required by law for all UK personnel who either sell or apply such chemicals
Blemish diseases	Diseases which cause marks on the skin of the tuber

Blindness in seed	Sprouts, usually affected by disease, which will not grow into shoots
Bloom	Reflective shine on tubers
BOD	Biological oxygen demand, mg/l. The BOD measures the polluting strength of dirty water and indicates the amount of oxygen needed by microorganisms to break down the organic wastes in a watercourse
Box	Container for potatoes, usually 1–2 t in capacity. Also known as a crate
BPC	Formerly British Potato Council, now Potato Council
Brock	Damaged or diseased potatoes unfit for human consumption
Bulk	Potatoes kept in one mass, supported on three sides by walls. Also called a pile
Chitting	Process of encouraging seed tubers to sprout prior to planting, to speed foliage development and increase yield
Chronological age	Age of a tuber measured in real time, i.e. days
Cleated-belt	Inclined conveyor belt fitted with rubber coated bars, or cleats, to prevent potatoes rolling back down the incline
COD	Chemical oxygen demand, mg/l. The COD measures the polluting strength of dirty water and indicates the amount of oxygen needed to chemically break down the organic wastes in a watercourse
Composite panel	Factory-made insulation panel, usually polyurethane injected between two metal skins. Jointed by gaskets and/or cam-lock mechanism
Condensation	The conversion of vapour in air to water when the air comes in contact with a surface that is below the dew-point temperature of the air
Condensation – subsurface	Condensation which occurs just under the surface of potatoes in a pile or box
Controller	Or control box. Controls operation of ambient-air cooling, louvres and refrigeration
Cooling coils	Heat exchanger (evaporator) of a refrigerator
Crop condensation	Condensation occurring on tubers, which can contribute to development of disease
Crop set-point	Target or desired crop temperature
Crop temperature	Average temperature of crop
Crop/ambient air differential	Differential between crop temperature and ambient air temperature
Crop/duct differential	Differential between crop temperature and duct temperature for systems with air blending
Crop/store air differential	Differential between crop temperature and store air temperature
Cull potatoes	As brock potatoes
Curing period	As wound healing period
Dead band	Tolerance in °C either side of a set point

Dew point	Temperature at which vapour in air will start to condense
Dew-point temperature	When air is cooled, it eventually reaches its dew point temperature when the air becomes saturated (100% RH); with further cooling, vapour in the air condenses on the coolest adjoining surface
Disease expression	Display of visible disease symptoms (e.g. lesions)
Disease infection	Entry into the flesh of the tuber by bacteria or fungus
Disease multiplication	Increase of disease that is already established
Dormancy	Period between tuber initiation in the soil and growth of first sprouts
Dormancy break	When 50% or more of tubers have sprouts of 3 mm or more in length
Dormancy, innate	Innate dormancy or 'potato rest' is a physiological stage during which tubers are not able to sprout, even under favourable conditions. Also known as 'natural dormancy' or 'endormancy'
Duct lower limit	Minimum allowable duct temperature
Duct temperature	Temperature of air in the main duct
EEP	Expanded extruded polystyrene (board insulation)
Emergence	Plant developmental stage, when first leaves of plant emerge from the soil
Evaporation	Loss of water from tuber via skin, due to the vapour pressure of air surrounding the tuber being less than the vapour pressure in skin cells. If air is at the same temperature as the tuber, it increases as RH falls below 97.8%
Evaporator	Heat exchanger (cooling coils) of a refrigerator
Fridge TD	Temperature differential between air flowing on to the cooling coil (evaporator) and evaporating gas temperature within the coil
HACCP	'Hazard Analysis and Critical Control Point' is a quality assurance system to ensure food from continuous production systems is safe to eat
Handheld thermometer	Electronic thermometer with a probe
Haulm	Stems of potato plants. Also knows as vines, shaws or tops
Heat	Energy in the form of heat. By-product of tuber metabolism
Heat exchanger	Device to allow transfer of heat from one fluid to another, e.g. the cooling coils of a fridge
Heat transfer	The process of heat transfer from one material to another, e.g. from tubers to air during cooling
Hot box	Insulated cabinet warmed to 32–36°C, and humidified with water, used for samples of potatoes to speed expression of bruising or to accelerate rotting
Humid ventilation system	Low-rate continuous ventilation of potatoes with humidified air

Hunting	Where fans or other electrical equipment switch on and off in quick succession due to rapid changes in air temperature. Avoided by use of time delays in switching equipment
Inoculum	Infective agents of disease
Interstitial condensation	Condensation forming within the building structure
Latent heat of evaporation	(Hidden) heat required to change water from liquid phase to vapour phase with no increase in its temperature (2.4 MJ/kg). Half of effective cooling of potatoes with air is due to heat removal by evaporation
Lateral	Secondary delivery duct off main duct
LED display	Light emitting diode digital display on controller or by store door
Lesion	An area of tissue damaged by disease
Low-rate tariff	Electricity sold at discounted rate to encourage use when demand is low
Main duct	Primary delivery duct for air coming from fan
Manometer	U-tube filled with coloured paraffin, used to compare pressure within a duct with atmospheric pressure
Metabolism	Biochemical processes of an organism's cells used to maintain life
Mummify	The process of drying a rotten tuber to a shrivelled mass that is dry and will not contaminate its neighbours with inoculum
Necrosis	Death of cells
Off-cycle defrost	Switching off a fridge compressor for a period of time to allow any ice on the evaporator to melt
Out-grades	Tubers too large or too small, or outside acceptable bands of disease, growth cracks, slug holes, etc.
Periderm	Corky outer skin layer of potato
Physiological age	Physiological age is the maturity of a growing, or stored, tuber and varies with the temperature of its surroundings, with warmer temperatures increasing maturity
Pile	A stored mass of potatoes, same as bulk
Pre-pack potatoes	Ware or table potatoes, usually sold in polythene bags
Pre-sprouting	As chitting
Provenance	Date of planting, emergence, tuber initiation, harvest and subsequent storage temperature for a particular stock of potatoes
PU	Polyurethane (insulation)
Pulp	Flesh of the tuber
Ramification	Infection
Recirculation	Ventilation with internal store air (no ambient air)
Relative humidity (RH)	Mass of vapour in air at a defined temperature compared with the maximum vapour it can hold at that temperature, expressed as a percentage

Respiration	Intake of oxygen and expression of carbon dioxide as part of the tuber's metabolism process
Roof space	Air space above the crop
Safe haven	BPC has a safe-haven scheme, which produces crops within a 'cordon sanitaire', so limiting the chance of disease entering a seed multiplication operation
Saturation humidity	Point where air contains the maximum mass of water vapour that can be held per unit mass of air at any given temperature
Sclerotia	Protected fungal body
Senescence	Point at which growing plant reaches the end of its life and dies; its cells stop reproducing and it goes yellow
Skin finish	The visual appearance of skin based on its reflectivity and presence of disease and defects
Skin set	As a tuber matures, its skin gains a strong adhesion to the tuber flesh below. The degree of skin set can be tested by rubbing the skin with the thumb
Soft rots	Bacterial wet rots
Split-grading	Separating tubers into two size fractions, usually at harvest
Stolon	Underground stem, the tip of which grows into a potato tuber
Storage diseases	Diseases that can multiply or infect in store but which may possibly be controlled by appropriate store management
Store air	Air above a pile or surrounding boxes of potatoes, as opposed to air within the voids between tubers
Store air temperature	Average temperature of air in body of store, surrounding boxes in box store or in roof space in bulk store
Store humidity	Relative humidity of the air in the body of the store
Structural condensation	Condensation forming on the inside of the building fabric or structure
Suberization	Laying down of the chemical suberin between damaged surface cells of tuber as the first part of periderm formation in the wound healing process
Temperature – dry bulb	Air temperature as measured by a thermometer
Temperature – wet bulb	Temperature reading in a moving air stream of a thermometer wrapped in a wet wick
Tipper	Or tippler. Machine for tipping boxes to empty them of potatoes
Top/bottom differential	Differential between crop temperature at top and base of a stack of boxes or pile
USDA	United States Department of Agriculture
U-value	Heat conductivity of a building material (W/m^2 °C). The lower the value, the better insulated the material
Vapour	Water in gaseous form
Vascular tissue	Tube-like structures in a plant or tuber involved in fluid and nutrient transport

Ventilation – airspace	Air circulated through the airspace surrounding boxes, but not forced through the potatoes themselves
Ventilation – positive	Air forced through tubers in boxes, using fans and ducts
Voidage	Voids or airspace between tubers
Web	Sieving area of potato harvester, made of a series of parallel bars, which holds potatoes but lets soil through
Wet rots	Tubers where the flesh has been invaded by disease organisms to form a liquid mass with little structural strength. The wet mass collapses on unaffected neighbouring tubers, providing inoculum and anaerobic conditions, which can lead to further rotting
Windrowing	Lifting a crop of potatoes with a harvester and returning it back to land in a windrow to allow the crop to dry. The windrow is lifted later in the day
Wound-healing period	Period for wounds on crop to heal, to form a barrier to prevent disease ingress to the flesh of the tuber

1   Physiology

Topics discussed in this chapter:

- Origin of potatoes.
- Where potatoes are grown worldwide.
- The developing tuber.
- Field conditions that affect tuber development.
- Physiology of the mature tuber.
- Tuber chronological and physiological age.
- Strategies for influencing physiological age.
- Physiology of the tuber in store.
- Tuber respiration and its effect on store climate.
- Tuber dehydration during storage.
- Optimization of tuber quality.

1.1   Introduction

Many problems postharvest originate during tuber growth. Store or packhouse managers must understand how tubers develop, so that they can distinguish between problems caused by physical conditions like flooding and nutritional deficiencies and those caused by disease. This allows managers to take appropriate remedial action in store and plan for changes in crop husbandry in future years.

A knowledge of the physiology of the living crop helps to explain the storage environment. Respiration heat emitted by tubers causes convection currents, dehydration of tubers, condensation on the crop and building structure and potential moisture-related disease. These consequences need to be either considered or prevented for successful store management. For other aspects of tuber physiology, which have little impact on storage, the reader must look elsewhere.

1.2 Origin of Potatoes

While there are hundreds of species of *Solanum* around the world, only about 200 produce tubers. Eight of these are cultivated on some scale. The potato, *Solanum tuberosum* L., is by far the most commonly grown and its exploitation and movement around the world is well documented (Hawkes, 1979). Cultivated originally in Peru it was brought to Spain and Portugal, from where it dispersed to other parts of Europe in the late 1500s. Over the next two centuries it was exported to North America, Australia, China and latterly elsewhere. Its universal adoption as a staple carbohydrate has not always been welcomed and has been affected by national propagation campaigns, such as in France in the 1760s, royal decrees in Russia, again in the 1760s, and stiff competition from other carbohydrates like rice and pasta along the way. None the less, the crop is now fourth in the world ranking of crops behind rice, maize and wheat, with over 300 million t produced annually (CIP, 2007). Its distribution across the world is shown in Fig. 1.1. Its value as a carbohydrate is due to the high amount of energy stored as starch within the tuber. Long starch molecules allow energy to be stored in a non-soluble form. When required, conversion of the starch into sugars allows energy transport around the growing plant.

At low temperatures, the tuber begins to convert insoluble starches into soluble sugars. This increase in concentration of ions in the form of sugar solution allows the cellular contents to remain liquid at low temperature (Wright and Diehl, 1927), reducing the tuber's freezing point, much as salting of the roads prevents ice forming in winter. Tuber cells can therefore function at lower temperatures, which may be a factor in helping them survive the Andean winter. This does not guarantee survival of intact tubers and we will see later how freezing damage can compromise storage. We will also see how this anti-freeze mechanism can cause fry colour problems in potatoes destined for processing. This is a major concern for store managers.

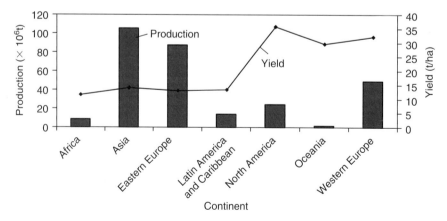

Fig. 1.1. World potato production and average yield by continent. (FAO, 2006.)

1.3 The Growing Tuber

1.3.1 Planting and subsequent crop husbandry

In the UK, prior to planting, stones are sieved from the soil and deposited in the bottom of furrows, usually 1830 mm apart. The seed tubers are then planted in a pair of secondary furrows, between 710 and 915 mm apart depending on the volume of stone in the stone-filled furrows. The secondary furrows are located equidistant between the rows of stones, and covered with a ridge of soil, which will eventually contain the daughter tubers. Planting date depends on final market, latitude, longitude and seasonal weather, with the earliest planting in January/February in the south-west and the latest in Ireland and Scotland in May/June. Tuber spacing along the furrow depends on variety, seed size and age, and the desired size of daughter tubers. If a small number of large ware tubers are required, seed tubers with apical dominance may be planted. If a large number of seed or salad-sized tubers are required, seed tubers that will produce multiple sprouts will be planted. Fertilizer is usually incorporated at planting, with a further application broadcast later if required. Weed control by sprayer wheels running on the separated stones may be required until the crop canopy closes. Thereafter the sprayer will be used to control blight, specific diseases or insects. Harvest of early potatoes will start in April/May, while main crop and seed crop will span September to November.

Knowledge of how tubers are formed can assist managers in discovering what went wrong when a crop fails its quality assessments.

1.3.2 Tuber development

Shortly after emergence, the stems of most potato plant varieties will begin to develop underground lateral stems called stolons. These can be subdivided (Fig. 1.2) into:

- Main stolons, borne directly off a main stem.
- Lateral stolons, arising from the axils between the main stem and the main stolon.
- Branch stolons, arising as branches from the main stolons.

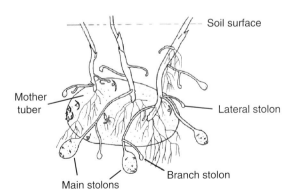

Fig. 1.2. Hierarchy of stolon emergence from plant stem. (H. Carnegie, Aberdeen, UK.)

These subdivisions are important as they reflect the hierarchy of main, lateral or branch stolon by which stored dry matter in the tubers will be converted back to sugar and reabsorbed by the growing plant in times of stress. Smaller tubers may also be 'reabsorbed' so that resources can be diverted elsewhere.

Following stolon formation, which occurs usually 14–30 days after the crop is visible above the soil, stolon tips will begin to swell (Fig. 1.3) as tuber initiation begins. New cells are created through cell division, starch is deposited after conversion from translocated sugars and the tuber periderm or skin is formed (Fig. 1.4). This process combines the laying down of stacked periderm cells with the deposition of suberin within and under these cells to form a protective barrier against disease and water loss. Dry matter accumulation and tuber expansion will then continue as the season progresses, although, as mentioned above, on occasions of stress this process can go into reverse.

The resulting tuber after harvest has a stolon scar at one end, where it was attached to the stolon, and the 'bud' or 'rose' end at the other (Fig. 1.5). At the bud end, the apical eye containing the bud is the last to be formed and contains the physiologically youngest bud (Rastovski and van Es, 1981). After a period of dormancy, one or more of the buds starts to sprout. The other buds or eyes are arranged spirally round the tuber, with the eyes nearest the stem being the first to be formed during tuber development. In the skin of the tuber are lenticels, tiny holes in the skin usually too small to be seen, which allow gas exchange between the cells within the tuber flesh and the surrounding soil atmosphere. There are about 250 of these in the skin of a 115 g tuber. The stolon or stem end, buds and

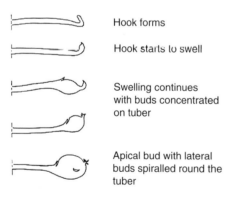

Hook forms

Hook starts to swell

Swelling continues
with buds concentrated
on tuber

Apical bud with lateral
buds spiralled round the
tuber

Fig. 1.3. Stages in the formation of a tuber. (H. Carnegie, Aberdeen, UK.)

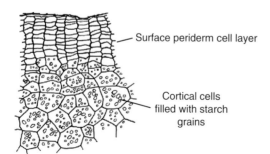

Surface periderm cell layer

Cortical cells
filled with starch
grains

Fig. 1.4. Periderm cells stacked one above the other with cortical cells below. (H. Carnegie, Aberdeen, UK.)

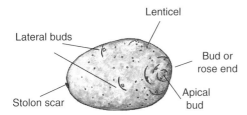

Lateral buds

Lenticel

Stolon scar

Bud or rose end

Apical bud

Fig. 1.5. Potato tuber with stolon scar to the left, bud or rose end to the right and apical and lateral buds in between. (H. Carnegie, Aberdeen, UK.)

Box 1.1. Hollow heart

The causes of hollow heart (Fig. B.1.1) are not fully understood but many researchers have proposed a link with water and temperature stress shortly after tuber initiation. Metabolites normally destined for the cells at the centre may be diverted elsewhere, causing some cells to die. Thereafter, when rapid growth resumes, due for example to a return to sufficient soil water status, any new growth will 'pull apart' the dead and dying area to leave an irregular-shaped cavity. In some situations, hollow heart observed early in the season will grow out if conditions allow for slow and steady growth.

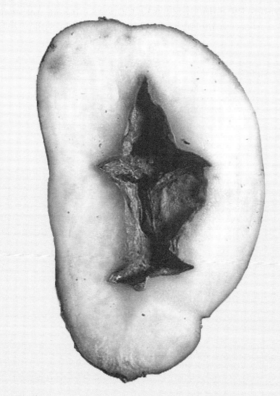

Fig. B.1.1. Hollow heart in a tuber. (Courtesy of Potato Council, Oxford, UK.)

lenticels are all important to the store manager as they are points of weakness in the tuber skin, where disease organisms can enter and multiply or wait for conditions that suit their development.

1.3.2 Chronological age

Research has shown that many physiological processes in the potato plant follow a predictable timetable for any given variety (Firman *et al.*, 1999). By repeated measurement of key growth stages in a range of varieties of potato crops, researchers have been able to state with some degree of confidence that:

- Tuber initiation follows crop emergence by a predictable length of time for any given variety.
- Dormancy break (or first sprouting) follows tuber initiation by a predictable period for any given variety.

Increasingly potatoes are being grown for a specific market, be it bakers, processing, pre-pack, salad potatoes or seed. Each has an optimum tuber size. The greater the proportion of the crop in the optimum size range the greater the potential profitability of the crop.

Firman *et al.* (2004) showed that the proportion of tubers within the optimum size range could be increased by extending or reducing the period between tuber initiation of the seed crop and the planting date of the resulting seed.

To produce many, but small, tubers, an extended period is required (Fig. 1.6a). In the UK, the seed crop is planted in mid-March, with tuber initiation in mid-May. The crop is harvested in early September, stored until the following April and planted late April/early May. The result is a crop with a prolific number of small tubers.

To produce few, but large, tubers, a short period between tuber initiation of the seed crop and planting of the resulting seed is required (Fig. 1.6b). The seed crop is planted in mid-July, with tuber initiation in mid-August. The crop is harvested in October, stored over winter and planted in mid-February or whenever soils are warm enough to allow growth.

This approach brings to the industry a degree of predictability it has not had before and would indicate that a number of unpredictable occurrences in the past, in terms of bulking rates, desired yield, etc., can be attributed to a lack of appreciation of the importance of seed provenance.

1.3.3 Physiological age

Many biochemical processes in plants are a function of heat and time. In potatoes this is best illustrated by sprout development. Under dark conditions, after dormancy break, sprouts will elongate much more quickly in crops that are stored warm than in those that are stored cold (Morris, 1966). Tubers that sprout in warm conditions will typically produce one dominant sprout, which results in low numbers of large daughter tubers.

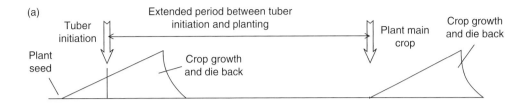

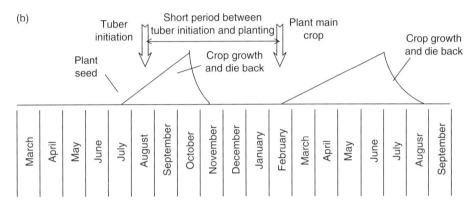

Fig. 1.6. Achieving a prolific or bold crop by changing the interval between 'seed crop tuber initiation' and 'main crop planting date': (a) system to produce many, but small, tubers; (b) system to produce few, but large, tubers.

Box 1.2. Chronological and physiological age

The tuber develops from a swollen stem into a fully mature tuber, which then goes into a natural, or innate, dormancy state for a period of time before it starts to sprout. The chronological age of the tuber is measured from tuber initiation and at dormancy break is around 6 months, but this varies for each variety. If the growing tuber is exposed to warm temperatures during growth or subsequent storage, its maturity or physiological age increases more rapidly and the tuber becomes physiologically older than it would have done at lower temperatures. Chronological age is the true age of the tuber in terms of months, while physiological age relates to its physiological state.

In ware crops stored for consumption, where sprouting is undesirable, tubers are ideally stored at temperatures of about 2.5–3.5°C to lengthen the dormancy period. If storage has to be at higher temperatures to prevent low-temperature sweetening, which is likely to be the case for processing crops, then suppression of sprouting may be necessary through chemical means.

In crops to be used for seed, advancing physiological age by keeping tubers warm under lights is often desirable where early emergence, early bulking and early harvest are required. For many producers the premium derived from physiological ageing has in the past outweighed the additional cost of the very high seed rates required. The apical dominance that results from this physiological ageing requires tubers to be planted closer together to establish an optimum stem density.

Managing physiological age allows the grower to produce a known quantity of stems and hence a relatively predictable sized crop. The crop has to be planted with extreme care so that sprouts are not broken off, which could result in blanking, or significantly delayed or variable emergence. The probability of sprouts being lost is greatly reduced by chitting, the process whereby sprouting tubers are exposed to natural or artificial light in a glasshouse or chitting shed (Ch10.5), to produce stronger, shorter and better-attached sprouts.

Physiological age is measured in day-degrees above 4°C from the time of first sprouting. Typically accumulations of 200–400 day-degrees are required to create a significant benefit for an early market. This can be achieved (Box 1.3) by warming seed to 10°C and storing it at this temperature for about a month in a well-lit chitting house prior to planting (Ch10.5).

While yields will seldom be as high as from un-aged seed grown for a full season, physiological ageing can have various benefits for a particular market:

- Early emergence.
- High early yield.
- Early senescence and bulking.
- Rapid skin set.

1.3.4   Influence of soil on skin finish

The visual appearance of skin, termed skin finish, based on its reflectivity and presence of disease and defects at harvest can be affected by soils in a number of ways, due to:

- Soil-borne diseases.
- Period since last crop of potatoes was grown.
- Un-decomposed plant residues.
- Soil type.

Box 1.3. Example of chitting temperature regime to achieve ~200 day-degrees

Physiological age = (10°C – 4°C) × 33 days = 198 day-degrees above 4°C

Table B.1.3. Time to achieve two physiological ages at a range of chitting store temperatures.

Temperature in chitting store (°C)	Time (days)	
	150 day-degrees	300 day-degrees
8	38	76
10	25	50
12	19	38
15	14	28

Where crops are grown in the same field more frequently than one year in seven, then disease incidence and severity on the crop increase. Various publications (Read *et al.*, 1995) recommend an interval of more than 6 years between potato crops if key diseases such as stem canker and black dot are to be controlled. In other reports, comparisons are made between crops grown more, or less, frequently than one year in six. More disease-related skin finish problems (e.g. black scurf and powdery scab) are reported with short rotations compared with long ones (Nolan *et al.*, 2000). In Europe storage diseases such as silver scurf, black dot and gangrene can also be exacerbated by short rotations. The concentration of potato disease inoculum, in the form of resting spores, sporangia or sclerotia (Ch4.5) depending on the type of fungal species, will be highest immediately after a potato crop and will then decline over a number of seasons. The level of the peak infestation will depend on the amount of trash left in the field after harvest, the size of the root mass and a range of pathogen specific conditions, the main ones being temperature and moisture. The subsequent decline in inoculum over the years may be caused by a number of processes working either alone or in combination. Spores can be leached away from the top layer of soil following rainfall, where other 'myco-phageous' fungi, bacteria or nematodes may destroy them. Alternatively they may simply die naturally, there being no opportunity to germinate and re-infect fresh potato tissue. In addition to these direct influences on decline, other factors such as tolerant weed hosts or potato volunteers may allow a particular pathogen to remain viable in the absence of a potato crop.

Time between cropping of potatoes can be reduced in some cases if pesticides are used to suppress diseases or pests. Such practices are coming under increased regulation to safeguard the environment, so husbandry measures or integrated pest management systems should be used wherever possible.

Soils support populations of suppressive organisms that have a role in disease reduction, as well as potential infective organisms (Elphinstone *et al.*, 2004). Attempts to replicate these have met with little success although various soil-improvement products that contain pathogen-suppressive organisms are starting to gain in popularity.

Careful management of the growing crop will minimize the inoculum and reduce the likelihood of producing crops with excessive amounts of disease. The role of storage is to minimize the further development of disease on the harvested crop by considering the disease triangle (Ch4.2).

As well as biological properties, the soil's physical properties have a direct influence on skin finish. Sandy soils are well known for producing a dull skin finish. Some would argue that a consistent skin finish within a given stock would be more important than a dull or bright finish because purchasers prefer a uniform product, while others would say that brighter skin sells better than duller skin and so ought to attract a premium. Recent electron-microscopy studies at the Horticultural Research Institute, Warwick, UK (Wiltshire *et al.*, 2005) have compared the skins of tubers grown in a range of soil types and have shown that the frequency of ruptured or collapsed cells within the epidermis is higher when crops are grown on sandy soils. An epidermis made of intact cells is better able to reflect light than one that contains damaged cells, so exhibits a brighter appearance.

Most sandy soils currently used for potato production produce processing crops where skin finish is less important than for washed and bagged crops.

1.3.5 Variety and skin finish

Some varieties produce a better skin finish than others even when grown in similar conditions. Between varieties there may be subtle differences in the way the epidermis and periderm are formed. In some varieties, epidermal cells appear very neatly stacked (Fig. 1.4). This stacking creates lines of weakness between cells, which can shear or grow apart if significant pressure is applied. While these shears are for the most part invisible to the naked eye, they can reflect and refract light in much the same way as damaged cells described previously. In other words, they can result in a dull skin finish. It has been proposed that where these lines of weakness are very deep then the resulting shears may be visible to the naked eye and appear as netting on the tuber surface, although this has been difficult to measure in practice (Wiltshire *et al.*, 2003).

For bright skins, varieties which have a cellular 'brick wall type' of stacking will contain fewer lines of weakness and therefore are less prone to microscopic cracking. Another cause of poor bloom is caused by the surface periderm cells collapsing (Fig. 1.7). The reason for this collapse is not fully understood, but appears to be influenced by maturity at harvest and evaporation during storage (Wiltshire *et al.*, 2005).

1.3.6 Dry matter

Discussion above referred to the process of sugar transport from plant to tubers and the formation of starch dry matter. Dry matter content can be measured in a number of ways, such as by comparing the weight of an oven-dried sample with its original fresh weight, but it is usually measured using a hydrometer, where a

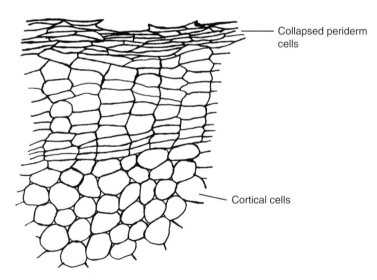

Fig. 1.7. Loss of bloom due to collapsed surface periderm cells. (H. Carnegie, Aberdeen, UK.)

known weight of potato attached to a float displaces a volume of water relative to its dry matter (Ch12.11).

The dry matter content of a tuber depends on:

- The potato variety.
- The amount of intercepted radiation during crop growth.
- Its health status, particularly that of its root hairs.
- The availability of water during tuber bulking.
- The position of the tuber within the tuber hierarchy.

These relationships are not clearly understood although the relationship between high dry matter content and water stress, induced using rain shelters, is relatively clear (Stalham, 1992). The complexity of the factors involved is reflected in the degree of variation in dry matter content typically found. This varies from plant to plant within a crop and also from tuber to tuber on a single plant. Variations as high as 8% are not uncommon (A. Veerman, The Netherlands, 2001, personal communication). Even within an individual tuber, dry matter content can vary by as much as 7% from location to location (Fig. 1.8). It is typically higher towards the edge of the tuber and lowest towards the centre, and typically highest near the stolon scar (Gaze *et al.*, 1998).

To ensure that a representative percentage dry matter figure is obtained for any sample, it is important to ensure that tubers are selected at random and that multiple samples are taken, whether they are taken from field or store.

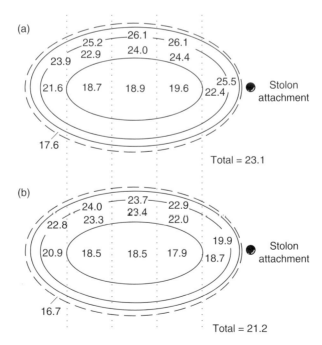

Fig. 1.8. Dry matter percentage distribution within 40–50-mm tubers of: (a) cv. Saturna; (b) cv. Russet Burbank. (Abbreviated and redrawn from Gaze *et al.*, 1998.)

1.4   Factors Affecting Growth

The store manager often has to identify the cause of a problem in a batch of potatoes. The origin of the problem may have occurred during the growing period.

   The phenomenon of skins losing adhesion to the tuber flesh below prior to, or
in the weeks after harvest, has been recognized by many store managers. Causes
are still not clearly understood although rapid changes in periderm dry matter
associated with the problem have been identified.

1.4.1   Growth problems which allow disease access to tubers

There are numerous examples where growth problems, often caused by stress on
the growing plant (see Box 1.4), allow disease to develop and subsequently cause
major problems in store. These are summarized in Table 1.1.

1.4.2   Influence of fertilizers on subsequent storage

Nitrogen nutrition can have a significant effect on a number of physiological processes in potatoes. It is included here because of its influence on crop senescence, skin
set and dry matter. Crops where nitrogen is over-applied, either initially or later
through top dressing, will take longer to reach 'maturity'. The haulm can be more
difficult to desiccate and it can also take longer to set tuber skins. Skin set can be
assessed during test digging at intervals prior to harvest. To do this, growers can use
their thumbnail on a random sample of tubers to check skin adhesion to get a rough
indication of skin set. For more accurate measurement, scuffing barrels (Wiltshire
et al., 2003) can be used to test samples of tubers selected at random from the growing crop. Where skin set has not been achieved, the crop should be left for harvesting later, unless unset skins are desired for the early, new potato market.

   On occasion, the skins will fail to set or will gradually become unset (Wiltshire
et al., 2003). This phenomenon is fortunately fairly rare but those reported cases
support routine sampling and testing for tuber skin set and dry matter content,
which can be used to adjust fertilizer applications in future years. Where skin set is
slower than anticipated, this is often due to over-application of nitrogen and fertilizer recommendations (e.g. SAC, 1990; DEFRA, 2000) should be followed more

---

Box 1.4. How stress can lead to disease

Stresses of various kinds can weaken potato plants' natural defences to such an
extent that disease development becomes easier. For example, extremely low temperatures (Wright, 1942) can lead to cellular membrane damage which can make
bacterial access much more likely and result in soft rotting. Higher temperatures
combined with water stress can lead to premature changes from starch to sugars
within tubers, providing a ready substrate (or nutrient source) for subsequent bacterial invasion (Ch1.4).

Table 1.1. Stress-related growth problems and potential for disease development.

Climatic effect	Result	Consequence
Lack of moisture at tuber initiation	Deep common scab	Difficult to dry – potential for soft rotting
Lack of moisture late in season	Jelly end rot (Fig. 1.9)	Easily ruptured source of moisture and soluble sugars, readily available to bacteria and leading to soft rotting
Excessive moisture through season	Anaerobic conditions allowing blackleg development	Potential to trigger soft rotting
Excessive moisture late in season	Lenticellular outgrowth (Fig. 1.10)	Damaged cells allow fungi and bacteria access to flesh of tubers
Extreme high temperatures	Heat-induced cell damage (necrosis) leading to localized disintegration of vascular tissue at bud (or rose) end	Potential for secondary dry rots and soft rotting
Moderate to high temperatures	Premature conversion of starch to sugars and membrane damage leading to 'glassiness'	Periderm easily ruptured, providing source of moisture and soluble sugars, leading to soft rotting
Extreme low temperatures	Freezing injury causing massive membrane disruption	Periderm easily ruptured, providing source of moisture and soluble sugars, leading to soft rotting

Fig. 1.9. Jelly end rot. (Courtesy of Potato Council, Oxford, UK.)

(a) Early formation of lenticel   (b) Suberized lenticel

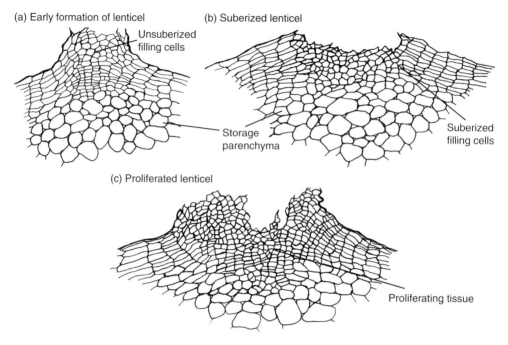

Fig 1.10. Lenticel formation and suberization; proliferation may occur in wet conditions. (From Adams, 1975. Redrawn by H. Carnegie, Aberdeen, UK.)

carefully in future crops. Over-application of nitrogen can also delay the crop in achieving the dry matter contents required by French fry and crisp processors.

While phosphates appear to have little effect on tuber quality, potash appears to have an influence in two areas. First, adequate potash nutrition can help in reducing susceptibility to bruise damage. While this is often debated, there is suffi-cient evidence to suggest that crops grown with inadequate potash are more likely to bruise (Fellows, 2004). Choice of potash fertilizer can also have an influence on tuber dry matter accumulation, as this tends to be higher where potassium sulfate is used instead of potassium chloride (Dickens *et al.*, 1962).

1.4.3 Drought

Water stress can affect the potato crop in a number of ways. It is particularly impor-tant from a quality perspective at tuber initiation and in the days and weeks that fol-low initiation. It is during this stage of growth that tuber periderm is not fully developed and common scab can infect tubers. Infection is most likely to occur in susceptible varieties where the soil zone in which tubers are formed is allowed to dry out during tuber initiation. Water would otherwise support a flora of bacteria antagonistic to *Streptomyces scabies*, the cause of common scab. When these are absent, common scab develops readily. Various computer programs are available to calculate optimal irriga-tion required during tuber initiation and range from simple water balance calculations based on rainfall, irrigation water evaporation and leaching, to more sophisticated models that take account of crop canopy and evapotranspiration, through to direct measurement of soil moisture status using various measuring devices.

Water status later on in crop development can also have an influence on tuber bulking rate, accumulation of dry matter (Gaze *et al.*, 1999) and susceptibility to bruising (Fellows, 2004). Rapid changes in soil moisture status, from very dry to very wet, towards the end of the growing season can cause splitting or cracking of tubers (Fig. 1.11). This is often more noticeable in certain varieties (e.g. 'Marfona' and 'Nicola') and following rapid mechanical destruction of haulm by machine flailing, where the ability of roots to take up water exceeds the rate of transpiration resulting in an overload of tubers with water. Providing cracks and splits are allowed to heal prior to lifting there should be little effect on storability, although market value is likely to decrease.

Rapid changes in soil water status from drought back to wet due to heavy rain can result in a number of tuber defects. If the drought is particularly stressful, then bulking will cease. Tuber growth will resume once the water supply resumes but this may follow discrete physiological changes in the tuber. This can result in the formation of 'dolly'-shaped tubers or chain tuberization (Fig. 1.12), where, in sequence, 'tuber–small section of stolon–smaller tuber' may develop on the same original stolon.

1.4.4 Flooding

Where flooding occurs and tubers encounter a period of time during which they stand in water, three things happen. First, there is an increased risk of disease. As described later (Ch4.3), most fungal and bacterial diseases develop rapidly where water is present. This is particularly so for bacterial diseases such as blackleg and soft rotting, but is also the case for a range of fungal pathogens. In areas where water sources are contaminated with brown rot bacteria, flooding may facilitate the spread of this disease. For *Erwinia* bacteria, which produce blackleg in the growing plant, their association with anaerobic conditions in store is well documented (Lund and Kelman, 1977; Pringle and Robinson, 1996). Similar anaerobic conditions in the soil occur during periods of flooding, allowing bacteria to develop and spread rapidly.

Fig. 1.11. Cracking in potatoes due to moisture variations in the soil during tuber growth. (Courtesy of Potato Council, Oxford, UK.)

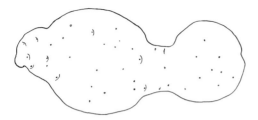

Fig. 1.12. Dolly-shaped tuber or chain tuberization. (H. Carnegie, Aberdeen, UK.)

Second, oxygen depletion and anaerobic conditions may also cause physiological changes within tubers. Blackheart and internal blackening are more commonly associated with oxygen depletion during storage due to crop respiration in very well-sealed stores. Low oxygen levels in the store atmosphere prevent oxygen from reaching into the tuber flesh, leaving central cells unable to respire and function properly. Over time the cells start to break down and go black. Similar effects can on occasion be observed in the soil following prolonged periods of flooding (Hooker, 1981).

Finally, prolonged exposure to water through flooding can cause outgrowth of lenticels. This can increase the chances of soft rotting occurring during storage, as bacterial access to tubers is made easier due to the damaged lenticels.

1.4.5 Frost

Whether occurring in the field through ambient low temperature or in store through insufficient insulation, the consequence of very low temperatures on potatoes is important. Low temperatures (Ch1.2) trigger a change in sugar concentrations in potato tubers and result in a deterioration of fry colours. These changes can occur either due to long term low-temperature storage or short-term exposure to very low temperatures.

Once temperatures dip below −2°C for more than 1 h, massive cellular breakdown can occur within a tuber (Wright and Diehl, 1927). The threshold will change slightly depending on planting depth and tuber sugar content. Cell contents freeze and expand, causing cell membranes to break. Upon thawing the contents are no longer confined and 'leakage' occurs. This is often seen as a blackening of tubers and is most noticeable in those tubers that have been directly affected by frost. Where tubers are partly exposed to mild frosts by being half in, half out of the soil ridge, then the frosted area will be clearly differentiated from the unfrosted, intact, buried part of the tuber.

1.4.6 Warm or cold growing seasons

Since duration of dormancy depends on the heat input in day-degrees above 4°C between tuber initiation and when the tuber starts to sprout (Fig. 1.13), warm summers, which result in higher average soil temperatures, will tend to reduce the period of dormancy while cold summers will tend to increase it. In extreme cases in the UK, potatoes have started to sprout while still in the ground. Examples are 'King Edward' crops in 2006. Such occasions are referred to as 'sprouty' years.

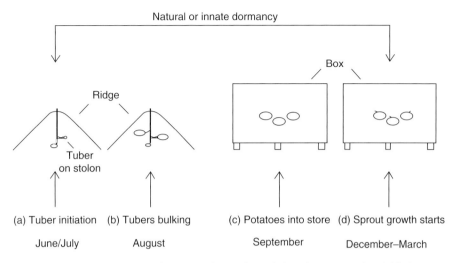

Natural or innate dormancy

Ridge

Box

Tuber
on stolon

(a) Tuber initiation   (b) Tubers bulking   (c) Potatoes into store (d) Sprout growth starts

June/July   August   September   December–March

Fig. 1.13. Natural or innate dormancy is number of days between tuber initiation and start of sprouting.

When managing crops after a particularly warm summer, the store manager should take the season into account when planning his cooling or sprout suppression practice.

Conversely, after a particularly cold summer, seed growers may have to ensure that the crop has been subjected to sufficient day-degrees of heat prior to planting, if the crop is to start growing immediately it is planted.

1.4.7 Location of tubers in the ridge

In store, it is common to see one or two tubers over a square metre of crop surface change from being a normal tuber to a bag of water contained by its skin. The rest of the crop can be quite sound and grade out well. In the same way, individual tubers tested for disease can give highly variable results. The likely cause of these variations is the position of the tubers within the ridge (Fig. 1.14). The likelihood of blight, soft rot, common scab, crushing due to wheel damage, greening, etc. will alter depending on their position. Sampling must take this variability into account, by ensuring large samples are taken and that these are replicated.

Other factors during harvest or in store may add to this variability, so this too has to be considered.

1.5 Physiology of Potatoes in Store

Potatoes are living organisms, which in store take in oxygen and give out carbon dioxide and heat. A thorough understanding of their physiology is necessary to understand the processes that occur in store.

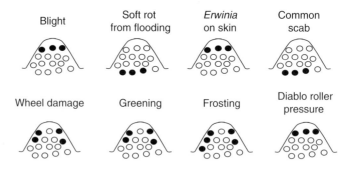

Fig. 1.14. Location in ridge will affect the likelihood of tubers having a particular problem.

1.5.1 Respiration

The biochemical processes by which energy is liberated from carbohydrates are complex and explained more fully in other publications. For this text it is sufficient to say that the processes require oxygen and are preceded by a conversion of starches to sugars. The mitochondria, the tuber cells' combustion chambers, oxidize glucose into nutrient that is required by the tuber to stay alive, and produce water, carbon dioxide and heat energy as by-products (Fig. 1.15). The water stays within the cells, while the carbon dioxide leaves the tubers via its lenticels along with a small amount of vapour, it being a wet gas. Heat is emitted from the tubers, mainly by conduction, and warms the air in the voids between the tubers.

The heat produced reduces the relative humidity (RH) of the air within the voids, increasing its water-holding capacity, and contributes to moisture loss through

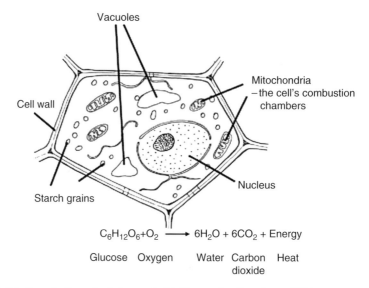

$$C_6H_{12}O_6 + O_2 \longrightarrow 6H_2O + 6CO_2 + Energy$$

Glucose Oxygen   Water Carbon Heat
                              dioxide

Fig. 1.15. Respiration of a tuber cell. (H. Carnegie, Aberdeen, UK.)

evaporation of water from the tuber skins. The warm air in the voids rises through the mass of potatoes, evaporating moisture from their skins as it goes. If the surface tubers are cooler than the tubers below, this warm, moist, upward flowing air will condense on these cool surface potatoes (Ch3.5). The moisture, so often visible near the surface of highly respiring potatoes, is therefore not given out by the tubers themselves as part of the respiration process, but is due to this up-current of warm moist air condensing on the cooler potatoes above. The resulting wet and warm potatoes provide ideal conditions for storage diseases to develop.

Oxygen consumption, carbon dioxide emission and heat generation are common measurements in the study of respiration by potato tubers. The store manager has to consider these processes as they have consequences on how a crop will store.

The rate of tuber respiration can be high in immature tubers (Fig. 1.16). Tubers lifted prior to senescence may still be increasing in size, with cell division proceeding at a rapid rate. If crops are harvested later when they are more mature, respiration will be considerably reduced. This high rate of respiration heat production associated with early harvesting has to be removed immediately after harvest if condensation on the layer of potatoes near the surface of the stored crop is to be prevented. Because immediate cooling of potatoes in addition to the removal of respiration heating (termed 'field heat') would require very large fridges and excessive energy use, high rates of ambient air are normally used to remove the respiration heat, allowing the crop to stay near ambient temperature. Keeping the crop warm allows wounds, the primary entry points to tubers by disease, to heal rapidly and prevent subsequent infection.

Respiration increases after tubers have been handled, transported over rough tracks by trailer or suffered damage in the form of bruises and cuts. The increase appears to be a combination of the actual movement and damage caused, together with the subsequent wound healing process (Meinl, 1972; Burton, 1989). Any handling operation will therefore be associated with increased respiration.

Fan- or wind-induced ventilation is required to minimize the rise in crop temperature from the respiration heat being generated. Moisture evaporation, which leads to tuber weight loss, can be high at this stage due to the poorly developed periderm.

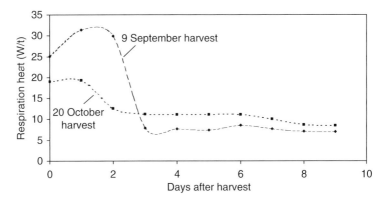

Fig. 1.16. Potato respiration in the days after harvest. (From Pringle *et al.*, 1997.)

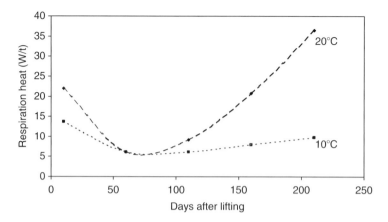

Fig. 1.17. Respiration of tubers when stored at 10°C and 20°C over a storage season. (Redrawn from Burton, 1973, assuming a respiratory quotient of 1.)

Respiration rate drops rapidly after harvesting and becomes relatively stable from senescence through to sprouting/dormancy break (Figs 1.16 and 1.17). It is less stable at high storage temperatures of 20°C, but storage temperatures in temperate climates are kept at 10°C or below, so in practice respiration over the storage period is relatively constant. Respiration rate with respect to temperature (Fig. 1.18) 1 month after harvest shows higher rates at elevated storage temperatures and a minimum at 5–6°C. As store temperature reduces below this temperature, respiration increases as part of the tuber's response to low temperatures (Ch1.2).

Respiration varies between varieties. In stores containing more than one variety, especially seed stores, where varieties are segregated into box rows (Fig. 9.12), one row may be respiring and generating more heat than another. These rows will therefore have higher convection airflows than the others, and may be more likely

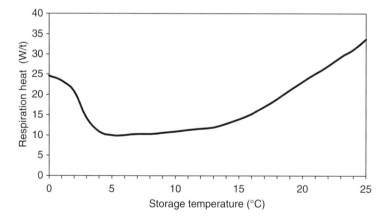

Fig. 1.18. Respiration of samples of potato tubers after about 1 month at various storage temperatures. (Redrawn from Burton *et al.*, 1955.)

to suffer subsurface condensation (Ch3.5). The presence of localized rotting in parts of the stored crop may increase the rate of respiration further, further increasing the likelihood of subsurface condensation.

Rapid increases in respiration rate occur at dormancy break and during sprout elongation (Ch1.3). This can lead to 'thermal runaway', where sprouting gives off additional heat, which in turn warms the crop, which speeds up sprouting, leading to increased respiration heat production. Eventually the cooling system can be overwhelmed and control of store temperature lost. This is why sprout suppression is more effective when performed before eyes are beginning to open and becomes more difficult once sprouts have developed.

1.5.2 Wound healing

Damage results from various harvesting and handling processes. Once damage has occurred the tuber will begin the process of wound healing during which a new waterproof layer of suberin is developed through the wound to prevent further moisture loss (Fig. 1.19a and b). Thereafter increased respiration fuels the process of cellular division in the tissues below and stacked corky cells develop to provide a thicker barrier against the invasion by pathogens. In time further deposits of suberin around the corky cells produce a more waterproofed finish. There are various factors involved but temperature is the most important and its effect on the process is summarized from numerous papers by Mitchell (2000) in Table 1.2.

After store loading, potatoes are usually held warm for a period of time to allow any wounds to heal. This is particularly important where damage levels are high and where disease organisms that can enter by wounds are known to be present. If crops have surface moisture on their skins, due to inadequate drying facilities or poorly designed ventilation, keeping the crop at warm temperatures can predispose the crop to rapid silver scurf, black dot or soft rot development. Cooling the crop rapidly would reduce this problem, but is likely to result in condensation forming on the crop (Ch3.5). The best strategy therefore is to have well-designed ventilation systems, which rapidly dry the crop and keep it dry, so that wound healing can be practised without fear of moisture-related diseases developing.

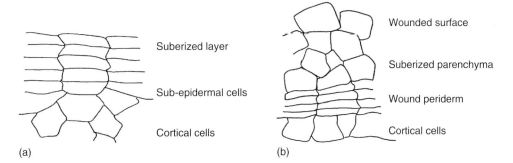

(a) (b)

Fig. 1.19. (a) Cross-section of tuber periderm; (b) suberization after wounding.

Table 1.2. Days to various stages of wound healing.

Temperature (°C)	Light suberization	Complete suberization	Start of periderm formation	Two layers of periderm formed
2.5–5	7–14	21–42	28	28–63
10	4	7–14	7–14	9–16
20	1–2	3–6	3–5	5–7

1.5.3 Dehydration of tubers

Ninety-eight per cent of the moisture that leaves a tuber during storage is lost through its skin by evaporation (Fig. 1.20). Only 2.4% leaves the tuber via the lenticels along with the carbon dioxide produced by respiration, it being a wet gas (Burton, 1989).

So long as the pressure within the cells of the tuber skins and the vapour pressure of the air in the voids surrounding the tubers are the same, no evaporation will take place. For this balance to occur, the RH of the air in the voids between the tubers has to be 97.8% (Hunter, 1985). This assumes that the temperature of the air is the same as that of the tuber.

Hunter determined this figure by blowing air humidified to various levels past a tuber (Fig. 1.21). The air had to be moving to remove the heat coming from the tuber, which would otherwise have heated the void air and reduced its RH. The graph shows that when the RH of the air is below 97.8%, moisture is evaporated from the tuber, and when the RH is higher than 97.8%, moisture in the form of vapour moves from the air into the tuber. At 97.8% vapour neither leaves nor enters the tuber.

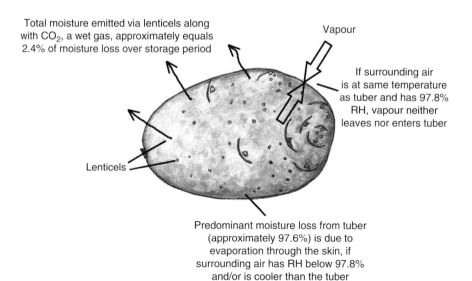

Total moisture emitted via lenticels along with CO$_2$, a wet gas, approximately equals 2.4% of moisture loss over storage period

Vapour

If surrounding air is at same temperature as tuber and has 97.8% RH, vapour neither leaves nor enters tuber

Lenticels

Predominant moisture loss from tuber (approximately 97.6%) is due to evaporation through the skin, if surrounding air has RH below 97.8% and/or is cooler than the tuber

Fig. 1.20. Vapour exchange between tuber and void air. (H. Carnegie, Aberdeen, UK.)

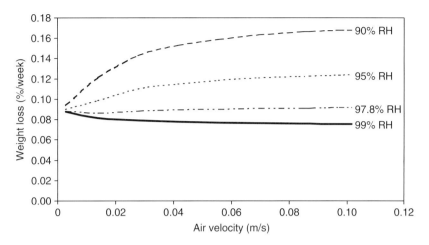

Fig. 1.21. Experiment to determine equilibrium relative humidity (RH) of air surrounding tubers.

1.5.4 Rate of moisture loss

During the storage period, the rate of moisture loss from the crop is proportional to the difference in water vapour pressure (WVP) within the cells of the potato skins and the water vapour pressure of the air in the voids. The difference is often referred to as the water pressure deficit (WVPD). Water vapour pressure is measured in kilopascals or millibars (0.1 kPa = 1 mbar; see Appendix 1 for metric–US imperial conversion tables), and varies with both air temperature and RH (Ch3.1). The greater the WVPD, the faster will moisture be lost from tubers.

The lower the RH of the ventilating air, the greater will be the WVPD. This confirms what we know that the drier the air, the more moisture loss though evaporation there will be.

But in addition, the colder the ventilating air compared with the tubers, the greater will be the WVPD. Ventilation of the crop with air cooler than the crop, no matter how humid, will always result in moisture loss through evaporation.

The cooler and drier the ventilating air, the greater will be the deficit between the water vapour pressures in the tuber skins and the void air and the more moisture loss there will be through evaporation.

1.5.5 Wound-healing period

The rate of loss of moisture through evaporation from a stored crop over a storage season tends to start high (Fig. 1.22), reduce during the dormant period and then increase if sprouting starts. Initial moisture loss is so high because:

- Evaporation from wound tissue is much higher than from intact skin. On a commercial crop, the moisture loss in the first week after harvest was three times that 2 weeks later (Burton and Hannan, 1957).

- Heat of respiration is highest for the 4 to 6 days after lifting (Peterson *et al.*, 1981; Pringle *et al.*, 1997), and this heat reduces the RH of the ventilating air.
- Respiration is increased following wounding (Burton *et al.*, 1992), contributing further to the reduction in RH of the ventilating air.

In intermittently ventilated systems (Ch3.7), extended ventilation is practised after store loading and during wound healing to dry the skins, remove respiration heat and to prevent subsurface condensation. For the majority of the time this reduces the RH in the voids between the tubers and increases moisture loss. With intermittent systems, once the wound-healing period is passed, extended ventilation stops, the RH in the voids rises and the rate of moisture loss from the crop decreases.

1.5.6 Cool down and cool storage periods

In systems which use intermittent ventilation (Ch3.7) without additional humidification, evaporation of moisture is minimized by only ventilating the crop when cooling is required. Some recirculation ventilation to prevent significant temperature differentials from developing is practised, but this is kept to a minimum. The rest of the time forced ventilation or recirculation of air is halted.

In systems where low-volume ventilation fans blowing humidified air run continuously (Ch3.7), the provision of near 98% RH air in the crop voids keeps evaporation to a minimum.

In either system, warm air leaking into store and heat gain through the store walls will increase the amount of cooling required, which will increase the evaporation associated with this additional cooling requirement. Well-sealed, well-insulated stores contribute to the minimization of crop moisture loss.

1.5.7 Fry colour measurements

Fry colour and the uniformity of colour within a batch of potatoes have a large influence on marketability of processing crops. It can be measured in a number of

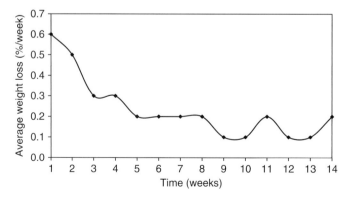

Fig. 1.22. Percentage weight loss per week over the first 14 weeks in an ambient-air cooled store. (From Pringle *et al.*, 1997.)

ways. Visual assessments on fried crop samples are possible by comparing their colour with charts published by various organizations such as the United States Department of Agriculture and the British Potato Council. While cheap and rapid, they are subjective and rely on specific lighting criteria and keeping colour charts in good condition.

More sophisticated, accurate and expensive methods are available; the Agtron system™ (Agtron Inc., Sparks, Nevada, USA) works by projecting a beam of light on to a French fry or crisp sample and measuring reflected light at a given range of wavelengths. The Hunter-L system™ (Hunter Associates, Reston, Virginia, USA) uses carefully controlled light emissions but expresses data as coordinates on red-to-green and blue-to-yellow axes. In each of the three cases, the ability to replicate prescribed fry conditions of oil type, heat and cooking duration, either by a particular purchaser or by other researchers, is critical if meaningful comparisons are to be made.

Preservation of fry colours is paramount when supplying the processing industry. Some varieties are better at keeping their fry colour than others and have been selected for use in processing companies. Fry colours are primarily related to reducing sugar content, so management of reducing sugars is critical. There are various influences on reducing sugar content that need to be considered when planning storage for processing.

Maturity

It is essential that crops are sufficiently mature at harvest if fry colour is to be preserved. While academics may argue about the definition of maturity, it is sufficient to say that fry colours can deteriorate rapidly if crops are harvested prematurely. Crops that initially produce a light golden fry colour may alter in store to produce darker fry colours if harvest occurs before senescence is complete.

Temperature

We have already seen that potatoes convert starch into sugars at low temperatures. Prolonged storage at low temperature will result in an increase in sugar content, which will result in darker fry colours indicated by low Agtron values (Fig. 1.23). A similar effect is sometimes seen when late harvested crops experience low temperatures in the field. Harvest should be started once senescence is complete but not so late that cold temperatures are experienced.

Duration in store

After prolonged storage, darker fry colours are sometimes experienced as the tubers begin their next yearly cycle and irreversible conversion of starch into sugars takes place to fuel springtime growth. This phenomenon is known as senescent sweetening and the degree to which this occurs varies from cultivar to cultivar. Senescent sweetening can be reduced, to some extent, by reducing the storage temperature slightly. This allows the total storage duration to be extended by a few weeks.

Carbon dioxide accumulation

While perhaps only a minor effect, deterioration in fry colour is sometimes associated with well-sealed stores where carbon dioxide levels have accumulated. This is

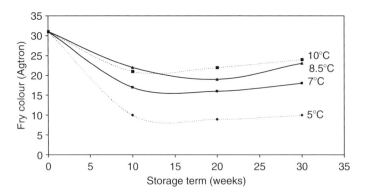

Fig. 1.23. Senescent sweetening over storage period at four storage temperatures.

easily avoided through either short bursts (e.g. 5 min/day) of flushing with fresh air that meets safe condensation-free criteria or the fitment of a fan just large enough to keep the store carbon dioxide at acceptable levels.

Hot fogging with chloropropham

This process uses the exhaust stream from a petrol-driven internal combustion engine to force heated chloropropham (CIPC) particles (Ch3.8) into a store. These engines are not particularly efficient and numerous contaminants are also present within the exhaust stream. Recently a link has been established between ethylene within the exhaust gasses and darker fry colours (Cunnington and Dowd, 2003). The effect can be significantly reduced by early ventilation of the store following CIPC application and by fixing catalytic converters to CIPC applicators. Other application systems for CIPC, either as cold particles or liquid sprays, are currently under investigation along with 'ethylene blockers'.

Achieving the appropriate compromise between each of the above factors is crucial, particularly for long-duration storage where temperatures 1–2°C lower than those for short-storage duration temperatures may cause a slight deterioration in fry colour but may ultimately reduce the need for CIPC applications and delay the onset of senescent sweetening.

1.5.8 Greening

Exposure of tubers to light can induce greening. This is caused by an increase in the number of chloroplasts in periderm cells. Consumers associate greening with the tubers being 'poisonous'. While chloroplasts and the chlorophyll within are not themselves poisonous, the increase in greening coincides with the production of nitrogenous steroidal triglycosides called glycoalkaloids. The two key compounds are solanine and chacocine and they are most likely present as a defence against consumption by both vertebrate and invertebrate pests. At low levels they enhance the flavour of potatoes but at the higher levels associated with greening they can induce a range of symptoms from stomach ache to dizziness and vomiting. There are well documented but rare cases where people have died after consuming large quantities of potato sprouts, where glycoalkaloid concentrations are many times higher than in tubers. Planting seed at sufficient depth and creating sufficient soil cover for daughter tubers by forming ridges usually prevents greening in the field. In the potato store, the prevention of greening is achieved by storing the crop in the dark, ensuring gaps in the store fabric and damaged louvres, which can allow light ingress, are repaired and switching off lights after access to the store. Greening may be more of a problem in retail polythene pre-packs where exposure to natural and artificial lighting in supermarket display cabinets can speed up the process.

Glycoalkaloids have been shown to be highly active against fungal and bacterial diseases (Percival and Bain, 1999). Attempts have been made to capitalize on this by generating aerial seed but with little commercial success. For the store manager who is storing seed only, with no likely possibility of consumption either by man or livestock, a small amount of greening is of little significance.

1.5.9   Dormancy

The biological and physiological definition of dormancy is not well understood and creates significant debate. Natural or innate dormancy is usually defined as 'sprouts being unable to develop even in conditions favourable to development'. All varieties exhibit a period of natural dormancy although this differs from variety to variety. The period during which sprouts become able to develop and start to sprout is known as dormancy break. This is indicated by the eyes visibly 'dilating' and is usually first recorded in the greenest tubers in the warmest part of a store. This is preceded by a shift from starch to sugars in associated cells, which then fuel the process of sprout development. This period is crucial to the store manager who should be prepared to undertake sprouting control if he is to effectively manage sprouting over the coming weeks and months. Imposed dormancy is the term used to describe any 'artificial' activity that stops sprouting after dormancy break. Tactics include:

- Low-temperature storage.
- Use of chemical sprout suppressants.
- Controlled atmospheres.

1.5.10   Sprout growth

Should sprouts start to develop, both respiration and moisture loss will increase rapidly (Burton *et al.*, 1955). This can lead to thermal runaway in store (Ch1.5.1) and a reduction in the value of the crop. The sprouts can, however, be knocked off if the crop is passed over a grader. The exposed flesh of the broken sprouts is vulnerable to invasion by disease, so if the crop is for use as seed, wound healing before dispatch is vital. Extensive sprouting leads to tubers shrivelling and becoming unmarketable. Boxes of such crops can on occasions of extensive sprouting fail to empty, in spite of being completely inverted, due to the intertwined nature of the sprouts.

1.6 Summary

This chapter reviewed the aspects of the physiology of the potato tuber that relate to harvesting, storage, grading and dispatch. This information provides an essential background for those interested in store management.

- Potato tubers are swollen stolons, which act as storage organs, attached to the stems of potato plants, with stomata that are sealed off by cellular growth, called lenticels, to provide gas exchange. Tubers produce sprouts and are capable of vegetative reproduction.
- Tubers receive food reserves in the form of sugars from the living plant and accumulate, then store them, as starch.
- Tubers' natural response to low temperatures is to convert some starch to sugar, which has considerable impact on fry colours and taste.
- The stages of development of a tuber are partly linked to period of growth, or chronological age, and partly to temperature-related physiological age.
- Warm growing seasons lead to early sprouting in store while cool growing seasons delay the break of dormancy.
- Seed can be manipulated to produce daughter tubers with multiple or single sprouts which, when grown on, produce small or large sized tubers, respectively.
- Soil conditions affect skin finish, with sands producing a dull skin and clays and silts a brighter, reflective bloom.
- Tuber dry matter content, of importance particularly with crops for processing, is linked to variety, sunshine hours, water availability and fertilizer rates.
- Most disease on tubers comes from either its seed or from infections developed during the crop's growth.
- While some diseases enter the tubers via its connection to the stem of the plant, others enter through the skin where its defences are weaker, such as lenticels, buds, pest holes or frost damage.
- As the primary defence of a tuber to disease is its skin, a principal aim in harvesting is to minimize tuber damage while the first task of storage is to encourage wound healing.
- Newly harvested crops should be dried using ventilation applied uniformly to the whole crop as it enters store.
- The removal of water by ventilation from wet harvested crops is of primary importance in preventing moisture-induced disease.
- Tuber heat production due to respiration of tubers in store results in convection currents, evaporation of moisture from the crop and the potential for condensation on potatoes just below the surface of the pile or box.

- Early harvested crops tend to have less disease than later harvested crops.
- Early harvested crops produce more heat than late harvested crops and require ventilation to prevent subsurface condensation and associated conditions conducive to infection by disease.
- Tuber dehydration during storage is minimized in intermittent ventilation systems by only ventilating the crop when necessary or by artificially humidifying the ventilating air in continuously ventilated stores.
- Keeping potatoes below 3–4°C in store can prolong dormancy of seed and pre-pack crops. If crops are not for seed, chemical sprout suppressants can be used instead.
- Greening of potato skins is prevented by storing potatoes in complete darkness and by minimizing the period under lights in retail display cabinets.

2 Harvesting and Store Loading Systems

Topics discussed in this chapter:

- Harvester functions.
- Field-to-store crop transfer systems.
- Into-store cleaning and separation systems.
- Damage minimization and tuber assessments.
- Pesticide application.
- Pre-storage drying.
- Loading into store.

2.1 Harvester

2.1.1 Description

Potato harvesters (Figs 2.1 and 2.2) combine a number of functions in the one machine. They differ in type and complexity but most have the following attributes. The harvester:

- Lifts potatoes and soil within the ridge from the ground, together with unwanted items such as clods, haulm, stones and rotting mother tubers.
- Sieves the lifted material as it travels up the inclined first and second webs, discharging the soil layer and small stones back to the field.
- Has steel fingers fixed above the webs, which align the haulm so that it is parallel to the direction of crop flow. This allows 'pinch' extractor rollers, located after the second main web, to grab the plant stems and return them to the field.
- Crushes clods and removes adhering soil using star wheels or axial rollers (see Ch2.2.3).

Fig. 2.1. Tractor with flail in front and harvester behind. (Courtesy of Grimme UK Ltd, Perth, UK.)

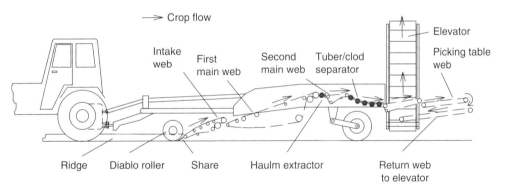

Fig. 2.2. Principal components of a potato harvester. (Based on a Grimme harvester.)

- May have picking staff aboard who manually remove any rots, remaining haulm, clods or stones from the conveyor picking table.
- Elevates the crop into a bulk trailer, or boxes located on a trailer; alternatively it may return the crop to the ground to form a windrow of potatoes to allow them to dry prior to being uplifted later in the day.

Where stones have not been separated from the soil in the field prior to planting, harvesters may be fitted with devices which mechanically separate stones from the potatoes. These separate out most of the stones from the crop, thereby allowing higher rates of harvesting than could be achieved when separation relies totally on picking staff standing on the harvester.

2.1.2 Performance requirements

The crop leaving the harvester should have a minimum of:

- Soil contamination, so that ventilation air can flow freely through the crop once in store.
- Decaying mother tubers or rots, which can result in clusters of rotting tubers in store through which ventilating air cannot pass; these then can develop into hot spots in store (Ch9.5, Box 9.1).
- Stones, which can damage neighbouring tubers as crop passes along conveyors and elevators, or which may rub against and damage tubers being transported in boxes or trailers.
- Haulm, which blocks air spaces between tubers and gets snagged on cleaning and separation equipment.
- Clods, which either harden during storage to cause mechanical damage later when potatoes are conveyed or alternatively disintegrate to form dust clouds during transport and grading.
- Damage to tuber skins, which can allow microorganisms in lifted soil and on the skin to develop on the moist unprotected flesh of the potato. Damage also spoils the appearance of tubers and may reduce their value.
- Internal bruising, which is almost impossible to detect on a picking line. This results in browning within the tuber and may cause a whole batch of potatoes to be rejected for a relatively small amount of bruise present.

A typical work rate for a two-row, trailed harvester lifting from stone-separated ground is 3.5 ha per 8-h day (SAC, 2005/6). If the yield is 45 t/ha, trailers or lorries must transport 160 t/day or 20 t/h. Into-store loading systems must be designed to receive crop at this rate. Harvesting rates will vary depending on the machine used, soil conditions, amount of discarded tubers present, number of pickers and harvesting organization.

2.1.3 Crop transfer to bulk trailer

The point where the greatest damage to potatoes commonly occurs during harvest is at the transfer between harvester and trailer (Maunder *et al.*, 1990). Keeping the harvester elevator discharge as near as possible to the surface potatoes in the trailer minimizes damage at this transfer point. When the trailer is empty, the elevator head is lowered close to the floor of the trailer where it is vulnerable to being hit by either end of the trailer, if the trailer tractor driver fails to keep station with the harvester or if the harvester stops suddenly. One collision between the elevator and the trailer can result in extensive damage to the elevator and in harvesting time being lost while carrying out what may be an expensive repair. Keeping a constant distance between trailer and harvester requires great driving skill, particularly when changing trailers.

To avoid collisions between the harvester unloading elevator and the trailer, the driver of the harvester may raise the elevator clear of the trailer sides and ends. Dropping potatoes from this height on to the hard floor of the trailer is then likely to damage the crop. Lining the floor of the trailer with foam material to cushion the first potatoes to be loaded minimizes this problem. Once potatoes cover the floor, they act as a cushion for the potatoes following. Improving the visibility through the front end of the trailer can further reduce damage, as the harvester driver can see the location of the elevator head in relation to the crop within the trailer.

Careful examination of washed tubers from the harvester and store intake can identify where damage is occurring (Ch12.11), so that any machine malfunction or damage points can be rectified before too much of the crop is spoilt.

2.1.4 Crop transfer to box

In the UK, 1-t boxes, with a plan area of $1.83\,\text{m} \times 1.22\,\text{m}$, are often carried on a trailer to transport potatoes from field to store. Preventing collisions between the harvester elevator and the boxes is even more difficult than when using a bulk trailer. The preferred solution is to fit a fall breaker to the top of the box (Fig. 2.3). The other option is to station an operator behind the box holding a straw-filled sack or cushion in the box, so that the first potatoes in are cushioned. This arrangement, though widely practised, is not recommended as the operator can be hit by the elevator, bombarded by stones accompanying the potatoes or can find his cushion trapped under potatoes if not constantly alert. The insulation material used for 100-mm pipes, when split to form an inverted 'U', can be fitted over the edges of the boxes to reduce tuber damage when moving from one box to the next.

Fig. 2.3. Loading boxes using a fall breaker to minimize tuber damage. (Courtesy of Eric Anderson, Scottish Agronomy, UK.)

2.2 Into-store Cleaning and Separation

A number of systems are available for transporting crop from field to store and for preparing material for storage (Fig. 2.4).

2.2.1 Reception hopper for trailers

Bulk trailers or lorries (trucks) either tip to empty their contents (Fig. 2.5) or are fitted with a horizontal, rubber belt, unloading conveyor (Fig. 2.6). The former needs a bulk hopper slightly wider than the trailer to empty into. The latter can discharge directly on to a belt conveyor or into the boot of a cleated belt elevator.

A common problem is a mismatch between the discharge height of the tipping trailer and the height of the reception hopper. The drop should be as small as possible to minimize damage to the crop. Attempting to correct height mismatches by the use of sloping ramps causes the trailer wheels, but not the trailer draw bar, to be raised. This causes the tipping angle of the trailer floor to be reduced, which can result in some crop or soil remaining when the trailer is tipped. Trailers and hoppers should be matched at purchase.

The bulk hopper usually consists of a wide horizontal rubber belt, with sloping steel sides (Fig. 2.7); it has a capacity allowing it to take the majority of the trailer contents. In larger hoppers the belt carries the potatoes horizontally for the first part of the hopper, rising at its far end where it discharges on to a belt or the boot of a cleated rubber belt elevator. Some bulk hoppers allow the trailer to be reversed over the hopper so that it starts to discharge its load near the discharge end of the hopper and completes emptying as it draws forward. This design allows for complete emptying of the load in one go and more rapid turnaround time for trailers.

The above form of bulk hopper with unrestricted discharge is preferable to ones conveying the crop through a hole in the hopper's rear bulkhead, which can subject potatoes to churning and associated damage as they are forced through the hole.

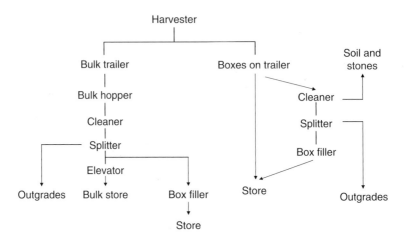

Fig. 2.4. Alternative into-store handling systems.

Fig. 2.5. Trailer tipping into bulk reception hopper. (Courtesy of RJ Herbert Engineering Ltd, Wisbech, Cambridgeshire, UK.)

Fig. 2.6. Self-unloading trailer. (Courtesy of RJ Herbert Engineering Ltd, Wisbech, Cambridgeshire, UK.)

Fig. 2.7. Bulk hopper. (Courtesy of Eric Anderson, Scottish Agronomy, UK.)

The flow of material from the trailer should be adjustable to provide an optimum rate of supply to the sizer or grader. This can be achieved using either a manually controlled variable frequency drive on the bulk hopper conveyor motor or an electronic proximity detector (Ch11.4, Fig.11.2) that senses the presence or absence of crop at the start of the grader and switches on the conveyor drive when more crop is required.

2.2.2 Options after reception

Once the potatoes are flowing in a stream, a number of operations can be carried out (Fig. 2.4).

These include:

- Removal of any remaining soil, clods, haulm and stones from the crop.
- Separation of crop into main crop and out-grades.
- Removal of any potatoes unsuitable for sale.
- Application of a pesticide.
- Conveyance to loading elevator or box filler.

Passing potatoes over cleaning and separation equipment, particularly at the high rates required to keep up with the harvester, will inevitably cause further damage. Most growers believe that it is better to do all the damage at one time, followed by an effective wound healing regime when tubers are warm, rather than cleaning and sizing later in the storage period, when the lower tuber temperatures will result in slower or incomplete wound healing.

Fig. 2.8. Potato cleaning and inspection prior to storage. (Courtesy of RJ Herbert Engineering Ltd, Wisbech, Cambridgeshire, UK.)

The importance of keeping falls to a minimum, conveyor belt speeds low and the use of soft materials to which soil cannot adhere has been known for many years (Vollbracht and Kuhnke, 1956). Cleaning and separation equipment should therefore have few and small drops, and have cleaning and separation areas of sufficient size so that the passage of potatoes across the machine is kept to a speed that causes a minimum of damage.

Cleaning the crop into store requires equipment capable of dealing with potatoes harvested from a range of conditions, from dry to wet sticky soils. Soil can build up inside spiral coils, on the sides of conveyors, the floor of chutes and on soft rubber flaps designed to cushion potato impact. This mud when damp may cushion tubers as they flow over the machine. When dry, it can turn to a sandpaper-type consistency, which may result in considerable scuffing of the crop. Excessive build-up of soil requires regular removal to keep the machine functioning effectively. PVC material should be used in preference to rubber for cushioning, as soil is less likely to adhere to it.

The benefits from putting the crop over a cleaning and separation system prior to storage (Fig. 2.8) include:

- The removal of any remaining soil, haulm, clods and stones ensures good airflow through the stored crop.
- Small quantities of oversized and undersized material can be removed from the main crop, so that they can be sold when prices are most favourable.

- Material unfit for sale is disposed of prior to storage, so storage space and electricity to run fans or fridges is not wasted storing unsaleable material.
- A bulk hopper and cleaning (and separation) system allows bulk trailers to be used for conveying potatoes from the field while using boxes purely for storage.
- Box life is enhanced if they are not used for transport.

The disadvantages of putting the crop over cleaning and separation systems are:

- Additional labour is required to staff the cleaning and separation system at a busy time of the year.
- The additional handling may cause extra damage and remove some of the bloom.
- If the soil is sticky, it will not fall off the tubers; it needs to dry first.
- The equipment has a high capital cost.

Packhouse buyers may allow suppliers who have modern, low-damage, into-store cleaning and separation equipment to clean their crops into store. If growers' separation and cleaning equipment is old and causes damage to crops, these suppliers will be encouraged to harvest crop directly into boxes and store 'as dug'.

2.2.3 Cleaning equipment

A wide range of cleaning equipment is now available for inclusion in store cleaning equipment. They are also to be found on harvesters and graders. They tend to suit different conditions (Herbert, 2006).

Spiral cleaner/grader
Spiral cleaner/graders (Fig. 2.9a) separate loose soil, clods and small potatoes from the main crop by the sideways action of the rotating helical coils. By altering the spacing between the coils, a rough grading up to 45 mm can be achieved. Steel coils can become scored by impact from hard angular stones and can trap small stones between the coils, so should be avoided where soils contain flints or granite. The coils can also become filled with moist soil and therefore in wet conditions need to be cleaned regularly. The rubbing action can damage the immature skins of first early potatoes.

Star wheel cleaners
Star cleaners (Fig. 2.9b) consist of a series of interlocking star wheels, which both carry and clean tubers as they pass over. The soft polymer fingers are backwards curving and raise the tubers as they push them on to the next bank of fingers. This action breaks up small clods and the rubbing action of the stars on the surface of the tubers gently removes lightly attached soil. Cleaning action can be adjusted by altering the speed that the stars rotate. Stars are used where considerable amounts of soil are present. If the potatoes are wet and muddy, the rubbing action may smear the tubers with a coating of mud, which may increase the likelihood of soft

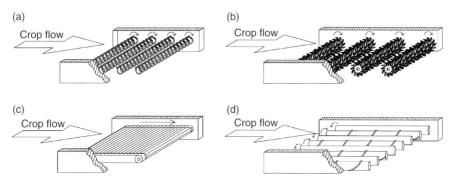

Fig. 2.9. Types of cleaning system incorporated in graders and potato harvesters: (a) spirals; (b) stars; (c) continental web; (d) axial rollers.

rot development in store. Like spirals, the rubbing action can harm immature skins and flints and granite stones can damage the polymer.

Continental web conveyor cleaners

The continental web conveyor cleaner (Fig. 2.9c) consists of a series of interlocking rods or rods fixed to narrow side belts to form an open web conveyor, such as that used on a potato harvester. The webs are normally covered with rubber (Huijsmans *et al.*, 1990) or hollow (McRae *et al.*, 1990) to reduce damage. The cleaning system relies on the web gap of 16–30 mm and agitation to achieve removal of soil, small stones or clods. This system will not break up clods but is useful as an initial cleaning system or pre-cleaner; it is used where soil readily falls off the tubers and is simple in construction and cheaper than other cleaning systems.

Axial rollers

Axial rollers (Fig. 2.9d) tend to be used mainly on potato harvesters when harvesting in wet, heavy soils; they are however sometimes used on separation equipment. The rollers work in pairs: one fluted roller rotates in one direction while a plain rubber roller rotates in the opposite direction. The rubbing action removes the soil from the potatoes and crushes clods. Different sizes of plain rubber roller can be fitted. The smaller the potatoes and the thicker the coating of soil, the thinner the rubber plain roller has to be. Damaged fluted rollers can damage crop and need replacing frequently. To minimize damage axial rollers should be removed from the cleaning line if potatoes are clean and dry.

Brush cleaners

Where only a small amount of soil is attached to tubers, brushes can be used to remove soil. Selecting stiffer brushes can increase the severity of abrasion. Brushes can work well to remove small quantities of soil but can become totally clogged with mud if large quantities of wet soil are present.

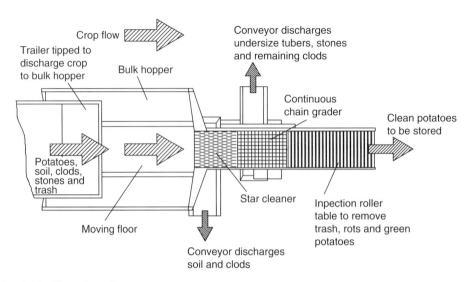

Fig. 2.10. Plan view of an into-store separator.

Certain models of brush or star cleaners will allow some of the banks to be rapidly removed for cleaning while another bank is installed. This allows the brushes or stars to be cleaned when removed from the machine, thereby minimizing cleaner downtime.

2.2.4 Separation equipment

Due to the large volume of potatoes going over the separator, size separation into store is not as precise at when grading out of store. Potatoes are likely to shrink in diameter by 2–3 mm during storage due to water loss by evaporation. However, sizing into store does allow a crop to be segregated into grades destined for different markets.

A typical separation system is shown in Fig. 2.8. The operation of into-store separation is shown in Fig. 2.10. The tipping trailer discharges the potatoes into a bulk hopper, fitted with a moving floor. The crop is conveyed on to a star cleaner, where soil and small potatoes are removed and clods are crushed. The crop then passes over a continuous chain grader, typically 45 mm in size, where undersize potatoes are removed. The remaining potatoes are put over an inspection table to allow staff to remove any remaining rots, haulm, clods, stones and unmarketable potatoes. The rollers continuously rotate tubers so that the entire surface of each tuber can be inspected. The resultant cleaned potatoes are then loaded into store.

2.3 Damage to the Crop

Damage during harvesting, cleaning, sizing and store loading should be minimized if the crop is not to be reduced in value. Damage results in:

- Whole tubers having to be discarded as unfit for sale.
- Either a reduced rate of crop throughput due to pickers being overloaded or an increase in the number of pickers required.
- Damaged flesh that is susceptible to disease establishment.
- Subsurface bruising, which is difficult to identify during out-of-store grading.
- Loss of raw material in the kitchen or at the processor through the extra peeling required to remove damaged tissue.

The degree and type of damage varies during harvesting and into-store grading and loading. On a well-maintained, well-adjusted modern harvester, damage levels of 1 of a scale of 1 to 7 (1 being lowest damage, 7 highest) can be achieved when the harvester is tested using an electronic damage sensor. The harvester-to-trailer transfer tends to be the point of greatest damage, which is why so much emphasis has been placed on the use of well-designed fall breakers. The main types of damage together with their main causes are listed in Table 2.1 (MAFF, 1982).

The potential loss from disease which may develop on wound tissue is difficult to quantify as it depends on the effectiveness of subsequent drying, ventilation, wound healing and store cooling. Agronomists who hand-lift trial plots with virtually no damage often remark that their stocks rarely experience the level of disease in store that machine-lifted crops do.

Table 2.1. Damage during harvesting.

	Causes		
Type of damage	Field	Harvester operation	Handling from harvester to store
Split	Pressure on ridge	Excessive speed of harvester webs	Drop into trailer Discharge into elevator hopper
Squash	Pressure on ridge		Discharge into elevator hopper
Slice		Disc/share setting	
Scuff	Stones and hard clods	Excessive speed of harvester webs	Mismatched hoppers and conveyors
Cut	Stones and hard clods	Excessive speed of harvester webs	Projections on machinery
Hole or indentation		Projections on machinery	Projections on machinery Levelling pile surface
Internal bruise	Pressure on ridge Stones and hard clods	Excessive speed of harvester webs	Drop into trailer or boxes Discharge into elevator hopper Drop on to pile; roll down face of pile

Abrasions and crush wounds are more likely to get infected than clean cuts (Adams and Griffith, 1978). Estimates have been made for the loss of material for sale. In the USA, 6.3% of the value of the crop was lost through damage (Preston and Glynn, 1995). A ware potato chain analysis in The Netherlands (Molema *et al.*, 2000) showed that 78% of the total amount of subcutaneous tissue discoloration was caused by impacts.

Damage can severely downgrade the value of potatoes. On one occasion in 2004 a sample of pre-pack potatoes was downgraded from a value of £165/t to £65/t due to damage alone. In a national survey (BPC, 2004a), during the years 2000–2002, 47% of respondents had loads rejected due to bruise damage while 60% had loads downgraded.

2.3.1   Damage assessment

The most popular damage assessment system used in the UK is the Damage Index (Robertson, 1970), which has merit in that it:

- Is simple to carry out in the field or store (Fig. 2.11).
- Takes 25, 50 or 100 tubers at random from a sample and divides them visually into four categories: severe damage >1.5 mm deep; peeler damage 0–1.5 mm deep; scuff damage to skin; and undamaged.
- Converts the percentage of tubers placed in the three damage categories into a single figure or 'index' suitable for use in experimental correlations with other parameters (e.g. disease development versus damage index).
- Gives an index for a sample, which is related to the weight of tissue (pulp) that has to be removed to obtain potatoes free from any marks when they are peeled.

Details of the system used are described in Ch12.11.

2.3.2   Bruise assessment

Bruises may not show up under normal conditions after harvest until 3–4 days later (Melrose and McRae, 1987). It is essential that bruising be identified as soon as

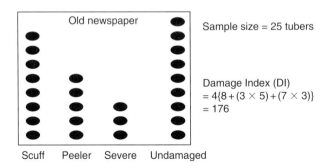

Old newspaper

Sample size = 25 tubers

Damage Index (DI)
$= 4\{8 + (3 \times 5) + (7 \times 3)\}$
$= 176$

Scuff   Peeler   Severe   Undamaged

Fig. 2.11. Schematic of a sample of tubers laid on a newspaper to assess damage.

possible after harvest so that either harvesting can be stopped or the cause of bruising identified. Methods of speeding up the identification of bruising are discussed in Ch12.11.

2.4 Pesticide Use

The best pesticide to prevent disease developing is air and plenty of it: air to remove heat produced by respiration; air to remove moisture on the potato skins; air to prevent subsurface condensation. Many growers, however, like the added security of applying a pesticide.

If synthetic pesticides are to be used they should be applied after the crop has passed over the cleaner of an into-store cleaning and sizing system. Applied liquid chemicals or dusts will not penetrate soil, so any soil should be removed before application. This is why application of chemicals on harvesters in the UK is rarely practised. If all fractions of crop are to be treated it should be applied before separation. If only one fraction is to be treated it should be applied following separation.

To accord with BASIS protocols (BASIS (Registration) Ltd, Ashbourne, UK), a chemical pesticide should be applied only if there is evidence that a disease is present. Justification is therefore required.

2.4.1 Chemicals allowed for application to tubers

A very restricted group of chemicals are suitable for applying to potatoes entering store. These are approved by the Pesticides Safety Directorate and listed in Table 2.2.

2.5 Pesticide Application

In the UK, 45% of ware and 39% of seed was treated with chemicals, into store, in store or prior to dispatch (Heywood *et al.*, 2006). The aim should be to maximize coverage to tubers with a minimum of chemical wastage and any related pollution.

Table 2.2 Chemicals available for treating potatoes into store (UK).

Target disease	Pesticide	Allowed in ware	Allowed in seed
Gangrene, dry rot, silver scurf, skin spot	Imazalil, thiabendazole	Yes	Yes
Sprout suppression	Chloropropham (CIPC)	Yes	No
Sprout suppression	Maleic hydrazide applied to growing crop	Yes	No

If it is possible to apply a chemical to seed prior to planting, this can be an effective way of avoiding chemical being lost to the surrounding soil in applications to the subsequent growing crop.

2.5.1 Hydraulic sprayer applicator

The most popular way of applying liquid chemical to harvested potatoes in the UK is to use a roller table, fitted with overhead hydraulic spray nozzles, with the spray assembly enclosed within a canopy (Fig. 2.12). The roller table both conveys the tubers under the spray, while at the same time rotating tubers as they pass. This ensures that their entire surface receives the spray. The speed of tuber movement across the rollers is controlled by the rate at which tubers are fed on to the roller table. The tubers behind keep the tubers in front moving forward. The canopy over the spray zone reduces potential spray drift, while a drip tray beneath the rollers collects any runoff.

2.5.2 Ultra low volume, spinning disc, controlled droplet spray applicators

If liquid chemical is dripped on to a fast rotating spinning disc, the liquid is broken up into a large number of fine droplets. A small amount of liquid will cover a large area of potatoes. As less water is applied than with hydraulic sprayers, the time

Fig. 2.12. Enclosed hydraulic sprayer applying chemical to potatoes on a roller table. (Courtesy of Eric Anderson, Scottish Agronomy, UK.)

taken to dry the tubers to prevent infection by disease is less. Application rates as low as 0.5 l/t can be used. This compares with 2 l/t with hydraulic sprayers. Since the spinning disc does not require much power, it can be driven from a car battery. The small droplet size makes low-volume sprayers very vulnerable to drift, so a canopy is essential. The labels on some chemicals state that they should not be applied at low volumes.

2.5.3 Electrostatic spray applicators

The use of electrostatic spray applicators for the postharvest application of chemicals to potato tubers was first demonstrated by Cayley *et al.* (1987). The applicators break up the liquid chemical into very small droplets in the same way as spinning disc machines, but then give the droplets an electric charge so that they are attracted to the tuber surfaces while repelling each other. Tubers pass below the sprayer on a roller table. The system can lead to very uniform application of chemical to tubers with reduced chemical runoff. As with the spinning disc technology, the application rates are lower than for a hydraulic sprayer.

2.5.4 Application of dusts and powders

Dust application involves a vibrating dispenser located above the flow of tubers passing across a roller table. As with hydraulic application, the rate of application is susceptible to variations in tuber size, cleanliness, throughput, tuber rotation and crop flow rate. Additionally, with dust application, the presence of moisture on the tubers can cause chemical application to be uneven, as the dust will selectively stick to wet tubers or wounds. A canopy enclosing the applicator can reduce upward and sideways drift of dust, but dust falling between the rollers tends to be dispersed into the atmosphere unless a well-designed fan extractor system is used.

2.5.5 Uniformity of chemical application

The rate of chemical applied to the tubers is dependent on:

- The rate of throughput of tubers – higher throughputs need higher chemical flow rates.
- The specified amount of active ingredient per tonne of crop.
- The amount of liquid required to effectively cover the whole tuber.
- Tuber size – the smaller the tubers, the more surface area there is to be covered per tonne.

The uniformity of chemical application in practice is very variable. In one trial (Bishop and Garlick, 1998), with the sprayer calibrated to apply chemical to potatoes passing over a roller table at 7.5 t/h, the rate of flow of potatoes varied between 2 and 15 t/h while the chemical applied to the actual tubers varied by ±42%. When the potato flow rate was low, chemical deposition reduced due to changes in spray pattern caused by the reduced number of tubers on the roller table. Furthermore, it was noted that tubers would often mesh together into a 'raft', which prevented rotation and uniform coverage. It has been recommended (Rodger-Brown *et al.*, 1999; Rollett *et al.*, 2001) that a buffer hopper be installed prior to the roller table and sprayer, to even the flow of tubers and so minimize the variation in application. Some form of agitation is required to prevent the rafting, though the less variation in size of the sample the less of a problem this is. Tubers need to be rotated 1.5 times for application to be effective (Bishop and Garlick, 1998). Soil on tuber skins prevents chemical reaching the skin of tubers. If the efficiency of application could be improved, the amount of chemical applied to tubers could be reduced (Hide *et al.*, 1994).

2.5.6 Calibration of applicators

Calibration of equipment is always important and should be carried out with each new chemical as well as at the beginning of season. In the UK, sprayer operator training and certification is organized by the National Proficiency Test Council (NPTC, Stoneleigh, UK). All operators have to have passed the appropriate Professional Accreditation (PA) modules required to operate the pesticide application equipment being used.

Developments in application systems are leading to systems that will automatically record the weight of tubers passing through so that spray volume is proportional to throughput. This should greatly assist with traceability requirements, where rate of chemical application has to be put on the shipment paperwork. An example calibration calculation for a spray applicator is presented in Box 2.1.

2.5.7 Fast drying of crop following liquid pesticide treatment

The application of water-based chemical may protect the crop from the target disease but the added moisture may increase the risk of infection from other diseases. This risk is minimized if the crop is ventilated to remove surface moisture immediately after treatment. This can be achieved by standing boxes in a windy location prior to loading into store, by blowing air through the crop immediately after treatment, or by fitting an air knife or fan over the flow of potatoes to remove moisture as the crop exits from the spray treatment chamber.

2.6 Transfer from Cleaner to Bulk Store Loading Elevator

For bulk stores the discharge from the cleaner/separator feeds into the intake of the loading elevator. The elevator needs a long reach to load the top of the pile without its wheels damaging potatoes at the base of the sloping face (Fig. 2.13). For a 4-m-high pile the reach required is 8 m. While the store fills with potatoes, the elevator is drawn back. As the cleaner/separator feeds the elevator, it too has to be moved back at regular intervals.

Box 2.1. Example calculation to calibrate a spray applicator

- Empty enough bags of a known weight of tubers similar in size to those to be treated, to evenly cover an area of the roller table.
- Select a point forward of the potatoes and then operate the table to measure the time taken for the potatoes to pass that given point.
- The 'spot rate' in t/h is calculated as:

$$\frac{\text{Number of bags} \times \text{weight per bag (kg)}}{\text{Times (s)}} \times \frac{3600}{1000} = \text{t/h}.$$

For example, if using four 25-kg bags and it takes 30 s to pass:

$$\text{Spot rate} = \frac{4 \times 25 \times 3600}{30 \times 1000} = 12 \text{ t/h}.$$

- The required flow rate in l/h is calculated as:

Application rate (l/t) × spot rate (t/h).

For example, if the application rate of spray is 2 l/t, at a spot rate of 12 t/h:

Required flow rate = 2 × 12 = 24 l/h.

- The flow rate on the sprayer is set to this value, the sprayer filled with clean water and the sprayer run for a timed period into a calibrated container. This checks that the flow rate is as indicated on the machine. If the actual rate is different from the expected rate, the flow rate should be increased or decreased as required and re-measured until the desired flow is achieved (Ingram and Storey, 1997).

Fig. 2.13. Long elevator loader required to reach the top of the pile. (Courtesy of G. Stroud, Potato Council, Oxford, UK.)

A tractor loader is normally unsuitable for loading a bulk store, since the length of loader frame required to prevent the front wheels crushing potatoes makes the tractor difficult to manoeuvre.

2.7 Transfer from Cleaner/Separator to Boxes

When storage is in boxes, a box filler is placed at the discharge from the cleaner/separator. The cleaner/separator can therefore be in a fixed location. At the start of loading a box, the end of the discharge conveyor is often lowered to minimize the drop on to the box floor (Fig. 2.14). A proximity sensor may be used to automatically raise the conveyor as the box fills. Previous designs of box fillers, which elevated the box to the discharge conveyor, have been discontinued due to the hazard posed to staff as the box is lowered to the ground.

2.8 Into-store Drying Systems

With bulk storage, there is no alternative but to load potatoes directly into store. The underfloor ventilation system present in almost all bulk stores provides a positive flow of air for drying the crop. So long as this is switched on to ventilate the crop immediately on loading, respiration heat removal and drying will start right away. Ventilated trailers (Fig. 2.15) can be used if there is likely to be a delay between harvest and loading into store.

Fig. 2.14. Box filler. (Courtesy of Haith Tickhill Group, Doncaster, South Yorkshire, UK.)

Fig. 2.15. Ventilated trailers. (Courtesy of Benny Jensen, BJ-Agro ApS, Hovborg, Denmark.)

With box storage, the potatoes can be loaded directly into store following harvesting or passed over an into-store cleaning and sizing system, if this is practised. Alternatively they can be pre-dried in batches either by leaving them in a windy location or placed on fan-ventilated drying systems (Ch6.8).

2.8.1 Batch drying using natural wind

The simplest of these systems is to stack boxes under a canopy and take advantage of the wind to ventilate boxes. This is possible only in exposed sites where windless days are rare.

The advantages of this system are:

- Low cost.
- Simple.
- No running costs.
- In good drying weather crops may dry in 2 days.

The disadvantages are:

- No control of crop temperature if nights are cool.
- If frost is forecast it may be necessary to move boxes rapidly indoors.
- The approach of warm weather fronts may result in condensation forming on the newly dried crop.

2.8.2 Batch-drying systems that use fans

Fan-driven ventilation will speed drying in sheltered areas or within a building. If the ventilation system is within a building, control equipment can be added to keep the batch temperature within a range suitable for wound healing and to prevent humid ventilating air from inducing condensation on the crop (Ch8.6). The drying system may be portable or fixed. A number of options are available.

Drying tents

Tents (Wedderspoon Processes Ltd, Forfar, UK) made of plastic fabric, which fit over a stack of 36 1-t boxes, laid out in a pattern of two boxes wide by three deep and stacked six boxes high, provide a portable box drying system (Fig. 2.16). A fan fitted to a frame on the top of the stack of boxes sucks air from the base of the stack, up through the six layers of boxes, a distance of 5.5 m. The fan is sized to cope with the resistance of six 0.8-m-deep layers of potatoes. Some tent users have fitted the tents over a fixed or mobile frame, or suspended them from the purlins of a building, to avoid the risk of staff falling when walking on the top boxes while putting the fan in place. Two alternative sizes of fan are used: a single-phase axial flow fan giving airflow rate of 2.4 m^3/s at 200 Pa or a three-phase axial flow fan giving 3.8 m^3/s at 200 Pa.

The tent will work on all types of boxes regardless of their design, but operators need to check that health and safety regulations for working at height are adhered to.

Letterbox

A letterbox duct is a large box, with holes in it to match the pallet apertures of boxes placed against it. It can be wall-mounted or made to be portable so that it can be moved by forklift to where it is required. For small batch-drying units, the fan or fans are usually mounted on the top of the duct (Fig. 2.17) and either blow or suck air through the boxes. To minimize the leakage of air from gaps between boxes,

Fig. 2.16. Tent moved from stack to stack to ventilate and dry newly stored crop. (Deveron Potato Growers Ltd, Boyndie, Aberdeenshire, UK.)

Fig. 2.17. Letterbox duct drying/ventilation system, for a 4×4 stack of boxes. (Courtesy of Tolsma Techniek, Emmeloord, The Netherlands.)

boxes preferably should be designed to suit letterbox ventilation systems (Ch6.8). While potatoes in traditionally designed boxes will still dry on this system, it will be slower than with purpose-designed boxes. The design of duct is in all other ways similar to that for a letterbox system, which caters for the whole stored crop.

Tapered fabric ducts

An alternative version of the letterbox duct is the Posi-vent™ (Clark & Sutherland Ltd, Aberdeen, UK) system of a fan and inflated tapered duct (Fig. 2.18). This hangs on the side of a stack of 40 boxes, four wide by five deep, two layers of boxes high. For four or six layers of boxes, respectively two or three sets of fans and ducts are required. The fan is an axial flow fan, which produces an airflow rate of $3.2\,\mathrm{m^3/s}$ at 120 Pa. The tapered duct helps distribute the air uniformly along the length of the stack. To install the system, the stack of boxes is built first and the fans are then hung on the end boxes by chains and hooks. The canvas ducts are clipped to the pallet apertures of the boxes by spring steel rods. As the ducts inflate they seal against the boxes.

Suction wall

A suction wall system can be made with two rows of boxes placed with a space between them (Fig. 2.19). The top and ends of the space are then sealed with canvas, and reinforced with timber or plastic battens, to form a plenum. If a fan is now

Fig. 2.18. Posi-vent™ duct drying/ventilation system. (Courtesy of A. Norrie, Slackadale, Aberdeenshire, UK.)

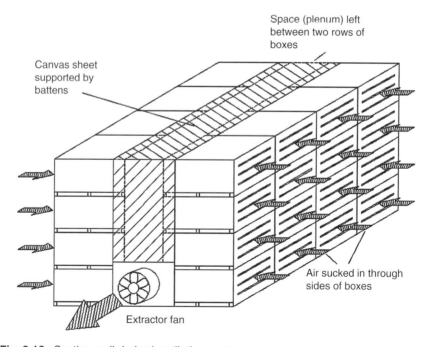

Fig. 2.19. Suction wall drying/ventilation system.

installed to extract air out from the plenum, the canvas is drawn inwards to seal itself against the boxes. Air is sucked horizontally into the sides of the rows of boxes, into the plenum, through the fan and back out into the air surrounding the boxes.

For best results the boxes should be designed to allow ventilation to flow horizontally through the potatoes in the boxes, rather than through the gaps between layers of boxes. A minimum airflow rate of $0.04\,\mathrm{m^3/s/t}$ is required to get the canvas to seal against the boxes.

Time taken to dry potatoes using these systems

The time taken to dry potatoes depends on how wet the ground is when they are harvested, whether condensation has already started to form on the tubers before ventilation starts, the airflow rate used to dry them, the proportion of air leaking between boxes and the RH of the ventilating air. In a sample of potatoes lifted in good conditions in late September in Scotland, ventilated at a rate of $0.037\,\mathrm{m^3/s/t}$, with $0.025\,\mathrm{m^3/s/t}$ actually flowing though the potatoes themselves, it took just over 5 days to dry all of the potatoes in the batch to achieve a skin resistance reading of over $1000\,\mathrm{k\Omega}$, where any soil present is in the form of dust (Pringle *et al.*, 1997). Ambient temperature varied between 9 and 18°C, while ambient RH averaged 76% and varied between 50 and 100%. Most growers using batch-drying systems do not dry their crops to this extent, ventilating their crops instead for a minimum of 2 days before transferring them to store.

2.9   Store Loading Equipment

2.9.1   Bulk

Bulk loading almost always uses elevator belts for the reasons given above. Management of the loading process is discussed in Ch9.5.

2.9.2   Boxes

Boxes are loaded into store using a forklift truck, usually fuelled by diesel or LPG, with solid tyres made from an aerated rubber. Forklifts with a 3.0–3.5-t lifting capacity are normally used for lifting 1-t boxes, in order to keep the forklift mast low enough to pass though doors. The last two boxes are often put up together (Fig. 2.20). This results in the load being 2.2 t in total as the boxes weigh approximately 100 kg each. Forklifts of this size are rated at what they can lift 500 mm from the fork uprights. Since UK boxes are usually 1830 mm long and 1220 mm wide, their centre of gravity is 610 mm forward of the fork uprights. The safe lifting capacity is reduced as a result and is further reduced the higher the last two boxes need to be raised.

Forklifts should be designed to minimize the vibration felt by the driver, in line with requirements in the UK under the Control of Vibration at Work Regulations (HSE, 2005a). Forklift exhausts should discharge vertically upwards, not down to the floor, as the jet stream from the downward discharge exhaust causes dust on the floor to become airborne.

Fig. 2.20. Forklift putting up two boxes together. (Slackadale, Aberdeenshire, UK.)

Forklift trucks fuelled by LPG rather than diesel give a cleaner exhaust but the fuel is more expensive, with spare bottles having to be stored in a secure compound. Electric trucks produce no exhaust, but usually only have the capacity to run for 8 h and therefore an extra set of batteries, costing up to £5000, is required to cope with extended workdays at harvest and grading.

Forklift trucks designed for uses in fields, fitted with agricultural tyres, are difficult to manoeuvre in the tight confines of a potato store and they also need a great deal of room to turn if boxes are to be stacked either side of a passageway. Use of these types of forklifts in store is therefore not recommended.

2.9.3 Sacks

Potatoes are usually only stored in jute or polypropylene sacks in countries with low labour costs (e.g. India). Sacks are normally loaded into position manually. If labour costs rise, the use of a sack elevator to reduce some of the burden of carrying sacks up multiple numbers of floors may be justified.

2.10   Cleaning and Separating Seed Following Initial Storage and Drying

Seed growers who harvest directly into boxes sometimes clean and size a portion or all of their crop in late autumn, after the potatoes have been in store for 3–6 weeks and are thoroughly dry. This requires boxes to be tipped on to a belt, passed over a cleaner and sizer, and returned to their boxes. Grading is therefore done in two stages, one in the late autumn and one prior to dispatch.

The benefits from this are:

- Attached soil has dried by then, so readily falls off the tubers.
- Grading can be carried out in a relaxed manner without the pressure of getting the crop into store.
- Low speed of throughput over the cleaner/sizer minimizes damage.
- Potatoes can be split into different size grades so that the quantities are known and are available for rapid dispatch.
- Grading time required in the spring pre-planting rush is reduced.

The disadvantages are:

- In order to minimize handling damage, the store is not normally cooled below 8°C until after cleaning and sizing is complete.
- Potential risk from a disease like gangrene, as wounds inflicted during cleaning and sizing at low temperatures can be slow to heal.

If cleaning and separation in this manner is practised, it should be done with a minimum of damage and after grading boxes should be well ventilated for 2–3 days to remove respiration heating and to dry any condensed moisture.

2.11 Summary

Harvesting and store loading are the most critical operations undertaken by growers if crops are to achieve their maximum sale price and disease development in store is to be avoided. The main points to be followed are listed below.

- Minimizing damage during harvest is more important than high harvesting rates.
- Regular sampling and inspection of crop leaving the harvester and entering store is vital to avoid large tonnages of potatoes being damaged before faulty equipment, incorrect harvester settings or poor harvesting practices are discovered.
- Potatoes entering store should be free from soil, clods, stones and haulm to allow free movement of ventilating air and the expense of storing waste material.
- Cleaning into store removes most foreign material but can cause damage and loss of bloom.
- Into-store cleaning equipment should be designed to suit the soil conditions of the area.
- Air is by far the best pesticide, to remove respiration heat, to dry the crop, to remove any condensation and maintain the crop dry in store.
- Any pesticide used should be applied to potatoes that are free from soil.

- Any liquid chemical applied to potatoes should be dried off using an air knife or ventilation by wind or fan.
- Whole store drying ventilation systems allow drying after harvest, and at any time thereafter, during the storage season.
- Batch-drying systems are cheaper to install but usually require double handling of boxes if drying is needed mid-storage.
- Seed cleaning 3–6 weeks after the end of harvest can reduce the time taken for out-of-store grading prior to dispatch, but care has to be taken to avoid damage and related disease development.

3 Store Climate

Topics discussed in this chapter:

- How tubers create their own, largely beneficial, microclimate.
- Crop temperature differences caused by tuber respiration.
- How sealed, insulated buildings provide optimum storage conditions.
- The need for within-store air recirculation.
- The use of ventilation with ambient air to cool potatoes.
- How potatoes cool.
- Causes of condensation on the crop and the building structure.
- Climatic zones across the world.
- Storage systems for different climatic zones.
- Stores open to ambient air.
- Assessment of different storage systems.
- Dormancy duration at different storage temperatures.
- Specific requirements for pre-pack, processing and seed storage.

3.1 Microclimate Surrounding Tubers

There are numerous influences that act on the stored crop. These are addressed in the following sections.

3.1.1 Tubers create their own microclimate

A single tuber left on its own in the grading shed will become shrivelled and have a soft, rubbery texture after a month. Had the same tuber been kept in the centre of a 1-t box, it would still have been firm. This is because potato tubers within a box, sack or pile create their own, largely beneficial, microclimate.

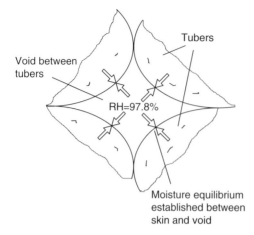

Void between
tubers

Tubers

RH=97.8%

Moisture equilibrium
established between
skin and void

Fig. 3.1. If air in the voids is at 97.8%
relative humidity (RH), moisture neither
leaves nor enters the tubers. This assumes
that metabolic heat is removed as soon as
it is produced.

If an airtight container is filled with potatoes and its lid closed, moisture will evaporate from the tubers and humidify the air in the voids between the tubers (Fig. 3.1). As this air approaches 97.8% RH* (Hunter, 1985), evaporation from the tubers will slow to almost zero and little weight loss will occur thereafter. A state of equilibrium is reached between the humid air surrounding the tubers and the tubers themselves. Storage in a sealed building with just enough air bled in to keep carbon dioxide and oxygen levels near ambient will minimize weight loss through evaporation and maintain skin appearance. Were it not necessary to remove respiration heat, supply oxygen and cool the crop after harvest, completely sealed storage would be the optimal storage system.

3.1.2   The effect of tuber respiration

Potato tubers are living organisms which consume a proportion of their food reserves to stay alive (Ch1.5). The respiration process results in oxygen being absorbed into the cells of the tubers, which is used in the oxidation of sugars to provide the substrates and energy required for cell maintenance. Two of the by-products of this oxidation process, carbon dioxide and heat energy, are released. A third by-product, water, remains within the cells (Fig. 1.15).

The respiration heat given off by the tubers warms the void space air, reducing its RH to below the optimum of 97.8%. Unless this heat is continuously removed, a steady-state RH of 97.8% can never be achieved, so some evaporation of moisture from the tubers will always take place. The heat generated by respiration is a maximum at harvest and varies with storage temperature, being lowest at 5°C and higher above and below this figure.

To avoid store carbon dioxide rising to a level that can affect the fry colour of the stored processing crop or cause blackening of the flesh, usually referred to as blackheart (Fig. 3.2), a very small amount of ambient air needs to be bled into store. If crops are respiring highly, as with early harvested crops, excessive carbon

---

*Other researchers (Cook and Papendick, 1978) suggest that the equilibrium RH is 99.3%, so
this figure may not be precise.

Fig. 3.2. Tuber with internal blackening due to elevated carbon dioxide level in voids surrounding the tuber. (Courtesy of Potato Council, Oxford, UK.)

dioxide levels can occur in a closed store within 24 h. In most stores, unless designed to be completely gas-tight, small gaps in doors and the building fabric provide sufficient air leakage to prevent problems with fry colour. Where stores are completely sealed, a small fan or control on the ventilation system should allow sufficient air to enter the store to maintain acceptable oxygen and carbon dioxide levels.

3.1.3 Recirculation of air within the sealed store

In an unventilated bulk pile of potatoes, the air movement within the pile is often likened to that occurring in a 'fire', where the heat produced in the respiration process causes air to rise up through the pile, drawing in air surrounding the pile as it does so (Fig. 3.3). While this natural convection current is much lower than that which would occur in a real fire where temperature differences would be so much higher, it is impossible to stop. The warm up-current of air evaporates moisture from tubers as it rises, until it meets the surface potatoes, which are often colder than the potatoes below. This results in condensation forming just under the surface of the top layers of potatoes, termed subsurface condensation (Box 3.1).

If a ventilation system is installed that allows air to be circulated though the crop, the respiration heat can be removed as it is being produced, thereby preventing this subsurface condensation, as well as minimizing temperature differences within the crop. When cooling is required, louvres can be opened to allow cool ambient air to be blown through the crop. Alternatively, a refrigeration system can be used to cool the air within the store, so it can be circulated to cool the crop.

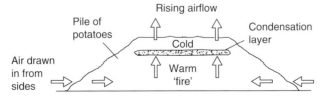

Fig. 3.3. In an unventilated pile, natural ventilation often results in subsurface condensation.

Box 3.1. Do potatoes sweat?

'Sweat' is the term normally used when humans exude moisture, when hot, to cool themselves. It is also used to describe the appearance of moisture on building sur-faces, but the former use is the more common.

As potatoes neither exude moisture (i.e. force moisture out through the skin) nor try to cool themselves by doing so, the term simply confuses the science and should therefore not be used.

If moisture does appear on potatoes that were previously dry, it is caused by moisture in the air condensing on potatoes whose skin temperature is below that of the dew-point temperature of the surrounding air (Ch3.5).

Where the cool surface of the pile is caused by heat loss through radiation to a cold headspace above, timely heating of the headspace air with electric heaters will prevent subsurface condensation (Ch3.5).

3.1.4 Response of sealed store to changes in ambient air temperature

While it might be thought that the atmosphere in a completely sealed store would be immune to changes in ambient air temperature, this is not the case. Just as the sealed store prevents air entry or escape, it also prevents vapour within the air from passing through the store fabric. In contrast, heat can enter or leave the store, with the amount determined by the insulation value of its fabric.

The RH in the headspace of intermittently ventilated, well-sealed potato stores varies between 90 and 98% (Pringle *et al.*, 1997), averaging about 94%.

On a warm sunny day, heat will enter the store and cause its air temperature to rise and its RH to fall. The RH in store will therefore fall to below 94% (Box 3.2).

Box 3.2. Temperature/relative humidity seesaw

For a mass of air enclosed in a sealed box or empty building, the relationship between temperature and relative humidity (RH) is like a child's seesaw (Fig. B.3.1). When the temperature of the air is raised as at a), due to heat entering the box, the air's RH will fall. If the temperature of the air falls as at b), due to heat leav-ing the box, the RH of the air will rise. This happens because warm air can hold more moisture than cold air.

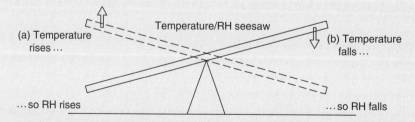

Fig. B.3.1. The temperature/relative humidity (RH) seesaw for an enclosed volume of air.

On a cold night, heat will escape from the store, causing the store air to reduce in temperature and its RH to rise. The RH in store will rise above 94%.

Changes in ambient air temperature will therefore result in oscillations in internal store temperature and RH; a rapid drop in outside temperature can result in condensation forming on the roof and on the coldest parts of the crop (Ch3.5).

3.1.5   Voidage within a crop

The voidage, or air spaces between the potatoes, allows free air movement through the crop. If these voids become blocked for any reason, air will not pass, so drying or cooling will either not take place or be slowed. Voids can become blocked due to:

- Large quantities of wet soil coming into store with the potatoes.
- Soil falling off potatoes being loaded into a pile and forming a soil cone.
- Small tubers separating out from larger ones as the potatoes fall from the elevator loading the pile.
- Haulm or weeds present in the crop.
- Rotten mother tubers or rots collapsing under the weight of neighbouring tubers, resulting in an impenetrable cluster of potatoes.

The smaller the tubers, the smaller the voids will be and the more likely they will become totally blocked. Seed potatoes, often grown on soils in cool, windy areas that regularly experience wet harvesting conditions, are most likely to encounter this problem. Blockage of voids can severely affect natural convection of air through crops and result in tubers staying moist for weeks, months or the entire storage period.

3.1.6   Effect of wet potatoes on store atmosphere

During very wet harvests, potatoes are sometimes harvested when there is free water in the soil. Such potatoes should be dried before loading them on to a bulk pile. If harvested into boxes, potatoes can be blown with a high volume of air to dry them out. They should not be put into a store un-dried. Damp potatoes and soil act in the same way as an air humidifier. Air moving through the damp material will approach a RH of 100%, so that store ceilings and potatoes may drip with moisture. These are ideal conditions for disease to flourish. The removal of this moisture will reduce the RH in store below 100% and prevent condensation forming on cold parts of the crop or roof. This will transform a store from one that drips water to one containing potatoes with dry skins and a headspace RH of about 94%.

3.1.7   Ventilation after harvest

If the potatoes are ventilated immediately after harvest, with air passing between each and every tuber, the respiration heat that causes subsurface condensation can be removed and the damp crop and soil dried rapidly. The respiration heat will warm the air slightly, increasing its water-carrying capacity and therefore its drying potential. Ventilation immediately after harvest will not only prevent subsurface condensation occurring but will dry any moisture present on skins and in the adhering soil.

Airflow sufficient to remove the high rate of respiration heat being produced is required for 4–5 days after harvest until the respiration heat produced has declined to near dormant levels (Fig. 1.16). Further ventilation is likely to be required to completely dry the crop, with either reduced rate, or intermittent, ventilation required to remove temperature differentials caused by respiration.

Thereafter ambient-air ventilation or refrigeration will be required to cool the potatoes to the desired long-term crop storage temperature.

3.1.8   Ventilation of crops with voids completely filled with soil

Since the soil pores of wet soil are filled with water, if air is forced through any remaining free voids, the soil pores will gradually dry out. If the potatoes are then put over a cleaner or grader, the soil will normally shatter and fall off. The clean potatoes will then have clear voids through which air can pass. Ventilation will remove any remaining surface moisture. This approach may not work in potatoes covered in soils with high clay content, so such soils should be avoided for potato production.

3.1.9   Temperature

Warm temperatures at harvest speed wound healing. Warm temperatures also favour the development of disease. Since the skin is the main defence against disease for potatoes, rapid wound healing is usually a priority. Once wounds are healed, cooling can start.

Warm storage temperatures shorten the dormancy period of potatoes and result in early sprouting. Any disease present will tend to multiply more rapidly at high temperature. Low temperatures of 3–4°C prolong dormancy and slow disease development, but can lead to the conversion of cellular starch to sugar (Ch1.5), which may result in dark fry colour of crisps or French fries. For long-term storage, potatoes are normally cooled using cool ambient or refrigerated air. The use of air for cooling always results in moisture loss from tubers (Ch.1.5). The need for cooling ventilation is minimized by ensuring that the store is well insulated to minimize the amount of heat entering the store, is well sealed to prevent warm ambient-air ingress and the control system ventilates the crop only when cooling is required.

3.1.10   Relative humidity of air surrounding tubers

As stated earlier, if no ventilation is taking place, the crop determines the RH of the air within the voids between the potatoes. With ventilation stopped, respiration heating will start warming the mass of potatoes and natural convection will occur. Other than by preventing leakage into store, the store manager has no control of RH within the box, bag or pile. Where a continuous, low-rate, high-humidity ventilation system is in use (Ch3.7), this maintains a high RH within the crop, which minimizes weight loss and constantly removes the heat of respiration by introducing cool ambient air or use of refrigeration if cool air is unavailable.

3.1.11   Optimal microclimate

Potatoes, kept together in a pile, box, bag or sack, create their own largely beneficial climate. Sealed stores keep store air RH high, minimize tuber weight loss and prevent the air from warm weather fronts leaking into store and condensing on the cool crop within. A small amount of ambient air is required to flush the store to keep levels of carbon dioxide low. Airflow through the crop, especially after harvest, is needed to remove respiration heat to prevent subsurface condensation. Air recirculation is required in intermittently ventilated systems to keep temperature differences in the crop to a minimum when no cooling from ambient air is taking place. For long-term storage and disease control, crops should be kept cool, but unless the air is humidified, ventilation and recirculation should be kept to a minimum to avoid excessive moisture loss from the crop.

3.2   Benefit from Sealed, Insulated Buildings

3.2.1   Sealing stores to minimize weight loss

Sealed crop stores minimize crop moisture loss by allowing the RH in the voids between tubers to approach 97.8%, at which point vapour in the air is at the same partial pressure as that within the skin of the potatoes. Air leaking into store, if at the same or lower temperature than the crop, will usually have an RH lower than this value (Fig. 3.4), so that the more gaps in the building fabric, the lower the store air RH will be and the greater the weight loss. Buildings that are exposed to wind will be more prone to such leakage than those on more sheltered sites. If ambient temperatures are higher than the crop temperature, air leakage may result in condensation (Ch3.5).

Cooling of potatoes requires ventilation to remove the heat. Cooling is achieved partly through evaporative cooling and partly through convective heat

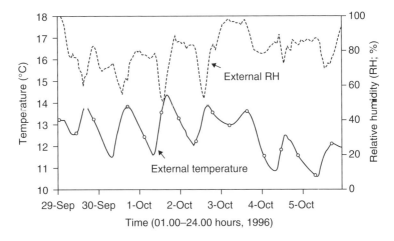

Fig. 3.4. Typical weather data for northern Britain.

transfer (Hylmö *et al.*, 1975a), the ratio of the two forms of heat transfer being approximately 55:45. As the cooling air warms as it passes through the warmer crop, its RH falls, causing it to evaporate moisture from the potatoes. Cooling therefore dehydrates the crop. The more cooling is required, the greater the moisture loss. Poorly sealed stores, in comparison to well-sealed stores, will require cooling systems to run for longer and will experience greater weight loss as a result.

3.2.2   Insulation to minimize weight loss

Insulation is vital in countries where ambient temperatures fall below freezing. Potatoes freeze when their temperature falls below between -1.0 and $-2.2°C$ (Wright and Diehl, 1927; Wright, 1942). Where ambient temperatures exceed the desired storage temperature, insulation slows the amount of heat conducted through the fabric of the building into the store. The lower the conductivity (*U*-value) of the walls or roof, the less heat will enter the store, and the less time the cooling system will need to run to remove this heat gain. This will both save on fan or refrigeration energy use and reduce the moisture evaporation loss associated with cooling. Well-insulated stores will therefore reduce moisture loss.

Insulation also reduces the likelihood of temperature differentials forming within the stored crop, which can have undesirable consequences. Potatoes stored in the cooler areas of the store can produce a dark colour when fried. This is particularly noticeable in bulk stores if their load-bearing walls are poorly insulated (Burton *et al.*, 1955). Condensation on the crop is also likely in these areas. The uprising convection current of warm air from the main mass of potatoes draws air downwards where the potatoes are cooler, which is by the walls when outside temperatures are low (Fig. 3.5). Since the downward-moving air has been warmed

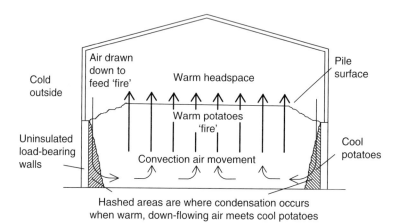

Fig. 3.5. Condensation may occur on cool potatoes near the sides of a pile.

and humidified by the potatoes, moisture in the air may condense on the cooler potatoes near the walls. This is sometimes visible when unloading is taking place.

3.2.3 Sealing to minimize condensation on the stored crop

If the stored crop temperature is below the dew-point temperature of the ambient air, a proportion of the vapour in the air leaking into the building may condense on the crop, leading to possible disease initiation and multiplication on the moist potatoes (Boxes 3.3 and 3.4).

3.2.4 Sealing to exclude vermin

With farm assurance schemes becoming increasingly important in the UK, keeping stored crops free from rats, mice and birds is increasingly important. This is most easily done if stores are sealed to exclude vermin. Bait is usually left in corners within the store to kill any vermin that do enter the store, but should be placed so that it cannot be touched by domestic animals or children.

3.3 Purpose of Ventilation

Ventilation is the term used to move air through the potato crop. It has two components:

- Recirculation ventilation, where the same air is circulated round the store and possibly through the potatoes themselves.
- Ambient-air ventilation, where outside air is blown or sucked into store, displacing the air that was previously there.

Recirculation ventilation is used with refrigeration systems, which cool the air within the store using a heat exchanger but do not replace the store air with fresh ambient air (Fig. 3.6a).

Ambient-air cooling/ventilation, usually coupled with recirculation, replaces the warm air within the store with cooler ambient air in order to cool the crop (Fig. 3.6b).

3.3.1 Removal of respiration heat

Harvest a dry crop of potatoes from dry soil, especially the earliest lifted crops which respire the most, and look at the top of the pile or box. If you do not recirculate air through the newly harvested potatoes within an hour or so after lifting, you will start to see condensation forming just under the surface of the top layer of tubers.

If the pile or box is ventilated immediately the crop is harvested, condensation due to respiration will not form. Ventilation to keep the crop free from this condensed moisture will minimize the chance of skin blemish disease developing or tubers starting to rot.

Box 3.3. Condensation formation

To understand how condensation occurs, imagine three 1-m cubes of air at a temperature of 4, 10 and 20°C, respectively (Fig. B.3.2). These can hold a maximum of 6.4, 9.5 and 17.5 g of water, in the form of vapour, respectively. These values are called their saturation moisture content or saturation humidity. The amount of water air can hold increases as the temperature of the air increases.

   If air at 10°C has less than 9.5 g of water, e.g. 8.5 g, in a cubic metre, the relative humidity (RH) of the air is calculated as the proportion this is of the saturation value:

   RH = 8.5 / 9.5 × 100 = 90%.

If the RH is 100%, this is equivalent to saturation; that is, the air can hold no more water at that temperature.

Warming the air
In some drying applications, air is warmed to increase its water-holding capacity, so that the rate of drying is increased. If the 10°C cube of air at 90% RH is electrically heated to 20°C, the amount of water in the air stays the same, but the air's water-holding capacity increases. The RH can again be calculated:

   RH = 8.5 / 17.5 × 100 = 49%.

Cooling the air
If the 10°C air at 90% RH is cooled to 4°C, the air at 4°C is capable of holding only a maximum of 6.4 g, so the balance of 8.5 − 6.4 = 2.1 g falls out of the air. In meteorology, where a warm front and cool air mass meet in the sky, this water falls as rain. In a potato store, the water the cold air cannot hold condenses on any surface which is below the dew-point temperature of the air.

   In this example, the temperature at which the cooled air reaches saturation is 8.4°C (CIBSE, 2006) and this temperature is called the dew-point temperature.

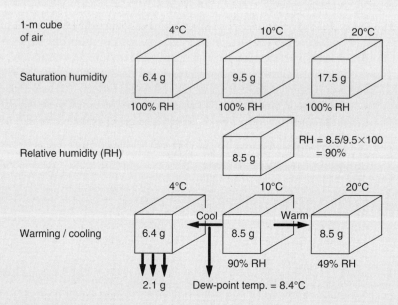

Fig. B.3.2. Effect of warming and cooling on air humidity.

Box 3.4. Psychrometric chart

The psychrometric chart (Fig. B.3.3) is a chart that allows the condition of air to be defined if any two properties out of a possible four are known. These are the dry-bulb temperature (°C), wet-bulb temperature* (°C), relative humidity (RH, %) and moisture content (kg moisture/kg) of the air. The chart can be used further to predict what will happen when the air is heated or cooled. Since moisture content is given on the vertical *y*-axis of the chart, if no moisture is added to or removed from the air, movement from one point to another on the chart will always be parallel to the horizontal *x*-axis.

It is instructive to plot the example used in Box 3.3 on the chart. In that example, moisture content was quoted in grams per cubic metre of air, whereas on the chart it is quoted in kilograms per kilogram of dry air. To convert from one to the other multiply or divide by the density of air at 20°C, which is ~1.23 kg/m³.

The starting point A is where the vertical 10°C dry-bulb temperature line and the curved 90% RH lines cross. If the air is heated from A using an electric heater (which does not change the moisture content of the air), the horizontal line AB on the chart shows this heating. Point B is where this line meets the vertical 20°C dry-bulb temperature line. At this point the new RH of the air will be found to be 47%.

If the air at A is cooled, the horizontal line AC shows the change in its condition until it meets the moisture saturation line at C. The temperature at C is the dew-point temperature 8.7°C, which is read off the dry-bulb temperature scale. On cooling the air further from C, the condition of the air moves down the saturation line until it reaches D at 4°C, losing moisture as it does so. The amount of moisture lost can be found by subtracting the moisture content at D from that at C, i.e. 0.002 kg/kg, on the vertical *y*-axis.

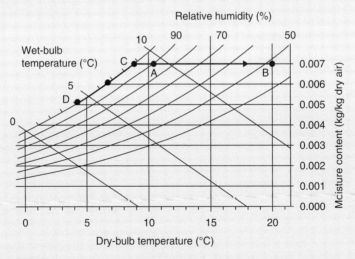

Fig. B.3.3. Psychrometric chart. (Section of chart redrawn with permission of the Chartered Institution of Building Services Engineers, www.cibse.org.)

---

*Temperature reading of a thermometer wrapped in a wet wick and rotated rapidly in the air to be measured to maximize the rate of evaporative cooling (see Box 3.5).

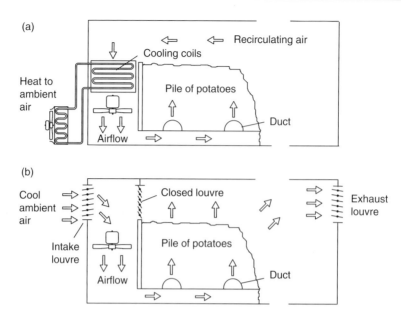

Fig. 3.6. Cooling of potatoes using: (a) recirculation ventilation with a heat exchanger (fridge) to remove heat from circulating air; and (b) ambient-air ventilation to remove heat by replacing the air within the building with cooler air from outside.

3.3.2 Drying of crop entering store

Potatoes are best harvested from soils that are slightly damp. Such soil will stick together and form a cushion on the primary web of the harvester, protecting the tubers from damage from the oscillating web bars.

The potatoes entering store will have soil still adhering to their skins. A third of the weight of the soil may be water. After significant amounts of rainfall, the proportion of soil moisture may be much higher. Ventilation immediately after harvest is therefore used not only to remove respiration heat, but also to dry out the adhering soil and remove any skin surface moisture.

Drying is usually achieved by blowing ambient air through the crop. This provides the lowest-cost method of removing the moisture. Drying may take between 3 and 20 days to achieve (Pringle *et al.*, 1997), depending on the rate of air passing between the tubers and the moisture-carrying capacity of the air.

Alternatively, drying can be achieved in a closed store using the evaporator of the refrigeration system to remove the moisture from the circulating air. This uses more electricity than ambient-air drying and will prematurely cool the crop, unless heat is added to maintain the crop temperature (Box 8.1). The addition of heat adds cost and produces carbon dioxide, the greenhouse gas. If the fridge is used for drying and is allowed to cool the crop as well, this may result in condensation occurring on the cooler potatoes when warmer, newly harvested potatoes are loaded into store (Ch3.5). During drying of the crop, therefore, every effort should be made to ensure that the crop already in store and the potatoes being loaded are at the same temperature.

3.3.3 Cooling the crop

Ventilation is used to cool the crop, either by bringing in cooler air from outside or by recirculating store air through the cooling coils of a refrigeration unit.

3.3.4 Minimizing crop temperature differentials

Temperature differences in the crop can occur for a number or reasons. They can result from:

- Stocks of potatoes entering store at different temperatures.
- Uneven cooling of boxes or piles.
- Lack of insulation in store walls or roof.
- Low headspace temperatures on cold nights due to poor insulation or air leakage.
- Surface cooling of potatoes by air distribution jets.
- Different respiration rates for crops of different varieties or maturities.
- Additional respiration heat output from microorganisms once disease has become established.

When fans are recirculating the air within the store, this forced air movement tends to dominate over any movement of air due to natural convection. Once fan-induced ventilation stops, natural convection currents will re-establish. Natural convection will take place between warm and cool parts of the crop. The greater the temperature difference between sections of the crop, the faster will the convection air movement become. When warm humid air from one part of the crop flows over cooler potatoes in another section of the crop, condensation on the cooler potatoes may result. To prevent disease development associated with condensation on the crop, these convection airflows should be minimized, and this is achieved by recirculating air to minimize temperature differentials within the crop.

In intermittent 'start–stop' ventilation systems, movement of air through the crop to minimize temperature differences will result in some crop moisture loss. The period of recirculation should therefore be just sufficient to minimize temperature differences to prevent condensation but no more.

3.3.5 Maintaining a low carbon dioxide level

In very well-sealed stores, carbon dioxide levels from crop respiration can rise to levels which can result in internal blackening, where the centre of the potato is asphyxiated. This is most likely to occur in processing stores held at warm temperatures, as crop respiration increases with temperature above 5°C (Ch1.5), or where newly harvested crops are rapidly loaded into store and the store immediately shut. In most stores, leakage between door and store, and gaps in the fabric, are usually sufficient to keep carbon dioxide levels below the 0.5% acceptable value. Where there is risk of carbon dioxide exceeding this value, a small fan can be fitted into the store fabric to give a supply of ambient air and so ensure carbon dioxide levels are kept to low. Carbon dioxide monitoring for well-sealed stores should be part of the monitoring equipment.

3.4   The Cooling Process

3.4.1   Natural convection

A pile of potatoes left to ventilate by natural convection alone increased 1.75°C in temperature for every metre of height (Burton *et al.*, 1955). This increase in temperature should have resulted in a base-to-surface temperature difference for a 3.7-m-high pile of potatoes of 6.5°C (Fig. 3.7). In practice the difference was only 2.8°C, as cool headspace air had penetrated the surface potatoes, removing the metabolic heat and reducing their temperature. The surface cooling caused the top potatoes to be 3.7°C cooler than they would have been had such cooling not taken place. The calculated rate of natural ventilation was 0.00064 m³/s/t (Hylmö *et al.*, 1975b), based on a measured respiration heat output of 12.8 W/t.

Temperature differentials of this magnitude are unacceptable in practice, when control of fry colour requires crop temperatures to be within a 1.0°C range. The cooling of the surface layer is also undesirable as it increases the likelihood of sub-surface condensation.

The increase in temperature up the pile was so large because the airflow rate was so low. A large amount of heat was warming a relatively small amount of air.

If the rate of airflow is increased by the use of fans, respiration heat output stays the same but the volume of ventilation air to remove the heat is increased. This reduces the temperature differential up the pile. Rates of airflow of 0.0035 and 0.0070 m³/s/t will reduce the temperature difference from 6.5°C to 1.3°C and 0.7°C, respectively (Fig. 3.7). Not only will this reduce the range of temperature differences within the pile, but it also prevents cool headspace air penetrating the

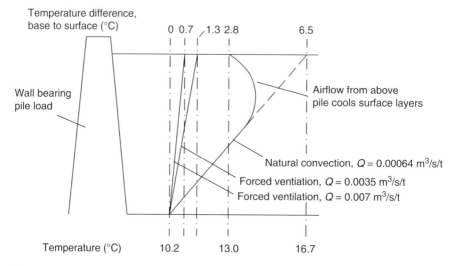

Fig. 3.7. Potato pile 3.7 m high showing how crop temperature increases with height under natural convection and two rates of forced ventilation.

surface of the pile and cooling the potatoes. The risk of subsurface condensation is therefore minimized.

If the fan-ventilated air is humidified to as near saturation as possible, the air within the voids will be kept close to 97.8% RH and so weight loss due to ventilation is minimized.

Constant ventilation with low rates of humidified air greatly reduces the temperature gradient within the pile, allowing the crop to be kept close to the desired set-point temperature. The humidified air delivered to the base of the pile minimizes weight loss in the base of the pile. As the air moves upwards, the respiration heat from the potatoes raises its temperature, reducing its RH fractionally as it does so. The rising air is therefore constantly drying the crop at every level up the pile, ensuring that condensation cannot occur and removing any skin moisture that may be present.

3.4.2 Cooling front

A cooling front in a potato pile is formed when cool air is blown upwards through warmer potatoes (Fig. 3.8). The potatoes at the base of the front are cooled to a temperature that is between the dry-bulb temperature of the cooling air and its wet-bulb temperature (Box 3.5), while those above the front remain at the same temperature as when cooling began. As cooling continues, this cooling front slowly moves up the pile of potatoes.

When ambient air is used intermittently for cooling a bulk pile or a box of potatoes, two cooling fronts form, the 'gradient cooling front' and the 'horizontal cooling front' (Hylmö *et al.*, 1975b).

The gradient cooling front
Prior to forced ventilation restarting, the temperature gradient in the pile will have been moving from the near-vertical gradient when the fans were on last, to the

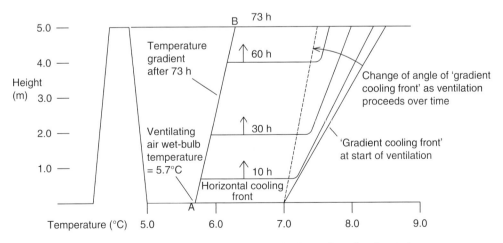

Fig. 3.8. An example of the progress of vertical and horizontal cooling fronts in an intermittently ventilated pile of potatoes.

Box 3.5. Cooling by evaporation of a wet-bulb thermometer

If unsaturated ventilating air is blown over a wet wick covering a thermometer (Fig. B.3.4), the thermometer will be cooled by evaporation to the wet-bulb temperature of the air. Since potatoes are about 80% moisture, they lose moisture in the same way as the wet wick, and are cooled by evaporation in a similar manner. Because the skins of potatoes prevent the release of moisture to the same extent as the wet wick, potatoes cool to somewhere between the air's dry-bulb temperature and its wet-bulb temperature.

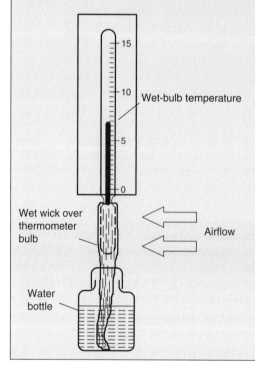

Wet-bulb temperature

Wet wick over thermometer bulb

Airflow

Water bottle

Fig. B.3.4. Thermometer and wet wick for measuring wet-bulb temperature.

---

inclined gradient that occurs when natural ventilation alone is taking place (Fig. 3.8). When the fans are switched on to cool the crop, the temperature gradient starts to move back to the near-vertical gradient associated with a high ventilation rate (Fig. 3.8). This cooling occurs simultaneously in all layers up the pile and serves not only to cool the crop, but also to reduce the base to top temperature differential.

The slope of the 'gradient cooling front' is determined by:

- The time since forced ventilation last occurred.
- The respiration heat output.
- The rate of ventilation.

The horizontal cooling front

Immediately ventilation cooling starts, not only does the angle of the gradient cooling front start changing to a more vertical orientation, but a horizontal cooling front starts to rise vertically up through the pile. This causes a temperature difference within the layer being cooled, which is illustrated in graph form in Fig. 3.8. In Hylmö's theoretical example, where the cooling air is a constant temperature and RH, the crop was at 7°C when cooling started, while the ventilating air was set at 6°C, 95% RH. As ventilation proceeds, the horizontal cooling front rises up the pile, while the angle of the gradient front becomes more vertical. After 73 h the horizontal cooling front reaches the surface of the pile and the gradient front lies along the line AB. At the end of cooling, therefore, there is still a temperature difference B – A°C between the base and surface of the pile, due to the respiration heat coming from the crop.

Height of horizontal cooling front
Cooling creates three zones (Fig. 3.9), in ascending order: a cooled region, a cooling front or zone, and a warm region (Lerew and Bakker-Arkema, 1976). The cooling front is a transition zone, with the potatoes at its base near the wet-bulb temperature of the ventilating air and the potatoes at its top near the temperature of the crop above. The rate at which the cooling front rises through the potatoes depends on:

- The airflow rate.
- The amount of evaporation taking place within the potato pile.
- The initial temperature difference between the cooling air and the potatoes.

The depth of the cooling zone depends on the airflow rate; high airflow rates resulting in deep cooling zones, low airflow rates resulting in shallow zones (Fig. 3.9).

As the cooling air is at the same temperature as the potatoes in the cooled region, it passes through this region almost unchanged, with only a small increase in temperature due to the heat being released by tuber respiration.

In the cooling zone, the warm potatoes 'see' a cool flow of air advancing towards them from below. The air, in contrast, 'sees' a warm mass of potatoes

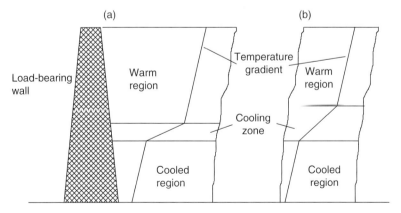

Fig. 3.9. Depth of the cooling front in a pile is related to ventilation rate: (a) shallow cooling zone associated with low airflow rates; (b) deep cooling zone associated with high airflow rates.

ahead of it. The potatoes are cooled in part by evaporation of moisture from the tubers and in part by heat being transferred from the warm potatoes to the cooler air (Hylmö *et al.*, 1975a). Approximately 55% of the cooling is through removal of latent heat from the crop (Box 3.6) and 45% by convection and conduction. The air receives heat from the potatoes and leaves the cooling zone at the temperature of the potatoes at the top of this zone. As the air rises through the warm zone, its temperature rises slightly due to the heat being released by tuber respiration. This increase in temperature reduces as the crop becomes more dormant.

In storage systems where low-rate ventilation with humidified air is continuous, the cooling front is virtually eliminated (Fig. 3.7).

Box 3.6. Change of phase from liquid to gas

When water changes phase, from a liquid to the higher-energy state of a gas, heat energy is required. The heat absorbed in the change of phase is hidden, termed latent, as it does not change the temperature of the water, only its state.

When water at 20°C evaporates to a vapour, it requires 2450 kJ of heat energy per kilogram of water. In contrast, only 4.18 kJ of heat energy is required to heat 1 kg of water by 1°C (Fig. B.3.5).

When cooling potatoes, the cooling airflow evaporates some moisture from the skin. Although the amount is small, as the latent heat of evaporation is so high, evaporation is estimated to account for over half the cooling effect of the air (Hylmö *et al.*, 1975a).

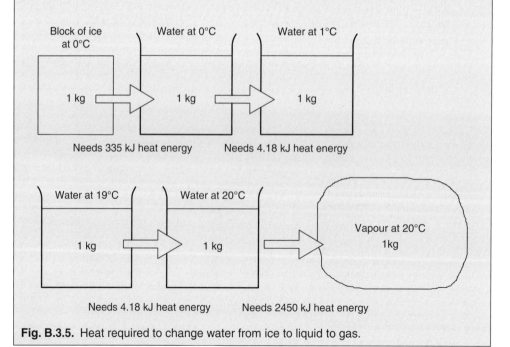

Fig. B.3.5. Heat required to change water from ice to liquid to gas.

3.4.3 Cooling using refrigerated air

Refrigeration systems cool the air within the sealed store using a heat exchanger, more commonly called cooling coils or an evaporator. During fridge operation, all louvres and doors in the store are tightly shut. In the direct expansion (DX) refrigeration systems used in most intermittent, 'stop–start' cooling systems, the fridge's evaporator cools the recirculating air by 2.5°C when the warm crop is first loaded into store, reducing to a 1.5°C temperature reduction when the crop approaches its storage temperature (Fig. 3.10). With DX refrigeration systems, therefore, the cooling air will never be cooler than the crop by more than 2.5°C.

Refrigeration systems used for continuous cooling using low rates of humidified air provide cooling proportional to the temperature reduction required. Since ventilation is continuous, cooling of only a fraction of a degree may be required.

3.4.4 Cooling using ambient air

Unlike cooling with refrigeration, ambient-air cooling uses air from outside the store to cool the crop. The temperature of this cooling air is dependent on the weather. Not only will the temperature difference between crop and ambient air vary from day to day, it may vary over the duration of the cooling period. The temperature difference between crop and incoming air could be as much as 10–15°C, if no limits are put on a minimum acceptable air temperature. This is much greater than the 2.5°C differential in a refrigerated store, and will, if allowed to occur, cause an excessive crop temperature difference (i.e. step in temperature) across the cooling front.

If ventilation is vertically upwards, as in a ventilated pile, warm potatoes will overlie the cooled potatoes so that when natural convection ventilation restarts after a period of cooling, cold air will meet only warmer potatoes, so condensation will not occur. If ventilation is downward, however, as in over-the-top

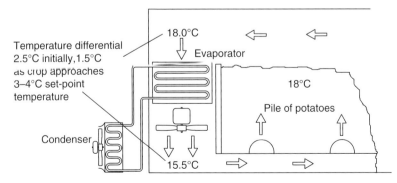

Fig. 3.10. Temperature differential between air entering the evaporator (air-on) and air leaving the evaporator (air-off).

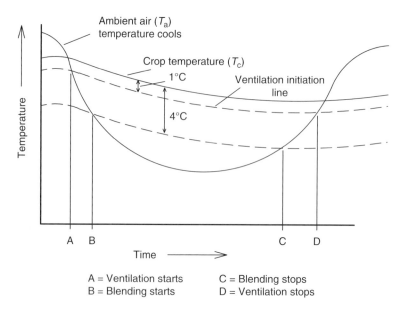

Fig. 3.11. Use of blending to prevent the inlet air temperature becoming too low.

ventilated box stores or in many letterbox ventilation systems, cool potatoes will overlie warm ones. When forced ventilation stops and natural ventilation re-establishes, warm air rising from the potatoes yet to be cooled may result in potential condensation on the cooler potatoes above. For ambient-air cooling, therefore, control software must be more sophisticated than for fridge units, if crop temperature differentials are to be kept within limits to prevent such condensation.

The controller software is usually set to allow cooling if the ambient air temperature is $1.0°C$ below the temperature of the crop (Fig. 3.11). Ventilation with the louvres full open will then take place, unless the ambient air becomes too cold (usually $4°C$ below the crop temperature). At this point a sensor in the supply duct will cause the inlet louvre to close and the recirculation louvre to open, so that the incoming air is never colder than the crop by more than the $4°C$ (Box 3.7). As the ambient air temperature rises again, blending will stop at the $4°C$ temperature differential and ventilation will stop when the potatoes are within $1°C$ of the ambient air temperature.

3.5 Causes of Condensation

3.5.1 Criteria for condensation to occur

Condensation on potatoes can occur if their skin surface temperature is below the dew-point temperature of the surrounding air.

Box 3.7. Maximum allowable crop temperature differentials

Subsurface condensation occurs on potatoes when the temperature of the tubers on the surface is below the dew-point temperature of air rising from warmer potatoes below (Ch3.5). If the RH of the rising air is assumed to be 96% and the crop temperature is 3°C, then the surface potatoes only need to be 0.6°C lower than the temperature of the potatoes below for condensation to occur. Temperature differentials in the top layers of potatoes in boxes or a potato pile should be kept to less than 0.5°C to provide a slight margin of safety.

Table B.3.1. Suggested maximum acceptable crop temperature differentials.

	Temperature differential (°C)			
Period	Between cooling air and crop	Between top and subsurface tubers in box or bulk	Between boxes in the same store	Between top and base of bulk
Store loading and wound healing	4.0	0.5	4.0	1.5
Cool storage period	4.0	0.5	1.0	1.0

Cool overhead ventilating air will reduce the temperature of the top layers of potatoes. In studies carried out (Pringle *et al.*, 1997), the temperature of the surface tubers never approached the temperature of the cooling air. The cooling air had to be very much cooler than the crop for condensation to occur. An estimated figure of 4°C temperature difference between cooling air and crop has therefore been chosen to prevent condensation occurring in the subsurface layer of the crop.

The recommended maximum temperature differences between stacks of boxes at store loading and during cool storage are to minimize convective ventilation, which may result in condensation. The differentials for bulk storage are taken from standard recommendations (Stark and Love, 2003).

This requirement can only arise if the surrounding air is warmer than the crop (Box 3.8).

The RH of the air due to moisture evaporation from the stored crop is so near 100% that even small temperature differences between groups of potatoes can cause condensation to occur.

3.5.2 Subsurface condensation due to a highly respiring crop

When potatoes are dormant, the respiration heat given out is small, less than 10 W/t (Burton, 1989). Just after harvest the heat produced can be many times this value (Ch1.5.1). This heat production causes natural convection to occur. The heat warms the tubers, which in turn heats the air within the voids, raising its temperature. This warm air is more buoyant than the rest of the air within the store and,

Box 3.8. Criteria for condensation to occur

Condensation on potatoes or a building surface will only occur if the temperature of the tuber or surface is below the dew-point temperature of the surrounding air. As the dew-point temperature of air at 100% relative humidity (RH) is the same as its dry-bulb temperature, it follows that the surrounding air must also be warmer than the tuber or surface. Table B.3.2 gives dew-point temperatures of air based on its temperature and RH (CIBSE, 2006).

Example 1
If the air's dry-bulb temperature is 17°C and its RH is 70%, the dew-point temperature is 11.6°C.

Example 2
To determine the risk as to whether ambient air would condense on potatoes in store when the store door is opened:

- Measure dry-bulb temperature of air and its RH, e.g. 14°C and 70% respectively.
- Get dew-point temperature of air from the table, i.e. 8.7°C.
- Check the temperature of the potatoes, e.g. 6.0°C.

Since crop temperature is below the dew-point temperature of the ambient air, condensation will occur on the stored crop if ambient air enters the store.

Table B.3.2. Dew-point temperature of air based on its temperature and relative humidity (RH).

Dry-bulb temperature (°C)	RH (%)													
	60	62	64	66	68	70	72	76	80	84	88	92	96	100
20	12.1	12.6	13.1	13.6	14.0	14.5	14.9	15.7	16.5	17.3	18.0	18.7	19.4	20.0
19	11.2	11.7	12.2	12.6	13.1	13.5	13.9	14.8	15.5	16.3	17.0	17.7	18.4	19.0
18	10.3	10.7	11.2	11.7	12.1	12.5	13.0	13.8	14.6	15.3	16.0	16.7	17.4	18.0
17	9.3	9.8	10.3	10.7	11.2	11.6	12.0	12.8	13.6	14.3	15.0	15.7	16.4	17.0
16	8.4	8.8	9.3	9.7	10.2	10.6	11.0	11.8	12.6	13.3	14.0	14.7	15.4	16.0
15	7.4	7.9	8.3	8.8	9.2	9.7	10.1	10.9	11.6	12.4	13.1	13.7	14.4	15.0
14	6.5	6.9	7.4	7.8	8.3	8.7	9.1	9.9	10.7	11.4	12.1	12.7	13.4	14.0
13	5.5	6.0	6.4	6.9	7.3	7.7	8.1	8.9	9.7	10.4	11.1	11.7	12.4	13.0
12	4.6	5.0	5.5	5.9	6.3	6.8	7.2	8.0	8.7	9.4	10.1	10.8	11.4	12.0
11	3.6	4.1	4.5	5.0	5.4	5.8	6.2	7.0	7.7	8.4	9.1	9.8	10.4	11.0
10	2.7	3.1	3.6	4.0	4.4	4.8	5.2	6.0	6.7	7.5	8.1	8.8	9.4	10.0
9	1.7	2.2	2.6	3.0	3.5	3.9	4.3	5.0	5.8	6.5	7.1	7.8	8.4	9.0
8	0.8	1.2	1.7	2.1	2.5	2.9	3.3	4.1	4.8	5.5	6.2	6.8	7.4	8.0
7	-0.1	0.3	0.7	1.1	1.5	2.0	2.3	3.1	3.8	4.5	5.2	5.8	6.4	7.0
6	-1.0	-0.6	-0.2	0.2	0.6	1.0	1.4	2.1	2.8	3.5	4.2	4.8	5.4	6.0
5	-1.8	-1.4	-1.0	-0.7	-0.3	0.0	0.4	1.2	1.9	2.5	3.2	3.8	4.4	5.0
4	-2.7	-2.3	-1.9	-1.5	-1.2	-0.8	-0.5	0.2	0.9	1.6	2.2	2.8	3.4	4.0
3	-3.5	-3.1	-2.7	-2.4	-2.0	-1.7	-1.3	-0.7	-0.1	0.6	1.2	1.8	2.4	3.0
2	-4.3	-4.0	-3.6	-3.2	-2.9	-2.5	-2.2	-1.5	-0.9	-0.3	0.2	0.8	1.4	2.0
1	-5.2	-4.8	-4.4	-4.1	-3.7	-3.4	-3.0	-2.4	-1.8	-1.2	-0.7	-0.1	0.4	1.0

being warmer, it can hold more moisture. It rises up the crop through the voids, evaporating moisture from potatoes in the process. If the skins of the crop above are colder than those in the layers below, the water in this warm humid air may condense on the cooler subsurface potatoes. This is similar to the situation that normally occurs in an unventilated potato pile (Fig. 3.3), but the high respiration associated with early harvesting and initial wound healing makes it more likely to occur. This condensed moisture becomes available to any microorganisms present. These can be on the skin, on sprout buds or within lenticels. The moisture can allow spores such as silver scurf to germinate, gangrene or skin spot to sporulate and re-infect, or anaerobic conditions to occur within the tuber. A complete film of condensed moisture surrounding the tuber can cause tubers at 10°C to become anaerobic within 6h (Burton and Wiggington, 1970), providing conditions that favour the development of soft rot and pectolytic *Clostridium* species.

3.5.3   Contribution of established disease to subsurface condensation

Normally the heat comes from tuber respiration, but increases in the host respiration and contributions from the pathogen's own respiration in infected plant tissues are common. Subsurface condensation in parts of the pile or in specific boxes may therefore be due to localized rotting occurring just below these areas.

3.5.4   Condensation due to excessive cooling of the crop surface

It is the difference in temperature between the potatoes lower down and those on the surface that results in subsurface condensation; thus condensation results from the combination of respiration heating and top-surface cooling. If air significantly colder than the crop is jetted across the top surface of the potatoes, some of this air penetrates the potatoes and cools the upper layers of the pile or box. When ventilation stops, the warmer air rising from below condenses on the cold potatoes near the surface (Fig. 3.12). This problem was traditionally minimized by the use of straw (Box 3.9) but its use in sealed stores is now rare.

3.5.5   Combination of highly respiring crop and excessive surface cooling

The greater the temperature difference between the potatoes below the subsurface layer and those within the cooled layer, the more likely that condensation will occur. Highly respiring, newly loaded crops are therefore more likely to experience such condensation than crops that have become dormant after a period in store.

3.5.6   Warm weather fronts affecting open stores

Fast-moving weather fronts (Fig. 3.13), particularly in maritime climates, can result in rapid changes in ambient air temperature and RH. Potatoes, in contrast,

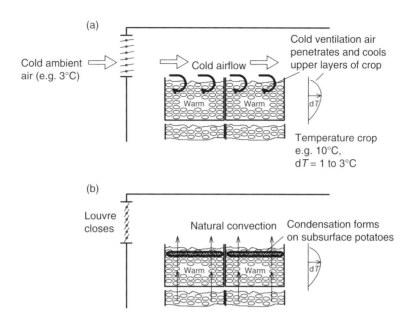

Fig. 3.12. Subsurface condensation due to over-the-top ventilation with air that is too cold: (a) ventilation with cold ambient air; (b) ventilation stops so natural convection dominates.

Box 3.9. Use of straw to keep potatoes near the crop surface warm

Traditionally, potato piles were covered with 0.3–0.4-m-deep loose straw or 0.36 m × 0.4 m × 0.8 m straw bales to keep the top surface of the pile warm and so minimize condensation or frost damage in the crop subsurface layer. While this is quite effective with bulk stores, in box stores only the top layer of boxes can be protected in this way. It is not practical to cover each box with straw as the stack is being built. As loose straw can clog up the sizing equipment when potatoes are being graded, it is preferable to keep the potato surface warm by use of sealed stores and sophisticated ventilation control.

---

when stored together in a large mass, will lag air temperature changes by a number of days. This lag will occur with potatoes stored outside. It will also occur if crops are stored inside with the doors left open, or if air leakage rates are high.

Should a warm front pass over and ambient air come into contact with crop with a temperature lower than the dew-point temperature of the ambient air, condensation will occur. This is a particular problem when stores are kept open for long periods for filling and is made worse if ventilating fans are switched to manual operation to encourage respiration heat removal and drying.

Warm crops are less likely to experience such condensation than cooler crops, as they are less likely to be below the dew-point temperature of the air. This

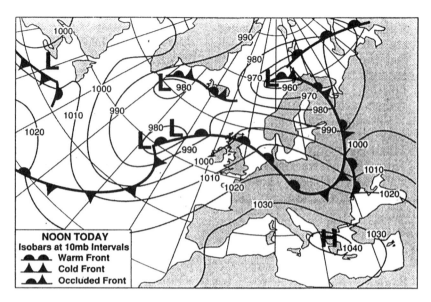

Fig. 3.13. Warm front passing over Scotland.

explains why crops in store should not be cooled until the store is full and the doors can be closed permanently.

3.5.7 Temperature differences in crops during store loading

If potatoes being loaded into store are at a different temperature to those already in store, convection currents will occur, which can lead to possible condensation on the cooler potatoes.

To illustrate this, suppose that two stacks of potatoes in boxes, one at 16°C and one at 10°C, are loaded into store, and the store door is shut (Fig. 3.14). The store air temperature will assume the average of the two stacks, i.e. 13°C. The void air between the tubers in the 16.0°C stack is warmer than the air in the store, so it will rise. The void air between the tubers in the 10.0°C stack is cooler than the store air, so it will move downwards. If the two stacks are close to each other, a convection-driven circulatory air movement will be established. If the RH of the store air is high, and the temperature of the cooler potatoes is below the dew-point temperature of the circulating air, condensation will occur on the top of the cooler stack.

3.5.8 Warm air entering store through leaks and doors

Potato crops are usually held below ambient temperature in the UK. If stores are leaky or if doors are left open, warm humid ambient air can enter the store and condense on the crop. In a box store, the top layers of boxes are most likely to experience condensation as the warm air leaking into store will tend to rise to the

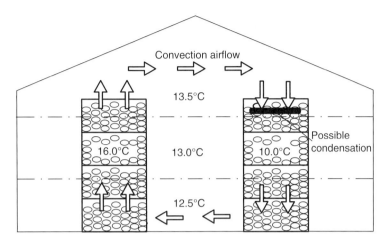

Fig. 3.14. Risk of condensation when boxes at two different temperatures are put into store.

roof space and fill it first (Fig. 3.15). If leakage is sustained, the whole upper part of the building will fill with warm air, with condensation occurring progressively down the layers of boxes.

An even worse scenario may occur if ventilation is started during these conditions. This can completely flood the store with warm air, which may result in condensation occurring on the entire crop.

3.5.9 Natural convection in a bulk pile

Subsurface condensation is not the only type of condensation in a pile caused by natural convection. If air is not free to circulate through ventilation ducts, the air

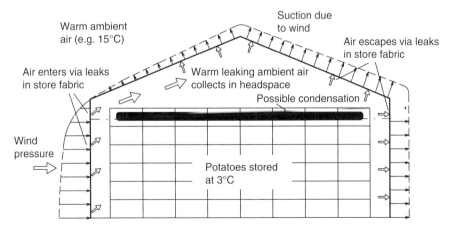

Fig. 3.15. Warm ambient air entering stores through leaks in the fabric can condense on the crop in store.

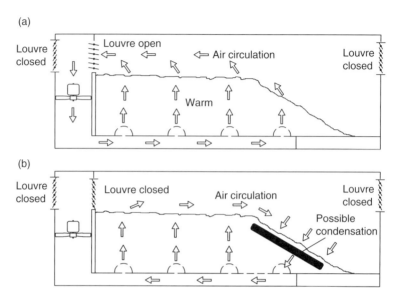

Fig. 3.16. Natural convection: (a) with recirculation louvre open and fan off, natural convection is anticlockwise; (b) with recirculation louvre closed and fan off, natural convection is clockwise.

drawn in by the 'fire' (Ch3.1) will enter the pile from areas where the heat being generated by respiration is less. Typically this is on the sloping front face of the pile or where potatoes are cooler, near the store walls (Burton *et al.*, 1955).

In warm weather, if warm humid air leaks into store, this warm air will be drawn into the sides and sloping front, to cause condensation (Fig. 3.16).

In freezing weather, if cold air leaks into store, the 'fire' is going to pull this freezing air in the sloping front and cold sides of the pile. This is where frost damage is likely to occur.

3.5.10 Recirculation ventilation in bulk stores

In bulk stores, it is common practice to recirculate air through the ventilation system to reduce the temperature gradient between the top and base of the pile. This is to ensure that all parts of the crop are as near as possible to the optimum storage temperature to ensure fry colours are uniformly light (Ch3.10). However, commonly warm humid air is allowed to leak into the headspace of these stores prior to recirculation taking place. If the potatoes at the base of the pile are below the dew-point temperature of the headspace air, recirculation ventilation may result in condensation around the ducts at the base of the pile (Fig. 3.17). At the warm temperature that processing crops are held, between 8 and 12°C, wetting the crop is likely to initiate rotting. The first signs the store manager sees that this is occurring are the surface potatoes slumping into hollows and ventilating ducts filling up with an odorous brown liquid from rotting potatoes.

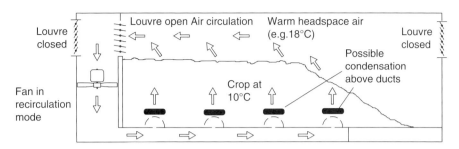

Fig. 3.17. Recirculation of warm headspace air can lead to condensation on crop above ducts.

3.5.11 Excessive spacing between ducts in bulk stores

Condensation just under the surface of intermittently ventilated bulk stores was observed (Hylmö *et al.*, 1976) in strips at the surface, half way between the lateral ducts spaced at 4m intervals (Fig. 3.18). Hylmö's evidence suggested that the longer air path between the duct and the surface where the condensation formed resulted in less cooling of the crop at this point. When a fall in ambient air temperature caused the air in the headspace above the potatoes to cool, the surface layer of the potatoes cooled too, resulting in cool potatoes overlying warmer potatoes below. Since the warmest potatoes lay between the ducts, condensation formed mid-way between each duct. Condensation could have been prevented by either reducing the distance between ducts or by installing roof-space heating. It is likely that continuous ventilation, which would have reduced the temperature gradient in the pile, would also have prevented its occurrence.

Modelling using computational fluid dynamics (Xu and Burfoot, 1999) confirmed this interpretation and suggested that, had ventilation stopped earlier, condensation could have occurred lower down the pile, out of sight of the store manager. The maximum thickness of condensation film was estimated to be 0.08 mm, well above the 0.03-mm film that caused anaerobic conditions within tubers to occur (Burton *et al.*, 1992).

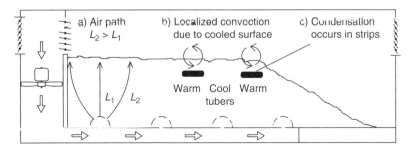

Fig. 3.18. Condensation can occur in strips between ducts.

3.5.12   Uneven airflows from ducts

Another investigation by the Hylmö team (Grähs *et al.*, 1977) found condensation in a number of 3000-t ventilated bulk stores, with temperatures varying from 6.5 to 9.0°C, leading to convection airflows and condensation-induced soft rotting (*Erwinia carotovora*) with secondary attacks of dry rot (*Fusarium*). The uneven temperatures were caused by uneven airflow rates varying from 0.0011 to 0.033 m³/s/t. Better design of the air distribution system or continuous ventilation with humid air would have prevented this problem. Conversion of old ventilation systems to continuous ventilation with low-rate humid air results in lower air speeds in existing ducts, which in itself results in more uniform air distribution.

3.5.13   Sudden reduction in ambient air temperature

While sealed stores are best for keeping the store atmosphere at a high RH to minimize dehydration, they are susceptible to condensation on the underside of the roof and on the coldest parts of the crop if ambient temperature falls rapidly, as often happens on clear, cloudless nights.

This can be illustrated by imagining a sealed, insulated store with the temperature of the crop, headspace air and ambient air all at the same 8°C, with a headspace RH of 96%. A steady-state situation will result where heat neither leaves nor enters the store (Fig. 3.19).

If the ambient temperature now rises (e.g. 14°C) heat will enter the store, raising the headspace temperature. If the headspace air temperature rises to 10°C, its RH will reduce to 80% (Box 3.2). Heat can pass through the insulation, but the vapour in the headspace air is trapped within the store. The warmed headspace will have little effect on the crop other than to increase weight loss through evaporation slightly.

If, instead of the ambient air temperature rising, it falls to 2°C, heat will leave the store and the headspace air temperature will fall. If the headspace temperature falls to 6°C, its RH will not only rise to 100%, but condensation will occur on the coldest surfaces within the store, usually the roof, where it can drip on to the potatoes. Condensation may also occur on the coldest parts of the crop. While this is a transient phenomenon, the moisture on the crop can allow spores to germinate and rotting to start.

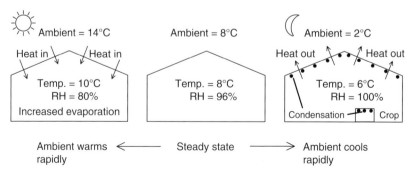

Fig. 3.19. Condensation can occur on underside of roof or on cold potatoes when ambient temperatures fall rapidly.

3.5.14 Venturi effect in base layers of box stores

This form of condensation is not common, but the author did find it in all five box stores on one farm 2 months after loading. Condensation occurred primarily in the bottom layer of boxes. Air speeds along the ducts formed by the pallet apertures between the bottom layer of boxes and the concrete floor were double those in the next pallet aperture up (Fig. 3.20). This was probably due partly to the resistance to airflow of the smooth concrete floor being less than that of the pallet apertures between the layers of boxes and partly because air from the fridge was cold and dropped to the floor. This differential air speed caused a Venturi effect, with air being drawn downwards in the base layer of boxes when the recirculation fans were operating. Once the fans stopped, natural convection dominated air movement and warm air from the warmer potatoes below rose up to condense on the cooler potatoes above.

3.5.15 Warming potatoes

As the criteria for condensation to occur require that ventilating air must be warmer than the crop and the crop must be below the air's dew-point temperature, warming potatoes, for example prior to grading following storage, is very likely to result in condensation. Only if the dew-point temperature of the warming air is below that of the crop will warming take place without condensation occurring.

3.5.16 Summary of causes of condensation

In summary, condensation on the crop will occur if the crop's temperature is below the dew-point temperature of any air with which it comes into contact. Air's dry-bulb and dew-point temperatures are the same at 100% RH, so by definition air must be

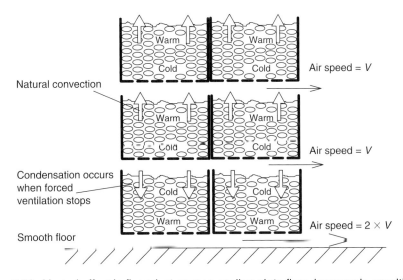

Fig. 3.20. Venturi effect in floor duct causes cooling air to flow downwards, resulting in cold potatoes overlying warm.

warmer than the crop if condensation is going to form. The lower the RH, the lower the dew-point temperature, the less likely that condensation on the crop will form. The likelihood of condensation can be predicted by looking up Table B.3.2 in Box 3.8.

For the reasons explained above, condensation on potatoes will occur when:

- Crop temperature in a store with its doors open is below the ambient air dew-point temperature when a warm weather front is passing or the crop has been cooled below the air's dew-point temperature.
- Excessive temperature differences develop within the stored crop, causing moisture in convection air currents from warmer potatoes to condense on cooler crop.
- Sudden reduction in ambient temperature causes headspace air to reach 100% RH.
- Potatoes are warmed with air having a dew-point temperature above the temperature of the crop.

All systems of storage are vulnerable to condensation when stores are being loaded and doors are open, or when no RH sensor is used to control ventilation during wound healing (Ch8.6). Once stores are shut and set to cool automatically, condensation should be prevented by well-designed ventilation controllers.

Temperature differences within the crop can be minimized by installing a continuous, low-rate, humidified ventilation system. Even these systems are prone to condensation on roof and crop from sudden reductions in ambient temperatures, so should be fitted with roof-space heating systems.

In intermittent ventilation systems, the problems of excessive temperature differentials and the associated condensation can largely be overcome by well-designed airflow distribution, coupled with precise, automatically controlled ventilation to minimize temperature differentials within the crop.

3.6 Climate Zones

Sample average maximum and minimum temperatures for the coldest and warmest months in the year, together with average RH, for different zones of the world are shown in Table 3.1 (BBC, 2007). More extensive data for the world are shown in Appendix 2.

When storing crops in different climatic areas, there is greater emphasis on some aspects than others. In continental climates where winter temperatures can reach −20°C and below, the prevention of freezing of the stored crop and minimizing crop desiccation due to ventilation with cold, dry air are the major priorities. In the maritime climate of the UK, the drying of crops and excluding warm humid air from entering the store and causing condensation on the potatoes is of greater concern. In tropical lowland climates where little air below 10°C is available for cooling, growers with simple stores depend on long-dormancy varieties, to allow a month or so of sprout-free storage (Ch3.8). For longer periods, total dependence has to be placed on refrigeration. In tropical highland climates, the altitude will determine how much cool night air is available for cooling; this will allow some ambient-air cooling to take place.

Table 3.1. Average minimum and maximum temperatures, and average relative humidity (RH), for the coldest and warmest months of the different world climatic zones.

Climatic zone	Coldest month			Warmest month			Main concerns during storage
	Temperature (°C)		RH (%)	Temperature (°C)		RH (%)	
	Min.	Max.		Min.	Max.		
Continental East Canada	−9 (−27)[a]	0 (14)	76	13 (4)	23(37)	73	Freezing in winter and weight loss
Maritime East UK	1 (−14)	6 (15)	89	12 (5)	21(31)	74	Crop condensation from warm air leaking into store
Continental/maritime The Netherlands	−1 (−25)	4 (13)	86	13 (4)	22 (34)	72	Freezing in winter and crop condensation from warm air leaking into store
Mediterranean Egypt	11 (3)	18 (28)	66	23 (18)	31 (41)	70	Sprouting and desiccation
Tropical lowlands Bangladesh	12 (7)	25 (31)	46	26 (22)	32 (36)	72	Sprouting and desiccation
Tropical highlands Peru	5 (−2)	21 (25)	60	8 (−4)	22 (26)	62	Sprouting and desiccation

[a]Figures in parentheses are the record minimum and maximum temperatures.

If ambient-air cooling is used, one author with experience of UK storage trials has found that storage temperatures tend to be a degree above average ambient temperatures. For the UK January temperature minimum/maximum of 1°C/6°C respectively in Table 3.1, the crop temperature would average 3.5 + 1.0 = 4.5°C. Another author with experience of tropical storage trials found that temperatures 1–3°C below the average could be obtained.

Since two crops may be grown per year in tropical areas, short-term, porous-walled, rustic storage for 1–2 months is common. Crops in such stores have in addition to risk from sprouting and desiccation the additional risk of attack by insects and animals.

Whether potatoes should be stored in the ground, stored for short periods in rustic buildings, kept in refrigerated buildings, or transported from one climatic region to another is discussed in Ch13.

3.7 Selection of Ventilation System Based on Climate Type

3.7.1 Introduction

In cool maritime and continental temperate climates with winters which allow ambient-air cooling to keep crops near 3–4°C, sprout-free storage for 6–7 months is usually possible. Storage at this temperature beyond 7 months requires refrigeration or sprout suppressants. Storage for 9 months is routinely achieved. Longer storage is possible but cumulative weight loss causes quality to deteriorate. Potatoes for processing are normally held at between 8 and 12°C for up to 9 months, with suppressants being used for sprout control.

In warm tropical climates, stores without refrigeration or sprout suppressants can hold potatoes only for the length of their innate dormancy period, a period less than 1 month (Wustman *et al.*, 1985). With sprout suppressants this can be extended to 4–5 months at 25°C or 1 month at 30°C. For a longer period of storage, refrigeration is required.

Storing potatoes in sealed, insulated, environmentally controlled stores minimizes the dehydration, disease and skin blemishes which can occur in unsealed stores and clamps. Disease-free, firm potatoes greatly increase their sale value.

Storage ventilation has evolved over time, with the system selected matched to the country's climate. These fall into four main groups, which are described below. They comprise systems of ambient-air cooling that are used worldwide, with refrigeration used where insufficient ambient air is available for cooling or in tropical regions where ambient air for cooling is not available. Simple systems based on storing potatoes in cool cellars or in open ventilated, timber structures are not mentioned here but are discussed in Ch5.1.

3.7.2 Intermittent ventilation on demand using medium airflow rates

The system predominantly used in the UK relies on ventilation on demand for removing heat of respiration, drying potatoes after lifting, cooling the crop following wound healing and recirculation of air through the crop to minimize temperature differences. The traditional airflow rate used is $0.02\,m^3/s/t$ (BPC, 2001a). This rate provided acceptable rates of drying, cooling and air distribution in bulk stores over many years, so when box storage was introduced the same rate was used. Experience has shown that so long as this airflow is distributed uniformly over the whole crop, whether in bulk or box, crops can be kept satisfactorily. However, for positive ventilation of potatoes in UK-style boxes (Ch6.8), where between half to three-quarters of the air (Pringle, 1989) may leak from gaps in and between boxes, this airflow rate may need to be multiplied by 2–4 times to achieve $0.02\,m^3/s/t$ through the potatoes themselves.

In bulk storage, uniformity of distribution is dependent on the design of the ducted ventilation system (Ch6.4).

In box storage systems, uniformity of airflow depends on (Ch6.6):

- The design of the air distribution system.
- The box stacking layout.
- The degree of precision with which boxes are stacked.
- Whether the store is airspace or positively ventilated.
- Whether the boxes are designed to suit the ventilation system.

Ventilation on demand has some potential problems:

- Excessive ventilation with non-humidified ambient air can result in up to 10% evaporative weight loss and associated pressure bruising, especially in the lower layers of potatoes in bulk stores.
- The cooling air in airspace ventilated box stores, where air is circulated round rather than through the boxes, may preferentially cool surface layers of potatoes, leaving layers below warmer. This can result in subsurface condensation when convective ventilation is re-established.
- Once forced ventilation stops, the low convective air movement will cause temperature gradients to develop, leading to a greater range in crop temperature and increased likelihood of subsurface condensation.

These disadvantages can all be minimized by sophisticated control of ventilation (Ch8.6). The system also depends on the UK's natural humidifiers, the Atlantic Ocean and the North Sea, to supply air that will not dehydrate the crop excessively.

While humidifiers can be put into the ventilation airstream, the high airflows used in the UK compared with the low-volume humidified air ventilation systems of the USA and Scandinavia will result in very large media humidifiers being used. In the UK's medium-rate airflow systems, humidification with spray jet humidifiers is the only realistic solution. These need to be fitted with effective droplet arrestors to stop the crop becoming wet through droplets landing on it.

Companies with a dehydration problem usually solve their bruising problems by installing more efficient fans, better sealing of stores and better system control.

3.7.3 Intermittent ventilation on demand using high airflow rates

The Dutch use ventilation on demand, but have pioneered the use of higher airflow rates of $0.042\,m^3/s/t$ (Scheer, 1998) to:

- Remove respiration heat from immature, early lifted crops.
- Rapidly dry crops lifted from wet soils.
- Minimize dehydration during ventilation to cool the crop.

The earlier crops are lifted, the less disease they tend to have. However, early lifted crops are prone to subsurface condensation due to the high respiration associated with immature crops (Fig. 1.17). The high airflow prevents respiration-induced subsurface condensation, providing the benefits from lifting crops early but without the problem of subsequent condensation and associated disease establishment and multiplication.

The emphasis in The Netherlands is therefore to prevent subsurface condensation by removing respiration heat, to rapidly dry the crop and keep it dry thereafter. High airflow rates allow this to be achieved.

Experiments show that when air is blown to cool potatoes, evaporation of moisture from the skin of the potato occurs during the process. As the air approach velocity (Box 3.10) increases to greater than 0.01 m/s (Rastovski and van Es, 1981), so the rate of heat transfer increases, but evaporation does not increase by the same proportion (Fig. 3.21). The skin is restricting the rate of moisture evaporation. By ventilating the crop for short periods with a high volume of air, a high rate of

---

Box 3.10. Air approach velocity

The air approach velocity is the speed of a column of air approaching a mass of potatoes. This is easier to measure than the air speed between tubers. The air velocity through the voids between the tubers is inversely related to the ratio of void, to the total space occupied by the potatoes. The air velocity therefore increases as it enters the crop.

If potatoes are stored 2.0 m deep and their density is 667 kg/m³, a column 1 m × 1 m in area will hold 1.33 t of potatoes (Fig. B.3.6). An approach velocity of 0.1 m/s will therefore be equivalent to a ventilation airflow rate of 0.075 m³/s/t, almost four times the rate (0.02 m³/s/t) used in UK stores.

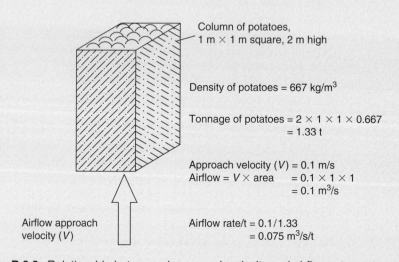

Column of potatoes, 1 m × 1 m square, 2 m high

Density of potatoes = 667 kg/m³

Tonnage of potatoes = 2 × 1 × 1 × 0.667
                                     = 1.33 t

Approach velocity (V) = 0.1 m/s
Airflow = V × area   = 0.1 × 1 × 1
                                     = 0.1 m³/s

Airflow approach velocity (V)

Airflow rate/t = 0.1/1.33
                        = 0.075 m³/s/t

Fig. B.3.6. Relationship between air approach velocity and airflow rate per tonne.

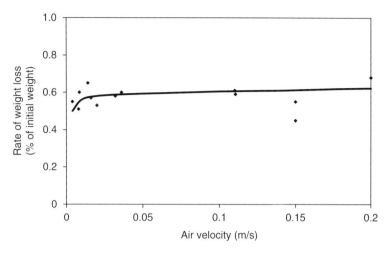

Fig. 3.21. Moisture loss versus ventilation air approach velocity.

cooling can be achieved without excessive moisture loss. High airflow rates therefore help to reduce dehydration.

To achieve these high airflow rates, larger fans than used conventionally in the UK are required. The success of the Dutch potato industry suggests that the additional cost of providing these high airflow rates is recouped by the quality obtained.

3.7.4 Continuous humidified-air ventilation systems – bulk storage

The system of continuously ventilating potatoes with a low rate of humid air resulted from a remarkable productive collaboration between mid-west American and Scandinavian technologists during the early 1970s.

By continuously ventilating bulk potatoes with air humidified to as near 97.8% RH as possible (Box 3.11), the following results are achieved:

- Crop weight loss is kept to a minimum by keeping the RH of the ventilating air as near to saturation as possible and by continuously removing the respiration heat.
- The temperature differential between top and base of the pile is kept to a minimum, with the difference declining as air speed rises.
- Subsurface condensation associated with natural convective ventilation is prevented.
- Sufficient drying of the crop occurs simultaneously at all levels up the pile to ensure tuber skins remain dry.
- Constant cooling allows the difference in temperature between the cooling air and the crop to be kept small, thereby minimizing the large step changes that can occur in intermittently ventilated systems.
- Continuous ventilation dominates natural convection, ensuring that only colder air meets warmer potatoes; this ensures that condensation on the crop cannot occur.

Box 3.11. How an evaporative humidifier works

Putting an evaporative, media type humidifier supplied by tap water in the main ventilation duct supplying the crop both humidifies and cools the ventilating air (Fig. B.3.7). Air leaving the fan at 10°C, 70% relative humidity (RH) will evaporate moisture from the humidifier and thereby cool the air to its wet-bulb temperature of 7.38°C and have an RH of 97% (Munters, 2007).

  Were the 10°C, 70% RH air supplied to the crop directly, the air would still be cooled to a value approaching the wet-bulb temperature, but the moisture would come from valuable potatoes rather than tap water. The humidifier therefore reduces moisture being absorbed by the ventilating air from the crop and so minimizes crop weight loss.

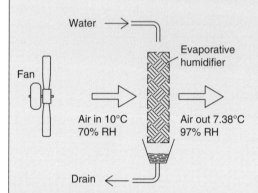

Fig. B.3.7. Increasing the relative humidity (RH) of ventilating air using an evaporative media humidifier.

In Scandinavia and the mid-west of the USA this type of storage now predominates (M.J. Frazier, Idaho, 2005, personal communication). The airflows selected are 0.0070 m³/s/t for initial storage after lifting and 0.0035 m³/s/t once the potatoes become dormant, although, increasingly, variable frequency drives are being fitted to slow the airflow further when crop respiration is low and so save electricity. These airflows are one-third of the UK airflow rates and so fans and ducts can be smaller and lower in cost. Low airflow rates for cooling can be used without causing significant crop weight loss due to the airflow being humidified.

Attempts to introduce this system for potato storage into the UK have been unsuccessful to date. This is due to a number of possible reasons:

- The high ambient humidity found in the UK allows potatoes to be stored with acceptable levels of weight loss using the established systems of ventilation.
- The fear that the low airflow may be insufficient to dry wet crops as rapidly as the higher airflow systems.
- An investigation (Potter, 2000) indicated that there was inadequate cool air available in the UK using ambient-air cooling alone to maintain a temperature of 4°C over the winter storage using this system. Additional backup refrigeration has therefore to be provided as standard.

3.7.5 Continuous humidified-air ventilation systems – box storage

The same continuous humidified-air ventilation systems have been applied to box stores. Cooling air is introduced at floor level while outlets are located as high up as possible in the store (Johansson, 1998). The cold air continually floods the store, rising up from the floor, displacing the more buoyant warm air coming from the crop as it rises (Fig. 3.22). This airflow movement can be likened to filling the store with water.

The rising cold air continually spills out the high-level louvre at the top of the store, like a river flows over a 'weir'. At the higher rate of $0.0070\,m^3/s/t$, the rate of air movement is 11 times that of the natural ventilation rate of $0.00064\,m^3/s/t$, based on the crop emitting 12.8 W/t (Hylmö *et al.*, 1975a).

The difference in sensation when entering a well-sealed potato store using intermittent ventilation compared with entering one using continuous ventilation with humid air is almost imperceptible. The potatoes in both have dry skins, as the heat they generate keeps them dry. This observation is verified by skin resistance readings (Pringle *et al.*, 1997). However, paper on a clipboard stays dry in the first but goes limp and moist in the humid store. The humidified ventilation system therefore keeps the skins of the stored tubers dry while limiting evaporation and weight loss by its high relative humidity (Box 3.12).

Boxes untreated with preservatives sometimes develop white penicillin moulds in such high-humidity storage as, unlike the potatoes, they do not generate the heat required to keep the wood dry (Pringle *et al.*, 1997).

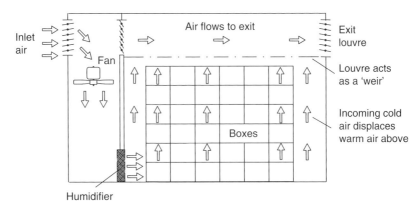

Fig. 3.22. Continuous ventilation of cool humidified air into store.

Box 3.12. Does high relative humidity in store exacerbate disease?

Continuous ventilation with humidified air reduces weight loss and minimizes condensation. Pathologists often say that high humidity encourages disease development. If so, continuous ventilation with humid air systems should result in increased disease. The total acceptance of this system in the USA, Scandinavia and elsewhere would not have occurred if this were the case. The likelihood is that in experimental work carried out by pathologists where humidities were high, unseen condensation was occurring. This free water, rather than high humidity, was therefore likely to be the key factor in disease development.

3.7.6   Seed storage and grading systems open to atmosphere

Potatoes in stores open to atmosphere are vulnerable to:

- Frost damage if ambient air is below $-2°C$.
- Condensation if ambient air dew-point temperature exceeds crop temperature.

While these two factors would suggest that all stores should be sealed, many seed stores are left open during the working day. This allows grading to take place without the necessity of opening and closing store doors. Such stores are usually only in areas where frost is infrequent. When a period of frost does occur, doors can be shut and heaters switched on to prevent potatoes from freezing.

The main risk in such installations is from condensation forming on the crop when warm weather fronts pass over the store (Ch3.5). The risk of such condensation can be significantly reduced if crops in store are cooled at a rate similar to the natural cooling of ambient temperature that occurs in late autumn and early winter. With a slow cool down, the crop will only rarely be below the dew-point temperature of the ambient air. This probably explains the Dutch recommendation that crops should be cooled slowly, so that they reach $4°C$ by early December (ZPC, The Netherlands, 1985, personal communication).

On the occasions where this strategy fails and crops do suffer condensation, positive ventilation for the whole stored crop, or very well distributed airspace ventilation, is required to rapidly dry the crop.

3.7.7   Storage using refrigeration alone

Refrigeration is used as part of the cooling equipment in all the systems discussed above where sprouting cannot be prevented using ambient-air cooling alone. However, refrigeration can be used on its own, with no ambient-air ventilation.

The use of refrigeration-only systems allows potatoes to be stored in tropical areas where cool ambient air is unavailable. This incurs a high energy cost. In temperate areas, refrigeration in theory uses approximately ten times more energy (Box 8.3) to cool potatoes compared with cooling using ambient air. The full benefit is not usually seen in practice, probably due to air leakage through the louvres required for ambient-air cooling. In tropical areas cool ambient air is not available. Where mountainous regions are nearby, it may be possible to store crops at altitude and transport them to markets when they are required (Ch13).

In temperate regions many growers are attracted to the apparent simplicity of using refrigeration alone to both dry and cool the crop, even though the energy costs may be higher. This can be done by using the fridge recirculation fans to recirculate the air in store as the crop is being loaded, and to rely on wind to change the air being recirculated. If stores are small, the high capital cost of installing ambient-air mixing and possibly humidification may make the overall costs of refrigeration-alone systems less than that of combined ambient-air cooling/refrigeration systems.

If the fridge is used to act as a dehumidifier, with the doors closed, it will tend to cool the crop during the drying process. The cooled, dried crop then may suffer condensation when the next load of warmer potatoes is loaded into store. Supplying heat using an oil heater, fitted with a flue to keep RH, carbon dioxide and ethylene levels in store down, can offset this cooling, but adds to the cost of drying and produces additional greenhouse gases compared with the use of refrigeration alone.

3.7.8 Comparative assessment of the four systems

The most 'natural' system of ventilation for potatoes is continuous low-rate ventilation of humid air, cooled by ambient air when available and refrigeration when not. It eliminates all the potential types of condensation with the exception of: (i) condensation on the roof and crop (and possibly boxes) caused by sudden cold spells of weather; and (ii) the problem of warm air entering the building through gaps and doors. The first can be prevented by the use of roof-space heating while ingress of air can be excluded by good sealing of the store and by ensuring that doors are kept shut.

Both the medium and high airflow UK and Dutch systems are potentially at risk from all the types of condensation listed above (Ch3.5). In these systems instrumentation and sophisticated controllers are required to ensure condensation on the crop will not occur. These systems can be made to work well so long as they are designed well. The high airflow rates are greatly valued to speed drying in wet harvests.

Seed stores with grading areas, which are kept open during the day, are appropriate where daytime winter temperatures range mostly between freezing and 10°C. Good drying facilities are needed to remove the occasional condensation that occurs on the crop. There is a danger with such stores that seed kept above 3–4°C may start to sprout prior to dispatch for planting. Varieties that sprout easily or are destined for later deliveries should therefore be kept within a closed sealed store at 3–4°C and opened just prior to grading and dispatch.

Stores that rely on refrigeration alone are expensive to run if cooling and drying are both carried out in the closed store using the fridge. In tropical countries the difficulty is in loading the stores without allowing ambient air to enter the store. This warm moist air will not only condense on cool stored potatoes, but also on the fridge cooling coils to rapidly form ice (Fig. 3.23), which has then to be melted to allow cooling to continue. Stores should be small enough to be loaded, wound healed and cooled as a single batch. If stores are large, then batches of potatoes should be loaded into an intake room, ventilated, wound healed and cooled prior to being loaded into the main store. In this way temperature differences between batches of potatoes can be minimized so that convective currents between batches do not occur.

3.8 Crop Storage Life

Potatoes are usually kept at harvesting temperature for 10–14 days to allow wounded skin tissue to heal, then cooled to a lower temperature for storage to minimize respiration, extend dormancy and slow the multiplication of any disease present (Ch1.5). Potatoes for seed and pre-pack are cooled to 3–4°C. Potatoes

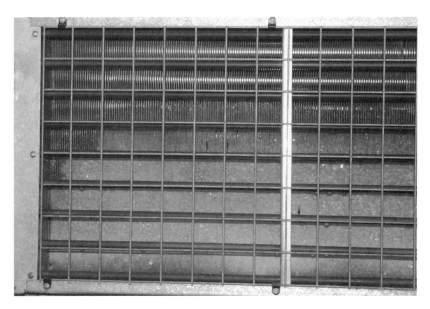

Fig. 3.23. Ice build-up on an evaporator. (Courtesy of G. Stroud, Potato Council, Oxford, UK.)

destined for processing are cooled to no lower than 8–12°C, to prevent the conversion of cell starch to sugar. Due to being stored warmer, dormancy break and associated sprouting will occur earlier in processing potatoes than pre-pack or seed, unless chemicals are used to control sprouting.

3.8.1   Length of dormancy versus store temperature

The length of time potatoes can be stored in an airspace-ventilated store varies between 1 and 12 months (Table 3.2). These data will vary considerably, depending on the variety. In cold continental climates which allow 5°C to be maintained over the winter using ambient-air cooling, crops can be kept for up to 8 months.

Table 3.2. Storability (in months) in naturally ventilated stores. (From Wustman *et al.*, 1985.)

Average ambient temperature (°C)	Seed potatoes		Ware potatoes	
	Under lights	Kept in dark	Sprout inhibitor	No sprout inhibitor
5	12	8	10	6
10	8–9	4	8	3–4
15	4	3	6–7	2–3
25	3	1–3	4–5	
30	2	1	1	

In warm tropical areas where temperatures average 30°C, storage of only 1 month is possible. Keeping tubers in trays exposed to bright artificial light or in a greenhouse can extend length of storage. This is suitable only for seed as greening of the skin takes place. Storage life can also be extended by the use of chemical sprout inhibitors.

3.8.2 Use of chemicals to extend dormancy

Potatoes for processing into crisps or French fries are stored at 8–12°C to minimize their sugar content so that, when fried, they have a light golden colour (Ch1.5). At this temperature, dormancy is short, and undesirable sprout growth soon starts. To extend dormancy, maleic hydrazide can be sprayed on to the growing crop prior to harvest, or CIPC (chloropropham) can be sprayed on to the crop at store loading. For longer-term storage in the UK, CIPC is applied as a fog to the crop once loaded into store (BPC, 2002a; Cunnington and Dowd, 2003).

CIPC
The active ingredient of CIPC is chlorpropham, which in the UK is usually dissolved in a solvent such as methanol or dicholoromethane. The chemical acts by preventing cell division in the tip of the sprout, preventing its growth. CIPC is injected into a hot airstream, produced by a fan, with petrol (gasoline) fed into the airstream and ignited using a spark plug (Fig. 3.24). The products of combustion enter the store along with the CIPC. In the USA, CIPC is usually applied as solid melted chemical with no solvent present. The store is fogged whenever tubers start to show signs of sprouts.

Recent research has focused on achieving a more uniform application of CIPC in box stores than has been achieved in the past, so that the maximum residue level of 10 ppm, set by the UK Pesticide Safety Directorate (EU, 1995), can be achieved. CIPC treatment is usually associated with an increase in sugars in the crop, which

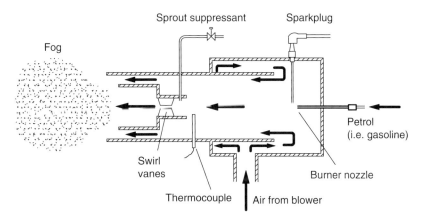

Fig. 3.24. Construction of a CIPC (chloropropham) fogger. (Redrawn from Mathews, 2000.)

adversely affects fry colour. This results from the production of ethylene from burning the petrol used to heat the hot airstream, which, being a plant hormone, initiates an increase in respiration and associated change from starch to sugar (Cunnington and Dowd, 2003). While sugar levels will subsequently reduce, the recommendation is to shorten the period that the stores are closed to 8 h if allowed by the label.

Other chemicals

Chemicals other than CIPC are used worldwide (BPC, 2002b). Diisopropylnaphthalene (DIPN) is used in combination with CIPC in the USA, while dimethylnaphthalene (DMN) is used in the USA and is undergoing registration trials in Europe. The naturally occurring caraway seed extract Carvone is sold in The Netherlands and Switzerland but is not registered for use in the UK at present. It is suitable for organic crops. Other potential sprout suppressants are clove oil, hydrogen peroxide and ethylene.

Ethylene enrichment of stores

Exposure of potatoes to ethylene for short periods has a dormancy breaking effect, while long-term, continuous exposure inhibits sprout elongation (Rylski *et al.*, 1974). Sprout suppression was reported with ethylene concentrations at 2 and 20 ppm. In stores where fry colour was not an important quality attribute, sprout suppression was accomplished through the continuous application of ethylene, maintaining a concentration of 4 ppm (Prange *et al.*, 1998). In commercial contract trials at BPC Sutton Bridge (Briddon, 2007), ethylene at 10 ppm controlled sprouting in three pre-pack varieties held at 3.5°C. It is not used in processing stores due to their high storage temperature and related high respiration level, requiring the store air to be refreshed twice daily to keep carbon dioxide levels down. It can however be used for seed.

Ethylene has approval as a commodity substance from the UK Pesticides Safety Directorate for use in potato stores as a plant growth regulator. The approval specifies a target ethylene concentration of 10 ppm in the store atmosphere and a minimum period between treatment and sale of 3 days. To maintain this level of concentration the store has to be very well sealed.

The ethylene generator (Restrain Company Ltd, Chatteris, UK; www.restrain.eu.com) uses a catalyst to convert alcohol to ethylene and a small amount of water, and is capable of maintaining an ethylene atmosphere at 10 ppm in a store capacity from 12 to 6000 t. Electronic controls and a computer program to operate the equipment come with the system. Remote data collection and alarms are also available features. Another system (Biofresh Ltd, Ripponden, UK) uses ethylene stored in bottles to maintain the ethylene-rich atmosphere.

3.8.3 Use of light to extend dormancy in seed

As illustrated by the data from Table 3.2, light is a very good sprout inhibitor. Potatoes have to be stored in thin layers, two or three tubers deep, usually on trays, so that the light can penetrate to the tubers' eyes without requiring the tubers to be turned. The light may be natural daylight or from fluorescent tubes. Under lights, skins rapidly become green, so extending storage life by this method is only suitable for seed.

3.9   Specific Requirements for Pre-packing

Supermarkets are demanding pre-pack potatoes to have:

- Bright shiny skins or bloom.
- Blemish-free skins.
- Uniform size.

To achieve these requirements, crops must be lifted early, ventilated to rapidly dry the tubers and remove respiration heat, and be kept free from condensation.

Bloom can be lost through excessive ventilation or recirculation of air through the crop, as tubers lose moisture and firmness. In intermittent ventilated systems, advanced control equipment should be installed to ensure that fans run only when required to cool the crop and to remove temperature gradients that could result in condensation.

Storage temperatures are commonly reduced to 2–5°C to prolong dormancy and slow any disease development.

3.10   Specific Requirements for Processing Potatoes

Crops must be mature at lifting as the fry colour of crops harvested when immature darkens rapidly in store. Skin finish and surface blemishes are relatively unimportant as skins are removed prior to processing.

Storage temperatures are commonly reduced to 8–12°C after harvest, the precise temperature depending on variety, to minimize the conversion of cell starch into sugar due to low-temperature sweetening (Ch1.5). Maintenance of uniform temperatures in store is vital to prevent cold areas of crop. This prevents crisps and French fries becoming too dark in colour when being fried.

At these temperatures rotting and disease development can be rapid and dormancy is reduced significantly. While disease development in pre-pack and seed potatoes can be slowed by the combination of keeping the crop both dry and cold, in potatoes for processing, disease prevention relies almost wholly on rapidly drying the crop and thereafter keeping it free from condensation.

Prolonging dormancy is achieved by the application of sprout suppressants, mainly CIPC, two or three times over a winter storage period.

3.11   Specific Requirements for Seed

Seed potatoes have traditionally been grown in colder, windy areas less attractive to aphids, which transmit virus diseases. Harvesting conditions in these areas can be wet, with considerable quantities of wet soil being harvested along with the crop. In bulk storage this can result in soil cones forming (Ch2.2) and ventilation airflows 'rat holing' through the voids surrounding the soil cone. Great care is required to minimize the amount of soil being lifted and continuous back and forth movement of store loading elevators is required to prevent soil cones forming.

In box storage systems, the presence of soil together with the small void space between seed-sized tubers results in reduced convective ventilation and airspace ventilation systems being less effective. In order to force air through these small passages, positive ventilation systems, where fans force air through the potatoes in the boxes, are commonly used.

Seed producers usually grow one to five varieties, of between one and three generations. However, some producers, especially those producing the highest-grade seed, will grow many more varieties, both to satisfy demand and to spread the risk of certain varieties becoming less popular. The requirement to segregate and store small amounts of these different varieties and generations was the main reason for the Scottish seed industry storing material in boxes rather than bulk.

In The Netherlands, where large cooperatives and companies handle the production of seed, individual growers are allocated a sufficiently large tonnage of one or two seed varieties that allows bulk storage to be used.

Seed stores in the UK may take 6 weeks or more to fill. An additional 2 weeks are then required to dry the last crop in and allow wounds to heal. During store loading the store temperature should track the temperature of the crop in the ground so that the temperatures of the crop entering store and already in store are similar (Ch3.5).

Unlike pre-pack or processing stores, which can be filled rapidly, closed and kept closed for the entire storage period, seed stores are continually being opened to pre-grade material, send off material early for export and respond to customer demands for seed to be delivered early for chitting in their own sheds. The risk of condensation forming on the crop by allowing humid ambient air to enter the building is very great. Seed stores should be designed with this in mind, and control and monitoring equipment should be incorporated to alert staff when opening doors may risk condensation forming on the crop inside.

3.12 Summary

Store climate should maintain potatoes firm, sprout-free, disease-free, and with uniformly low sugar levels if destined for crisping or use as French fries. The main factors that influence storage conditions include the following.

- Tubers stored together in a well-sealed building create their own, largely beneficial, microclimate.
- Some bleeding of air into store is required to prevent oxygen starvation and to ensure carbon dioxide remains below an average level of 0.5%.
- Ventilation immediately after harvest is required to dry damp harvested crops, and to remove respiration heat that would otherwise warm the crop and cause subsurface condensation.
- Ambient ventilation air varies both in temperature and RH due to changes in weather and so can cause wetting of crops, as well as drying during wound healing and crop drying if ventilation is uncontrolled.

- Ventilation is required after wound healing to cool the crop with cool ambient air when available. If refrigeration is fitted, it can be used when no cool ambient air is available.
- The temperature of ambient air used for cooling is likely to vary during the period of ventilation.
- Ambient air used to cool the crop also removes moisture. In non-humidified storage systems ventilation duration should be kept to the minimum required for temperature control.
- Cooling of potatoes results in a cooling front developing, which then rises slowly up through the pile or box.
- Evaporative cooling accounts for approximately 55% of cooling; convection and some conduction accounts for the remaining 45%.
- The temperature difference between top and base of a pile is a maximum in naturally ventilated piles and reduces as ventilation rate increases.
- Temperature differences between the base and top of a pile, moisture loss during cooling and formation of cooling fronts can all be minimized by the use of continuous low-volume ventilation of crops with humidified air.
- At the high relative humidities produced by potatoes when stored together, condensation on the crop is an ever-present risk.
- Condensation can form on potatoes whenever the temperature of the skins of tubers is below the dew-point temperature of the surrounding air.
- While intact skins of tubers are the main barrier to disease infection, keeping skins dry is the second.
- Where it is not possible to avoid subsurface condensation by ventilation and headspace heaters, straw can be applied as an insulating layer to keep the pile surface warm.
- While low-temperature storage is normally used to prolong dormancy in potatoes, chemicals can be used instead where low temperatures would otherwise cause sugar accumulation and dark fry colours.

4 Disease Control in Store

Topics discussed in this chapter:

- The host–pathogen–environment disease triangle.
- How diseases become established in the growing crop.
- Storage strategy modified by condition of crop at harvest.
- Influence of moisture, temperature and hygiene on bacterial diseases.
- Assessing the risk of fungal disease development in store.
- Influence of moisture, temperature and hygiene on fungal diseases.
- How disease gains access to the tuber flesh.
- Avoidance of soft rotting during early storage.
- Disease ecosystems and competition.
- Hygiene on intake and cleaning equipment and out-of-store graders.
- Holistic approach to disease control.

4.1 Introduction

In seeking to limit the development of disease on tubers in a potato store, store managers must realize their limitations as to what they can do:

- In a store with a single airspace, individual batches of potatoes cannot be held at different temperatures to limit specific diseases; the whole stored mass has to be kept at a compromise optimum temperature.
- Even in stores fitted with refrigeration, it will take 30 days to cool a crop by 15°C, e.g. from 18°C to 3°C. It is therefore not always possible to cool a crop rapidly to minimize disease development; cool-down will always be slow.
- The crop itself primarily influences the RH of air in the voids between the tubers. The manager therefore has limited control over RH.

- Sections of the crop that were harvested wet, or have become wet through condensation, can be dried by directing air towards the wet areas or by ventilating the whole stored crop.
- Ventilation of the crop can be taken a stage further, by using prolonged ventilation to desiccate (mummify) rots and prevent the pectolytic oozes from digesting the skins of neighbouring tubers.

In summary, store managers have to select one single compromise temperature for the store airspace. They have little control over the RH within the crop voids, but can selectively dry wet sections of crop without necessarily desiccating the main mass of potatoes.

4.2   Host–Disease–Environment Triangle

The 'disease triangle' proposed by Vanderplank (1963) helps to visualize contributions to disease development of three components: (i) the pathogen present; (ii) the host tuber; and (iii) the surrounding micro-environment (Fig. 4.1). The 'triangle' helps us understand how disease may develop and how disease development may be slowed by various control strategies. In potato stores, the host is the potato tuber; the pathogen a range of bacteria and fungi; and the environment primarily temperature, free surface moisture on the tuber skin surface, and the RH and carbon dioxide concentration of the void air. The store manager's ability to 'interfere' with any leg of the triangle will affect the amount of disease that develops.

Were store managers free to select highly resistant varieties, grow and store them in the absence of pathogens and choose conditions that prevent disease development, they could then concentrate solely on the other crop quality factors such as fry colour or sprouting. However, this dream scenario rarely occurs!

4.2.1   The host

The potato tuber's ability to resist disease development is dependent on a number of genetic, physiological and physical factors. Genetic resistance can include a tuber's ability to:

- Resist disease through production of a thick and well-structured epidermis or periderm.
- Produce biochemicals which interfere with disease infection.
- Produce suberin between cells under damaged areas to protect the flesh from invasion by pathogens (Fig. 4.2).

The magnitude of each of these factors can vary from variety to variety. For most common diseases, standard tests and assessments have been developed and varieties are routinely evaluated for disease so that the grower can select varieties appropriately, or at least understand the disease risks associated with any prescribed variety (e.g. Independent Variety Trials in UK: BPC, 2006a). Disease resistance scores on a basis of 1–9 are usually provided in recommended variety leaflets.

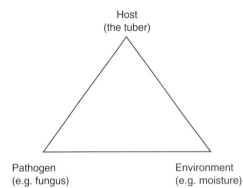

Host
(the tuber)

Pathogen   Environment   **Fig. 4.1.** The disease triangle. (After
(e.g. fungus)   (e.g. moisture)   Vanderplank, 1963.)

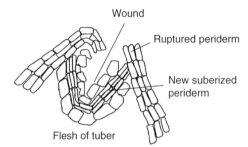

Wound

Ruptured periderm

New suberized
periderm

Flesh of tuber

Fig. 4.2. Formation of new suberized
periderm beneath a wound.

The physical condition of the host can include factors such as damage incurred during lifting and grading operations, which provides access for certain pathogens. While such damage could be assigned to the environmental leg of the disease triangle, it is included here to allow us to concentrate primarily on store conditions when considering environment.

4.2.2 The pathogen

Most potato storage pathogens originate either from the crop's seed tubers or from infections that invade the plant during its growing phase. Pathogens, be they bacteria or fungi, have a wide range of different life cycles, and infect their host in different ways. Bacteria tend to be limited in the ways they can multiply, move, and spread and survive. Fungi in contrast may have life cycles that allow over-wintering and survival in the absence of a host, spore dispersal over long distances, or sequential spore dispersal. As such, survival, spread and development can be different for a bacterium compared with a fungus. For most pathogens, disease life cycles (Agrios, 1988; Bissonnette, 1993) have been explored and are well known (Fig. 4.3). Many pathogens exhibit diversity within the species, so that different subspecies (or strains) may have different optimal conditions for their development and react to sprayed chemicals in different ways. The objective for most store managers is to reduce disease inoculum levels as much as is practicable. Tactics

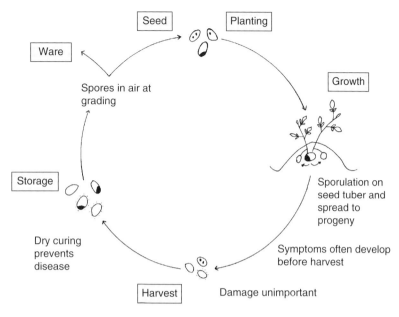

Fig. 4.3. Life cycle of silver scurf (*Helminthosporium solani*). (Source: British Crop Protection Council, modified from BPC Store Hygiene CD.)

may include selecting seed with low disease levels, employing agronomic techniques that prevent disease development, and rigorously cleaning stores and equipment prior to store loading or handling the crop.

Store managers should assess the amount of disease on tubers coming into the store so that management can be tailored to particular needs. This may include early sale of crops already starting to rot, rapid drying or omission of wound healing where silver scurf is the predominant risk. For most diseases it is impossible to improve quality by storage (an exception may be soft rotting, where rapid and sustained drying can 'mummify' rotted tubers so they do not infect neighbouring tubers) and disease levels normally increase during storage. To evaluate how well the storage environment minimizes subsequent disease development, the crop should be sampled at intake, periodically during storage and at grading, to assess any decline in quality over the storage period. A low disease increase may suggest a satisfactory storage regime.

4.2.3 The environment

Many pathogens will die, slow their development or enter a static, resting stage if their environment becomes hostile. Their life cycle therefore fails. Environmental conditions that induce such failure include some that can be influenced by the store manager such as temperature and moisture on the tuber surface. A key contribution to the shape (and size) of a disease triangle comes from the application of fungicides that kill or prevent further development of a proportion of the pathogenic organisms present. The application and efficacy of these chemicals are discussed in more detail in Ch2.5.

4.2.4   Interaction between host, pathogen and environment

In theory, a disease triangle will collapse if any one of the three components meets criteria where disease development is impossible. For example, where a variety expresses complete immunity to a given pathogen, then modifying the environment or removing the pathogen will be irrelevant for disease control. Likewise, if a pathogen is completely absent from a growing region then changing variety or storage conditions will have no influence on development of that disease. More often than not, however, it is difficult to reach these absolute conditions for a given disease and so the role of the store manager is to understand that modifying one point of the triangle will simultaneously influence both of the others.

For example, *Polyscytalum pustulans*, the causal agent of skin spot, grows most rapidly on an agar plate in the laboratory at 15–18°C. However, suberin deposition and wound healing of the potato tuber is most rapid at this temperature range. At 15–18°C, the tuber's wound healing ability allows formation of a barrier to the pathogen before it can access the tuber flesh (Fig. 4.2). In practice skin spot tends to develop most rapidly on tubers in store at 3–5°C, where the rate of wound healing is slow, while the fungus can still grow rapidly enough to access the flesh of the tuber (Box 4.1).

Box 4.1. Slow rate of disease expression at low store temperatures

Due to the low temperatures at which potatoes are stored, disease multiplication may take weeks or even months to develop visible lesions. This makes it difficult to identify what caused the problem and, without a control, there is no certainty that any particular explanation is correct. In a storage experiment (Pringle *et al.*, 1997), a quarter of a stock of 'Desiree' potatoes was stored in an experimental, ambient-air cooled, 45-t box store, while the remainder was stored in the farmer's ambient-air cooled box store. The farmer noticed condensation on the potatoes in his store in October and circulated air round the boxes to dry off any condensation. In January he observed some skin spot, *Polyscytalum pustulans*, in this stock, and by April the disease had developed to the extent that the entire stock had to be sold for stock feed. The 45 t of potatoes stored in the experimental store, which had skin resistance sensors fitted to monitor condensation, were kept free from condensation and suffered only mildly from skin spot (Table B.4.1). The 45 t were sold for seed. The length of time between the likely cause and the subsequent expression of the disease, together with the absence of a control, would normally mean that the two events would not have been linked.

Table B.4.1. Comparative infection on tubers from farm and experimental store on 7 April 1997.

	Skin spot infection		Silver scurf infection	
Store	Proportion of tubers (%)	Surface area (%)	Proportion of tubers (%)	Surface area (%)
Farm store	100	14±16	100	9±9
Experimental store	28	1±2	24	0.5±1

4.3   Storage Strategy Dependent on Crop Condition at Harvest as well as Storage Principles

While this chapter concentrates on storage diseases it is crucial to remember that the disease triangle is greatly influenced by what happens prior to store loading. The store manager should endeavour to apply the knowledge derived from the previous chapters when deciding on the preferred disease control strategies in store. For example, a crop of a dry rot-susceptible variety, grown from infected seed and lifted early in the growing season, will exhibit extremely poor prospects for each point of the disease triangle (Wale and Clayton, 1999). Such a crop will be at risk as:

- Spore concentrations of the pathogen in the soil may be at their peak before spores die or are leached away from the rooting zone.
- The host tubers may not have developed physiologically to a point where wound healing is sufficiently rapid.
- The environmental conditions may be so warm that attempts to cool the crop to a temperature that will reduce the multiplication of the fungus may be just too expensive on refrigeration energy.
- Cooling will inevitably take too long.

Disease development in this case will be influenced more by the environmental conditions at harvest rather than environmental conditions following store loading.

4.4   Bacteria

4.4.1   Avoidance

Diseases caused by bacteria, such as soft rot, blackleg, brown rot, ring rot, etc., can cause significant losses during storage if they are not controlled. Methods of access to the flesh of the tuber can vary. With soft rotting, caused by *Erwinia carotovora* subsp. *carotovora*, access is via damage during lifting, through lenticels where these have burst or become more prominent and exposed due to the growing tubers standing in waterlogged soils, or to a combination of fungal pathogens, pest damage and bacteria. Where a fungal pathogen is present, soft rotting typically develops as a secondary infection.

Store managers can therefore improve the chances of storage without rotting, even for susceptible varieties, by avoiding crops:

- From waterlogged sites.
- Infected with other primary diseases such as blight.
- With heavy pest infestation such as slugs.

Grading-out defects prior to storage may provide sufficient 'salvageable' crop to pay for the additional work involved.

Where it is impossible to avoid crops that are waterlogged, diseased or damaged, an understanding of the life cycle of the bacteria can help when devising disease limitation strategies. The bacteria that cause soft rotting develop most rapidly in susceptible varieties and at high temperatures; the presence of free surface moisture, which creates anaerobic conditions, increases their rate of development. Surface moisture can either be due to wet lifting conditions or generated through condensation during storage (Ch3.5). In either case careful removal of surface moisture, through either windrowing in the field or ventilating the crop during wound healing, can ameliorate the disease risk.

4.4.2 Influence of moisture

Clayton and Cunnington (1996) used stocks inoculated with antibiotic-resistant strains of *E. carotovora* to measure the effects of surface moisture on the development of bacterial populations throughout storage. By preventing condensation during curing they achieved a 50% decrease in bacterial levels compared with crops where condensation had not been prevented.

While moisture is clearly a major factor in bacterial multiplication, the rate of multiplication is more rapid under warmer conditions and less rapid under cooler temperatures. As the temperature of storage for crops destined for fresh use tends to be lower than for crops destined for processing, tuber surface moisture is a more significant risk in the latter.

Bacterial rotting results from the breakdown of cell structures caused by pectolytic enzymes produced by bacteria. This allows rapid access to nutrients and substrates within potato cells and further multiplication of the bacteria.

4.4.3 Influence of temperature

The metabolic rate of bacteria, and consequently their ability to multiply, decreases as temperature decreases. Gray and Robinson (1988) demonstrated that bacterial multiplication was considerably slower below 10°C than at 20°C, taking over 12 times as long to multiply tenfold (Fig. 4.4). For most potato pathogenic bacteria,

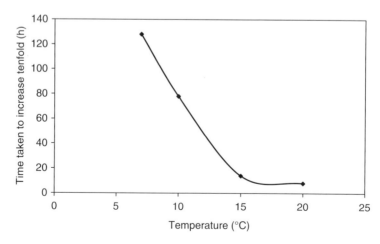

Fig. 4.4. Time taken to achieve tenfold multiplication of bacteria (*Erwinia*) at different temperatures. (After Gray and Robinson, 1988.)

their optimum temperature for development is usually higher than normal storage temperatures. For *Erwinia* species these optima lie in the low 30s Celsius, with subtle differences between the two subspecies that are prevalent in the UK. For example, subsp. *carotovora*, the cause of soft rotting, has an optimum temperature marginally higher than subsp. *atroseptica*, the cause of blackleg, and so will predominate in hotter growing seasons.

Knowing that the rate of bacterial multiplication decreases with temperature would suggest that newly harvested crops with a high bacterial count should be cooled immediately after harvest, rather than being allowed to stay at harvest temperatures to speed wound healing. But not only would such cooling be slow, a maximum of 0.5°C/day, it would also reduce the rate of periderm formation, the primary defence to disease entry. In addition, cooling in a store still being loaded is very likely to result in condensation forming on the crop due to temperature differences between the crop in store and the new crop being loaded (Ch3.5). The order of importance of management actions is therefore to:

- Ensure wounds are healed faster than disease can gain entry.
- Any tuber surface moisture at harvest should be dried and subsequent condensation avoided by good distribution of ventilating air.
- Once wounds are healed, the store can be closed and the crop cooled. During cooling, temperature differentials within the crop should not exceed 0.5°C or condensation, followed by possible bacterial multiplication, is likely to occur.

Where bacteria have entered into the flesh of the tubers, through stolon ends for example, wound healing will have little positive benefit, so cooling to slow bacterial disease development could be justified. As such crops are very likely to break down in store whatever action the store manager takes, their immediate sale as ware or stock feed is probably the best strategy.

4.4.4 Prevention of contamination by resting bacteria and bacterial slimes

The ability of bacteria to survive long term in the absence of the host depends on a species' ability to avoid desiccation through the production of polysaccharide slimes that bind with soil particles and provide protection (e.g. brown rot). Where this mechanism is absent, as is the case for soft rotting *Erwinia*, then long-term survival is rare except in larger clusters of soil and rotting debris. Store managers can avoid such contamination by ensuring stores are cleaned of debris and that any remaining soil particles are allowed to dry between seasons to remove any remaining inoculum.

Some store managers choose to use a disinfectant after cleaning to ensure that the bacterial risk is minimized. This may be good routine practice in areas where brown rot and ring rot are fully established and is a crucial precautionary step in very high-health, safe-haven schemes where absence of key pathogens within the seed multiplication chain is essential. For other crops, however, it has not proved necessary in the UK to use disinfectants against soft rotting so long as routines for cleaning stores and grading and handling equipment are in place.

Typically these routines include:

- Removing large items of debris from boxes and graders.
- Steam cleaning/power washing boxes where rotting has taken place.
- Cleaning and washing graders and allowing their surfaces to dry before re-use.

A particular problem with crops contaminated with bacterial disease occurs at harvest and grading. The presence of rotting mother tubers or early rotted daughter tubers can result in the production of a 'bacterial soup' (Fig. 4.5). This soup or slime can smear the neighbouring sound tubers with a film of inoculum. Since harvesting and grading inevitably cause damage to the skins of sound tubers, the damaged areas become covered with the slime. There is then a race between wound periderm developing and bacteria multiplying rapidly enough to infect the tubers (Fig. 4.4). This creates a high-risk or large disease triangle situation, which can result in hot spots particularly in bulk storage of processing potatoes (Box 9.1). Putting pickers on the harvester to remove any rotting tubers before they contaminate sound tubers can reduce smearing. If smearing does occur, high rate ventilation to rapidly dry the inoculum is the best remedy.

In extreme cases, where the presence of rots has been sufficient to contaminate the grader with a soft, muddy slime of inoculum, it may be necessary to thoroughly clean the grader prior to grading the next crop or a disease-free stock of the same crop. While store loading has been the topic considered up until now, similar attention to hygiene ought to be standard when grading seed crops in the spring. Where such rotting is present, it is often better to dispose of the crop rather than compromise the reputation of the producer, should ware or seed subsequently start to break down in delivered material.

Fig. 4.5. Soft rotted potatoes prior to disposal.

4.5 Fungi

4.5.1 Removing the pathogen

Fungal spores can vary in size from <10 μm (*P. pustulans*, skin spot) to >50 μm (*Helminthosporium solani*, silver scurf). Their structure also differs; in particular, cell wall thickness can vary from >5 μm for *Fusarium* dry rot spores down to <2 μm for *P. pustulans* (Fig. 4.6). As well as presenting these facts for interest, they are included to help develop an understanding of how to influence the pathogen component of the disease triangle. For example, *Helminthosporium* spores are easily captured by the filter of vacuum cleaners used for store hygiene, whereas *Polyscytalum* spores would pass through most filters. Store hygiene strategies for silver scurf should therefore be based on vacuuming stores and grading areas to remove spores rather than attempting to destroy them with disinfectants. Similarly, *Polyscytalum* spores will be more prone to desiccation in store than those of *Fusarium* species. Providing there is adequate time and conditions to allow spores to dry out, a store manager might be less concerned with store hygiene following a skin spot problem than following a dry rot problem.

4.5.2 Evaluating risk prior to storage

As well as store hygiene, the obvious way to influence the pathogen component of the disease triangle is to try to prevent disease development in the growing crop. While this is not always possible, the store manager can at least develop an understanding of the factors that contribute to disease risk and use crop histories to anticipate potential problems in store. He can then take the necessary remedial actions should the market desire low disease levels. Various advisory tools have been published to guide the store manager using this approach (BPC, 2006b).

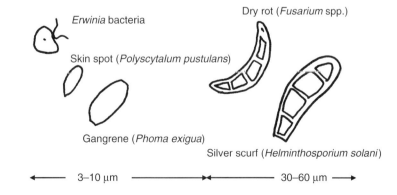

Fig. 4.6. Size of fungal spores and bacteria (not to scale).

In the previous section, an example of bacterial rotting was illustrated. The store manager could create a picture of risk by considering the impact from fungal infections, varietal susceptibility, waterlogging of the crop in the field and damage at loading. For common fungal diseases a store manager can combine similar pre-disposing factors to pre-judge disease risk. With all of these factors, the presence of moisture on the tuber surface can provide conditions that allow the pathogen to overwhelm its host. A number of examples of predisposing factors follow.

Gangrene

- History of infection on the farm.
- Poor or uncertified seed used.
- Seed untreated or treatment inappropriately applied.
- Late, cold harvesting conditions.
- High damage levels at harvest.
- Susceptible variety grown.

Dry rot

- History of infection on the farm.
- Poor or uncertified seed used.
- Dry rot present in seed.
- Unhealed damage after grading.
- Seed untreated or treatment inappropriately applied.
- Early harvest under warm conditions.
- High damage levels at harvest.
- Susceptible variety grown.

Skin spot

- History of infection on farm.
- Poor or uncertified seed used.
- Seed untreated or treatment inappropriately applied.
- Late, cold harvesting conditions.
- High damage levels or scuffing at harvest.
- Susceptible variety grown.

Silver scurf

- Poor or uncertified seed used.
- Seed untreated or treatment inappropriately applied.

Black dot

- Poor or uncertified seed used.
- Seed untreated or treatment inappropriately applied.
- Soil contaminated/infected by short rotation.
- Long period between emergence and lifting.

Blight

- Susceptible variety; note that foliar and tuber resistance differs.
- Conditions conducive to development of blight in the field.
- Poor burn-off/haulm destruction prior to lifting.

Further evidence of these predisposing factors can be found by undertaking the following risk assessments.

Undertaking washed seed inspections prior to purchase
Washing seed prior to purchase will provide information on visible coverage by most key diseases (Box 10.2). These data can be improved by undertaking microscopic examination of incubated excised eye-plugs removed from the infected tubers. More recently these methods have been complemented with new molecular diagnostic techniques for disease detection. It is too early to say how these might contribute to the day-to-day management of a potato store, but some of the larger potato supply companies are already using these tests to identify and eliminate problematic stocks during seed multiplication or to sell ware crops that may be difficult to store for immediate packing or processing.

Undertaking washed quality assessment prior to storage
Prior to storage some diseases such as silver scurf or black dot may not be fully expressed, but with a little practice store managers or their agronomists can pick out problems in their earlier phases. The place to look is often determined by the route the disease takes to gain entry (Ch4.6) to the flesh of the tuber. For pathogens that can travel by digesting the stolon (e.g. black dot) then the store manager might start with tubers where the stolon was easily detached during test digging and examine the few millimetres around the stolon-scar for disease development. Likewise some bacterial rots and fungal blemishes, whose life cycles are driven by waterlogging in the field, may be more prevalent on the underside of tubers, as they were positioned in the soil.

Frequent monitoring of crop disease status once in store
The more informed the store manager is about any changes in disease status, the better able he is to manage any change in store management. This may include removal of parts of the crop for early sale before the disease becomes excessive, altering the store temperature or applying additional ventilation to a section of the store that has become damp.

4.5.3 Resistance of the host

Choice of variety is often beyond the remit of the store manager. However, an understanding of resistance and susceptibility to key diseases will help in undertaking the risk assessments described previously. For some diseases, such as gangrene and blight, methods for determining host resistance are well documented (e.g. Gray and Paterson, 1971; Dowley *et al.*, 1999). For others like skin spot and dry rot, various assessment methods have been published (Kerr and Parrish, 2005) but not all varieties have been evaluated. Each country and producer group will have some relevant data that identify high-risk varieties. For the group of diseases that affect visual quality, such as silver scurf and black dot, genetic differences in the host have been studied but as yet they have little practical significance for the store manager (Hilton *et al.*, 2001).

4.5.4   The role of free surface water, relative humidity and temperature

Like humans, disease organisms can only access water in its liquid state. Water in the form of vapour in air is inaccessible.

The RH of air within the voids of a mass of potatoes in a well-sealed store is controlled largely by the potatoes themselves, at or around 94–97%. The crop alters the condition of any air passing through it. In an experiment by Grähs *et al.* (1978), air being blown into the base of a pile, albeit at a low volume rate of $0.0044\,m^3/s/t$, with a temperature of 7.3°C and RH of 87%, both cooled and increased in RH as shown in Fig. 4.7. At the 'turn', the air is at its wet-bulb temperature. Above the 'turn' the air temperature rises, due to the heat of respiration emitted from the potatoes, but stays at the air's wet-bulb temperature. Even though store managers are able to control when ventilation takes place, they cannot control the RH of the air within the voids between the tubers.

As the condition of the ambient air used for ventilation is totally dependent on the weather, the manager has even less control over the RH of ambient air, unless a humidifier is used, other than to stop ventilation when weather conditions are unsuitable.[†]

Air with high RH is often associated with disease development because, at values of 94–97%, even a small temperature difference of 1°C within the potatoes will cause warm humid air to condense on the cooler potatoes above. This condensed moisture may then become available to disease organisms.

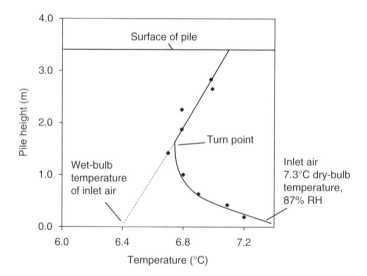

Fig. 4.7. Condition of ventilating air as it rises up a bulk pile. (Redrawn from Grähs *et al.*, 1978.)

[†]The term 'dry curing', used to define the process of ventilating the crop to dry it, is a dubious term as it suggests the quality of air is controlled.

For most fungal diseases water is crucial to some or all stages of growth. Usually it is moisture in combination with warm temperature that creates conditions that allow disease development.

4.5.5   Silver scurf, a case study

Various descriptions and illustrations of the life cycle of the silver scurf pathogen have been published. The one reproduced above (Fig. 4.3) was published by the British Crop Protection Council and illustrates that heat and moisture in combination create opportunities for disease development. There is evidence that each of the key stages in the life cycle – infection, lesion expansion and sporulation – are influenced by water, temperature or a combination of the two (Box 4.2).

Infection
Clayton *et al.* (1998) devised a set of four categories for measuring phases in the infection process (Fig. 4.8):

- Germ tube production to a length two times the diameter of the spore.
- Penetration to a depth equal to or greater than one potato epidermal cell.
- Ramification of germ tubes into new fungal mycelium and curved or proliferated to surround and invade potato epidermal cells.
- No development observed.

The frequency of spores within each category was measured over time on inoculated tubers under temperatures of 5, 10 and 15°C and under moisture regimes of 1 h to 4 days. Progress from penetration to ramification (infection) was observed much more quickly at 15°C than at the lower temperatures, and reached >5% of all spores present after only 6 h of surface moisture being present (Fig. 4.9). This was sufficient to cause substantial surface infection, which was confirmed on replicate samples that were inoculated, treated and stored long term prior to visual assessment. The lesson for store managers is that if silver scurf spores are present on the

---

Box 4.2. Quick guide to fungi

This box describes some of the terms used when describing fungal diseases to help store managers with limited knowledge of the subject.

Germination: term used to describe growth phase where spores produce germ tubes (which become hyphae) allowing growth and plant invasion to continue. Usually has a set of specific triggers (temperature and moisture).
Hyphae: individual strands of a fungus that grow in length, often through plant tissue.
Lesion: term used to describe a discrete visible symptom of disease.
Mycelium: collective term used to describe a matrix of hyphae.
Spore: reproductive cell (or cells) that allows a fungus to disperse, survive adverse conditions and multiply. Usually has a thick protective wall (sporangium).
Scletorium: tightly packed mass of fungal mycelium. Allows over-wintering and survival.

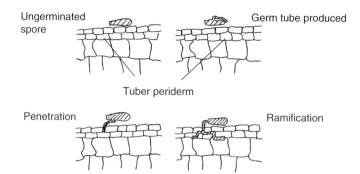

Fig. 4.8. Stages of fungal infection. (After Clayton *et al.*, 1998.)

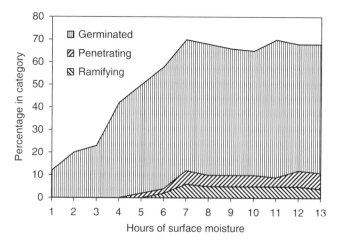

Fig. 4.9. Percentage of silver scurf spores in each development stage after exposure to different durations of surface moisture at 15°C. (After Clayton *et al.*, 1998.)

newly harvested crop or as fresh spores from in-store sporulation, then surface moisture from lifting from wet soils or through condensation must be removed within a few hours to prevent infection. Condensation on tubers can occur in minutes rather than hours if air with a dew-point temperature exceeding the temperature of the crop enters via a door or through the store ventilation system.

Lesion expansion

Hardy *et al.* (1997) induced timed condensation on store-infected or inoculated potatoes for different periods of time at a range of temperatures. After a period of storage, tubers were transferred to high RH conditions at 15°C to induce sporulation. Spores were collected by washing tubers and filtering the wash water and were used to indicate the rate of disease development (Fig. 4.10). Similar to results for infection, they found the combination of high temperatures, common during store loading and wound healing, together with presence of surface moisture to be important in speeding up disease development. At lower temperatures, the presence of free surface moisture allowed an infection to become established, but was less likely to trigger massive disease expansion.

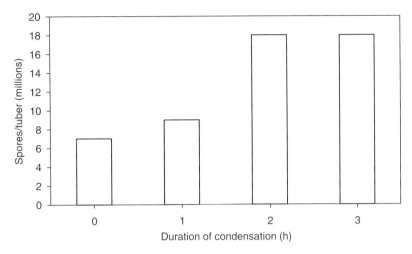

Fig. 4.10. Spores harvested from stored tubers infected with silver scurf following different durations of condensation simulated after store loading. (From Hardy *et al.*, 1997.)

Sporulation

Most pathologists incubate tubers at relatively high temperatures (15–20°C) and at high RH, by adding a water-soaked paper wick to their incubation chamber, when confirming a diagnosis of silver scurf. The conditions created are suitable to allow sporulation and the spores are used for anatomical confirmation of silver scurf. It is therefore safe to assume that both temperature and RH have a role to play in sporulation. The influence of free surface moisture, however, is less well understood. Anecdotally, many store managers and consultants are able to recall the charcoal-like appearance of silver scurf-infected tubers following condensation in-store. The charcoal-like appearance is created by the production of fungal sporangiophores (Fig. 4.11), which support a fresh 'crop' of sporangia of silver scurf and

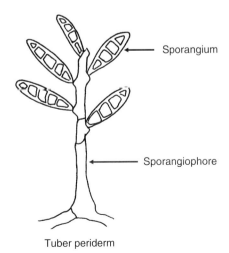

Fig. 4.11. Sporangiophore and sporangia of *Heliminthosporium solani* growing on the tuber periderm.

can occur even at low storage temperatures. Indeed, some retailers may be able to recall a similar appearance after a crop has been washed and bagged and presented for sale.

The lesson for the store manager is that crops already infected with silver scurf can deteriorate in quality and provide fresh inoculum to infect otherwise healthy tubers if surface moisture is allowed to develop. By combining the findings from various experiments, certain 'rules of thumb' can be devised to help the store manager predict quality post-storage.

At 5°C

- 1 h of condensation = twice the disease.
- 2 h of condensation = twice the disease.
- 3 h of condensation = twice the disease.

At 10°C

- 1 h of condensation = twice the disease.
- 2 h of condensation = twice the disease.
- 3 h of condensation = four times the disease.

At 15°C

- 1 h of condensation = five times the disease.
- 2 h of condensation = six times the disease.
- 3 h of condensation = eight times the disease.

So far, the link between temperature and moisture has been considered. Many experimenters report an increase in silver scurf as a result solely of increased temperature. In many of these cases, condensation and subtle changes in surface moisture were not measured.

4.5.6   So how important is condensation?

In combination with temperature, free surface moisture resulting from condensation has so far been shown to influence the rate and frequency of infection by gangrene and skin spot and also influences the rate of multiplication of soft rotting and blackleg bacteria (Clayton and Blackwood, 2001). In all three cases the length of time over which free surface moisture is present appears to be crucial: the longer the period, the more likely disease will develop. Dry rot however appears to behave differently from other fungi. In experiments similar to those described for silver scurf, induced condensation events were shown to reduce the incidence of dry rot infection (Clayton and Blackwood, 2001). In a range of experiments to control dust within potato stores, water applied to graded potatoes was also shown to reduce the development of dry rot. There are a number of explanations for why this happens, including:

- The moisture may dilute the wound exudates so that germinating dry rot spores are less able to 'find their way' into wounds.
- Dry rot spores may not be as 'sticky' as the spores of other fungi so they may just 'float away' from the wound tissue.

- The moisture may help maintain a wide population of naturally occurring antagonistic organisms that may prevent infection by dry rot (this is speculative).
- Any free surface moisture present during the grading process will act as a 'lubricant' as tubers move along the grader, resulting in less damaged or torn skin and less opportunity for disease to enter.

In any case, dry rot has been included here to demonstrate that once in a while something will come along that can 'buck the trend', so the store manager needs to be aware of the potential for particular control strategies for one disease to compromise those of another disease. On balance, however, the recommendation to prevent or remove condensation by appropriate ventilation control strategies should be followed.

4.5.7 And how important is temperature?

Warm crop temperatures favour the development of most storage fungi. Fungi can be classified into two groups: (i) those that develop more during curing (wound healing) and high temperature storage, such as dry rot, silver scurf, black dot and blight; and (ii) those that develop at lower temperatures, such as skin spot and gangrene (Fig. 4.12). While the interaction with moisture is probably ever-present, moisture may persist at undetectable levels so is discounted in this section.

The objective of good store management is to use the risk criteria identified in Ch4.4 and modify storage accordingly if certain diseases are to be controlled.

So, for example:

- A pre-pack crop that has already been designated high risk for skin spot may have its holding temperature raised by 0.5–1.0°C to escape low temperatures where the disease develops most rapidly. While full disease control is not achieved, many businesses use this approach to keep a stock within specification for sale.
- A processing crop that meets high-risk criteria for dry rot might be stored at a slightly lower temperature to avoid disease development. This may require the period it is stored to be reduced to meet target fry colours.
- A pre-pack crop meeting high-risk criteria for silver scurf or black dot may have its curing period reduced or omitted to reduce the period of high temperature during which each disease can develop.

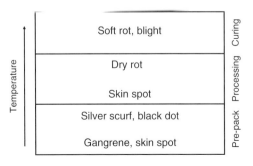

Fig. 4.12. Disease development associated with a range of storage temperatures.

4.6   How Disease Gains Access to the Tuber Flesh

Our theme in this section has been to 'remove the pathogen' from our disease triangle through good hygiene or to limit its growth through chemical applied to the tuber. If we know where or how the pathogen enters the tuber, we can decide what, if any, adjustments to store climate will minimize further infection or multiplication.

4.6.1   Bacteria

Pathogens can enter tubers via a number of routes. Some bacteria (e.g. ring rot) can persist within the vascular tissue of the growing plant. The disease can therefore move from seed to daughter tubers without ever leaving the plant. The same bacteria can also persist for some time away from a host plant and can enter a crop through damaged tissue, so good hygiene practices combined with careful, low damage potato handling are necessary. It is this ability of a pathogen to enter a crop via a number of routes, combined with the loss in crop value due to the infection, that make closed systems for seed multiplication so important. Closed systems, such as the UK British Potato Council's (BPC) safe-haven scheme, produce crops within a 'cordon sanitaire', so limiting the chance of disease entering a seed multiplication operation.

Other bacteria also travel within the vascular tissue (e.g. blackleg-causing *Erwinia*), which explains the classic blackleg tuber symptoms of rotted tissue around the point of stolon attachment. Bacteria can also enter the tuber through lenticels and wounds, sometimes mixed with fungal pathogens such as blight (*Phytophthora infestans*).

Infection by *Erwinia* via lenticels follows the breakdown, or complete rotting, of mother tubers, which exude bacteria that are then washed through the soil to coat daughter tubers. Where soils are wet then lenticels can sit in an 'open' position. Bacteria can then continue multiplication, if conditions are right, and cause rotting either in the field or early on in the duration of storage (Fig. 4.13). This occurs both for *Erwinia* soft rotting and for brown rot (*Ralstonia solani*); in both cases the avoidance of irrigation and the bacteria that irrigation water may contain can greatly increase the long-term storability of the crop. Bacteria may on entry, or soon afterward, enter a latent phase whereby they are 'sheltered' by the tuber's formation of a protective corky layer over the open lenticel. While bacteria in this phase do not cause visible symptoms they can survive until appropriate conditions for development are triggered, so they remain a potential risk to storability. In some cases an

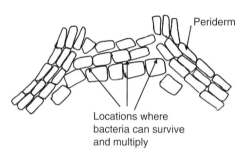

Locations where
bacteria can survive
and multiply

Periderm

Fig. 4.13. Open lenticel allowing access by bacteria.

intermediate stage may be seen where bacteria have started to multiply and affect tissue around the lenticel, but at some point environmental conditions, such as lack of tuber surface moisture or low temperatures, become less conducive to disease development. In these cases, the rot becomes dry and localized and is often referred to as a bacterial hard rot. Entry of bacteria through wounds can similarly be halted, with the resting organisms remaining a potential source of infection should environmental conditions become favourable to disease development.

Bacteria may be present on equipment, on grading lines, on tonne boxes, or in the exudates from decaying mother tubers. When tubers are handled and damage occurs then the bacteria can develop within the damaged areas easily and quickly (Fig. 4.2). Wound healing and deposition of suberin may help the tuber avoid excessive weight loss but will not prevent bacteria that have already entered from multiplying and developing.

Blight is usually regarded globally as the most devastating disease to affect potatoes and massive reductions in crop productivity can and do result from foliar infection during the growing season. In many cases though, it is the role of the invading *P. infestans* blight spores as a carrier of bacteria that makes blight even more devastating in storage. Bacteria are commonly found within laboratory cultures of *P. infestans* and are difficult to remove. Once tubers become affected by blight then two disease triangles, one for the fungus and one for the bacteria, coincide. Both *P. infestans* and the soft rotting bacteria grow most rapidly at higher storage temperatures and very significant losses during storage can result, particularly in processing crops.

Having an understanding and appreciation of these entry points allows the store manager to develop the risk assessments described previously (Ch4.4) and follow 'risk avoidance strategies' to minimize disease development. One for soft rotting is shown below.

Risk avoidance strategy to minimize soft rotting in store

- Avoid wet crops or sites prone to waterlogging.
- Make sure harvesting/grading equipment is clean/dry before use.
- Avoid lifting crops where mother tubers are still breaking down.
- Remove any rotten mother tubers there are from the harvester picking table.
- Minimize damage during harvesting and grading.
- Avoid storing crops that are infected with blight.

4.6.2 Fungi

Fungal diseases can be divided into those associated with tuber damage and those that are not.

For gangrene and dry rot, prevention of damage is paramount for disease control, as these pathogens usually require a wound to gain entry. Damage minimization is required during both harvest and loading and also during the grading process. Wound healing after loading or grading is an important part of the control strategy for these diseases. This is especially so for dry rot in graded seed. The rate of suberin deposition and development of wound periderm is at its slowest

5–6 months after harvest, which coincides with a peak in seed grading activities. As well as careful handling, grader hygiene is essential to reduce disease development. Diseases such as skin spot can develop after superficial damage to the tuber skins by sand particles on machinery and neighbouring tubers. Again, adequate wound healing is an essential part of the control strategy, and applications of CIPC, if used, need to be timed carefully as they can slow the development of periderm cells forming over wound tissue, predisposing the tubers to more severe infection. Diseases that do not require damage for entry are either carried into the store as latent infections (e.g. black dot) or, like silver scurf, can germinate and penetrate the tuber periderm unaided (Ch4.5).

4.7 Ecosystems

An understanding of how one pathogen interacts with another can be useful for decision making prior to, and during, storage. Read and Hide (1984) showed how control of silver scurf could result in more severe symptoms of black dot. The causal organism of black dot (*Colletotrichum coccodes*) competes for the same ecological niche, the potato epidermis, as does that of silver scurf (*H. solani*). In various experiments Read and Hide (1984) demonstrated that where silver scurf was targeted for control then black dot developed readily in stored crops. This may not be particularly exciting to most store managers and could in fact sound depressing. The store manager, however, might begin to consider the risks of each disease developing in the presence or the absence of the other. This should become easier as diagnostics improve. For example, where silver scurf-free seed is purchased and the washed fresh sector market is targeted, then soil with black dot ought to be avoided to reduce the risk of disease development. Similarly, for a crop already heavily infected with black dot, fungicide treatment to reduce silver scurf would be unnecessary since this disease is unlikely to out-compete the black dot. The crop would also be unlikely to reach a premium market given the high disease levels in the first place!

Using one disease as protection against another sounds like a lose–lose situation. What is more promising for the future is the potential of non-infectious organisms being used as biological control agents either in soils, as potatoes develop, or in store, as tuber surface treatments (e.g. various proprietary blends on *Trichoderma* fungi). Various contenders exist within each category, although this may change in time as the rigorous checks on environmental and consumer impact currently applied to agricultural chemicals are implemented for biological control agents. This may or may not preclude their long-term use.

4.8 Hygiene

Store hygiene, or removing the 'pathogen' from the disease triangle in store, is covered later (Ch9.2). What interests us in this chapter is the need for the store manager to extend the principles of good hygiene to all other activities that take place in or around stores. We have seen from the previous section on mode of disease entry that access to tubers becomes much easier once tubers are wounded.

It therefore follows that removing or reducing the levels of pathogens from equipment or processes that can cause damage is equally important. Such operations include the transfer of crop to and from hoppers and the use of graders. The need for hygiene measures and their cost–benefit require considerable judgement. The store manager should be able to make a common-sense risk assessment, similar to that described previously for disease brought in from the field, and might consider the following before grading:

- Are conveyors and riddles well-maintained, low-damage machines?
- If they are covered in dried, hard, sandpaper-like soil deposits, the following stock will damage easily.
- Has the grader recently handled a diseased stock? If so, the grader can be considered dirty.
- Will the disease(s) in question affect saleability of the stock about to be graded?
- Will less well-controlled conditions during post-grading storage and dispatch affect disease development?
- How long does the stock have to 'hold out' before market? If it is to be consumed within a few days then cleaning the grader before handling might prove more costly than beneficial.

The store manager should also consider the 'host' and how that influences the likely need for cleaning a grader:

- How big and what shape are the tubers? Where they are large and long then they will damage more easily.
- How tightly will they need to be graded? If the end product needs a very narrow size distribution and is to be graded from a stock with a wide size distribution, then expect a lot of damage.

4.8.1   Hygiene cost–benefit evaluation

With all store management and grading operations it is important to consider the cost and the benefit. There are many occasions where the benefits of cleaning a grader will be small; for example, when grading a disease-resistant variety after a very clean stock. Likewise there will be cases where over-zealous cleaning will come at a cost beyond the benefit; for example, where disinfectants/detergents remove protective paint from equipment and shorten its working life or where the labour requirement for spray washing each and every box outweighs the benefit in removal of possible pathogens. There may therefore be occasions where a programme of frequent grader cleaning will not pay for itself. Exceptions might apply to high-grade seed production or import/export operations where hygiene is paramount.

4.8.2   So when does cleaning a grader pay for itself?

If the store manager has worked through a risk assessment and has found that the grader has already handled a stock with a particular disease, the next stock to be handled is susceptible to that disease and/or disease development will affect marketability, then cleaning and disinfection could be advised. In a series of studies undertaken for the BPC, Clayton *et al.* (2000) considered four distinct and different diseases: (i) gangrene (caused by *Phoma foveata*); (ii) dry rot (*Fusarium* spp.); (iii) blackleg (*E. carotovora*); and (iv) silver scurf (*H. solani*). In the cases of gangrene and dry rot, the grader was 'contaminated' by first grading infected stocks. For blackleg it was contaminated using pulped infected potatoes. For silver scurf, spores were mixed with dry soil and either sprinkled over the grader or pasted on as a moist soil. Grader cleaning routines were then tested by undertaking a cleaning process and, at each stage, passing healthy tubers over the grader. Success of the treatments was scored depending on contamination or infection showing up either in the healthy test tubers or, in the case of silver scurf, in a daughter crop grown from these.

Figure 4.14 shows that contamination of the grader by blackleg bacteria was reduced to relatively safe levels simply by washing. A further step of allowing the grader to dry nearly eliminated the contamination entirely. The next step in the process, of disinfecting the grader, provided no further improvement in crop health and can therefore be considered unnecessary in this case.

Figures 4.15 and 4.16 show a similar set of cleaning routines for gangrene and dry rot, but which included the additional treatment of first using sacrifice tubers passed over the table to clean its sieves, belts and rollers. In both experiments, where no cleaning was undertaken, denoted by the 'contaminated' bar, disease levels were excessive and sacrifice tubers did not sufficiently reduce infection. In each case full cleaning routines were required to reduce the incidence of disease, albeit to still alarming levels. In such cases the use of disinfectants in the wash water could provide a benefit.

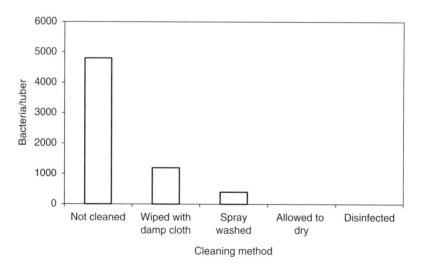

Fig. 4.14. Bacterial numbers (*Erwinia* per tuber) recovered from tubers after their passage over a contaminated grading line cleaned in different ways. (From Clayton *et al.*, 2000.)

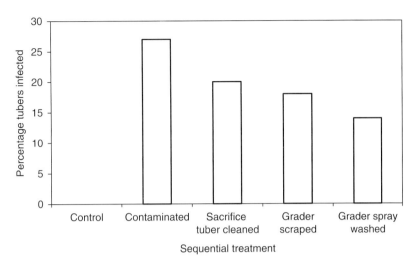

Fig. 4.15. Percentage of tubers infected by gangrene after their passage over a contaminated grading line cleaned in different ways. (From Clayton *et al.*, 2000.)

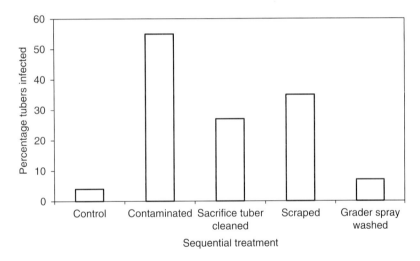

Fig. 4.16. Percentage of tubers infected by dry rot after passing over an inoculated grading line cleaned in different ways. (From Clayton *et al.*, 2000.)

For silver scurf, the grader was contaminated by dust containing spores (Fig. 4.17) and by a wet slurry containing spores (Fig. 4.18). In both cases, clean tubers passed over the uncleaned grader (control) became heavily contaminated. Where inoculum was dry and dusty (Fig. 4.17), sweeping or vacuuming had only a moderate effect on reducing contamination of clean stocks. Where it was caked on to the grader as wet slurry (Fig. 4.18), attempts to remove the dry cake (by scraping with a section of semi-circular guttering chosen to match the size of grader rollers) resulted in high levels of infection, probably because the hardened cake became broken and crumb-like and could stick to tubers more easily.

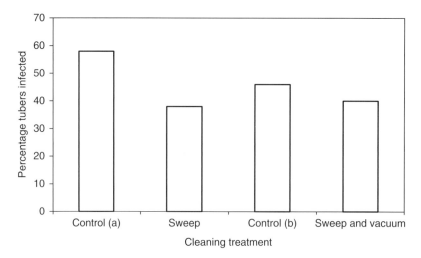

Fig. 4.17. Percentage of infection in progeny tubers after seed tubers were passed over a grading line that was contaminated with dust containing silver scurf and then cleaned by different methods. (From Clayton *et al.*, 2000.)

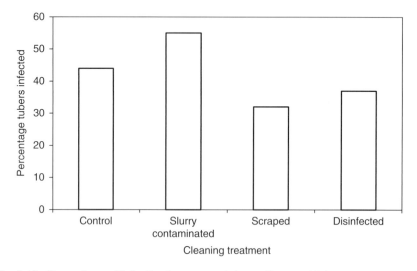

Fig. 4.18. Percentage of infection in progeny tubers after seed tubers were passed over a grading line that was contaminated with wet slurry containing silver scurf and then cleaned by different methods. (From Clayton *et al.*, 2000.)

Whenever there are concerns about the true costs and benefits of hygiene, then a few common-sense questions should allow for a safe, cost-efficient and effective cleaning programme to be devised. These would include:

- Will development of a certain disease affect marketability?
- Is the building designed for easy cleaning?
- Have crops been stored without disease developing in the past?

4.9   Holistic Approach

This chapter has dealt with the predisposing factors that allow diseases to develop, their mode of entry, disease development and some mitigating actions a store manager might take. In this final section the predisposing factors and possible control solutions are brought together in Table 4.1 so that a store manager can find the possible causes and appropriate solutions to a given disease problem.

There are some unavoidable agronomic omissions in the table; field factors such as volunteer control and adequate length of rotation have been excluded to save repetition. These are requisites for good control of most of the diseases listed.

Table 4.1. Predisposing factors that contribute to disease development.

Disease	Predisposing factor	Possible action by store manager
Soft rotting	Susceptible variety	High air flow
	Poor seed quality	Limited curing
	Wet in field	Prevention of condensation
	Wet at harvest	Low-temperature storage
	Poor damage control	Good store and grader hygiene
	Wet during and after loading	Sale of stock as soon after harvest as possible
	High-temperature storage	Regular sampling and temperature monitoring
	Poor air flow	
Skin spot	Susceptible variety	Better seed selection
	Late harvest	Good skin set at harvest
	Poor fungicide application or none applied	Full curing regime
	High damage and scuffing at harvest	Good store hygiene policy
	Sandy soil	Prevention of condensation
	Inadequate curing and wound healing	Elevated storage temperature
	Condensation in store	Fungicide treatment if permitted by market
	Low-temperature storage	
	Premature CIPC application	
Dry rot	Susceptible variety	Better seed selection
	Early harvest	Minimize damage at harvest
	High damage levels	Later lifting
	Intense grading to achieve narrow size band	Good wound healing followed by lower-temperature storage
	Inadequate wound healing	Ventilation and wound healing after grading
	High-temperature storage post-curing	Warm prior to dispatch

Continued

Table 4.1. Continued.

Disease	Predisposing factor	Possible action by store manager
Silver scurf	Poor seed selection	Hygiene in store and during grading
	Poor fungicide application or none applied	Prevention of condensation
	Late harvest	Rapid pull-down after loading
	Condensation postharvest	Low-temperature storage
	High temperature in store	Fungicide treatment if permitted by market
Black dot	Poor seed selection	Low-temperature storage, including rapid pull-down after loading
	Field history of infection	
	Long interval between haulm destruction and harvest	
	Long duration in field between emergence to lifting	
	High-temperature storage	
Gangrene	Susceptible variety	Good hygiene policy
	Poor seed selection	Thorough curing regime
	Poor fungicide application or none applied	Prevention of condensation
	Very late harvest	
	Wet and cold harvest	
	High damage levels	
	Poor wound healing	
	Condensation	
	Low temperature during storage	

4.10 Summary

Potatoes inevitably enter store along with a range of disease microorganisms either already infecting the crop or posing a potential threat. This chapter has focused on the following issues.

- As stores comprise a single airspace, at-risk stocks cannot be treated differently from those in the rest of the store, apart from subjecting them to additional ventilation.
- The crop itself largely dictates the RH in the voids between the tubers, so there is little a manager can do other than ensure the crop is dry.
- The concept of the 'disease triangle' helps visualize the pathogen, host tuber and micro-environment interaction.
- This interaction is well illustrated by the disease skin spot, which poses most risk below the disease's optimum growing temperature, as this is when the host tuber's resistance is low and allows the disease to gain entry.

- Most disease in store has its origins in the growing crop.
- Field conditions like flooding and the presence of blight and other infections provide early warning of potential problems in store.
- Moisture on the crop, be it from wet adhering soil or condensation, is a major factor in disease development.
- Disease development tends to be faster at warm storage temperatures than at cool temperatures.
- The potato's main barrier to infection is an intact, well-healed skin.
- The rotting of blighted or soft rotted tubers can cause bacterial slimes to ooze from the affected tubers and contaminate and digest neighbouring sound tubers in a domino-like manner.
- The removal of potential pathogens may be achieved by vacuuming spore-laden soil from store, removing last year's detritus from boxes and cleaning handling equipment before use.
- For each crop entering store a disease risk assessment should be undertaken and an appropriate management plan put in place; this may include sale or even disposal of a crop of little value.
- Monitoring of stocks in store provides feedback for selecting suitable markets as well as an evaluation of the store and its management.
- Microorganisms can only access water in its liquid state; air with high RH on its own is not enough. Warm temperatures speed their multiplication.
- Silver scurf development is influenced by both surface moisture and temperature.
- While many diseases are triggered by moisture, dry rot infections seem to be reduced by the presence of moisture on tuber skins.
- Bacteria gain access to tuber flesh through the stem attachment of growing crops, via wounds, lenticels, pest holes and the lesions of other diseases.
- A fungus like silver scurf can penetrate the skin unaided, but the majority of fungi need wounds, abrasions or other disease lesions to penetrate the flesh.
- Diseases like silver scurf and black dot, which occupy the same ecological niche, may each develop at the expense of the other if one disease is targeted by a chemical; they should therefore be considered together.
- Hygiene on grading lines is important to minimize contamination, but should be cost-effective and appropriate.
- Disease control should be holistic and consider the risk of disease from field to final grading at dispatch, with appropriate management steps being taken at each stage to minimize risk of disease development.

5   Store Design and Structure

Topics discussed in this chapter:

- Low-cost storage structures.
- Building types and complexes.
- Building components and materials.
- Hygiene considerations.
- Insulation materials and their selection.
- Minimization of air leakage.
- Health and safety.

5.1   Low-cost Structures

Since potatoes arrived in Europe in the 16th century (Burton, 1989), they have been stored in simple low-cost structures, some lasting the period of storage only and some permanent. The basic objectives were to prevent greening, frost damage, being eaten by rodents or birds, and theft. Storage involved either the use of cellars or caves, or protection from the weather using straw and soil. Caves or cellars were particularly good for storage as they reduced the extremes of temperature. In tropical countries, where frost is not a problem, lightweight timber structures, designed to keep out insects and rodents, are common.

5.1.1   Clamps

Above-ground piles or clamps of potatoes were the traditional method of potato storage in Europe and are still used in various forms in many countries (Wilson and Boyd, 1945; Burton, 1989). They comprise simply a pile of potatoes, which is covered with straw and soil to give some protection against the weather and

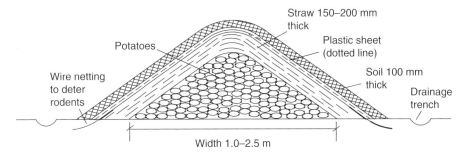

Fig. 5.1. Potato clamp in-field storage.

rodents. They have a low capital cost but high labour requirement. Although there are a number of different types, they all require the potatoes to be heaped into a long narrow pile typically 1.0–2.5 m in width. Height is dictated by the angle of repose of the tubers, approximately 35°, giving a corresponding height of 0.35–0.88 m (Fig. 5.1). A covering of straw, 150–200 mm thick when compressed, is laid over the top of the tubers. This stops greening, insulates the crop from hot or freezing weather and minimizes subsurface condensation on the stored crop. Ditches are dug into the soil on either side of the pile to reduce the risk to the crop from flooding and wire mesh is inserted in the ground to deter rodents. After 2 weeks, once the initial high level of respiration following harvest has subsided, the straw is covered with two plastic sheets which overlap at the apex of the clamp to prevent rain penetration but have a gap between them to allow for ventilation. The plastic sheet is then covered with friable soil 100 mm thick to prevent the plastic being blown away by strong winds and to prevent freezing winds from entering the clamp. In very cold climates a second layer of straw followed by more earth is applied.

A more sophisticated version of the clamp is the 'Dickie Pie' in which one or two 'A' ducts are placed parallel to the long sides (Fig. 5.2) and bales of straw, placed around the perimeter, allow a greater bulk of potatoes to be stored. These 'A' ducts are open at each end and can provide some ventilation of the tubers, allowing the size of the clamp to be increased to between 4.0 and 5.0 m in width. Blocking the ducts with a bale of straw or equivalent can close the 'A' ducts. The cross-sectional area of the duct should be a minimum of 0.013 m^2 per every 10 t stored.

5.1.2 Lightweight tropical structures

Low-cost tropical potato storage is quite different from storage in more temperate climates. Two harvests a year are common in some areas, so storage may only be needed for 1–3 months. This allows potatoes to be stored in simple stores at temperatures that would result in sprout growth if kept longer (Ch3.8).

Most low-cost storage aims to protect the potatoes from the sun, high air temperatures and low RH while keeping predators and insects out. Air movement is

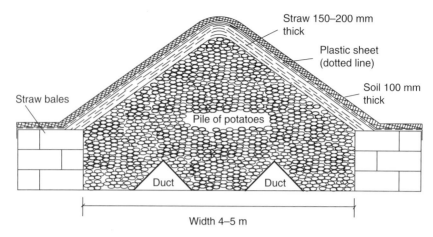

Fig. 5.2. 'Dickie Pie' in-field storage.

encouraged to keep the potatoes as cool as ambient conditions will allow. The bulk of potatoes is kept small to prevent self-heating and a maximum depth of 1.2 m is traditionally suggested. Storage lots should be reduced in size if the temperature exceeds 25°C (Hunt, 1982).

The most basic store is a woven basket, covered over with straw, kept in the residential house. The structure of the house, which provides a stabilized environment for the residents, and sometimes animals, against extremes in ambient temperatures, also benefits the storage of potatoes. The weave of the basket allows for some natural ventilation and thus reduces the likelihood of hot spots or localized deterioration.

Where a free-standing low-cost store is used, it should preferably utilize night-air cooling when the temperature is at its coolest and the RH is at its highest. Ventilation relies on natural convection, using the tuber respiration heat to draw cool ambient air through the crop. The materials used will be what is locally available, but should ensure that all surfaces in contact with the potatoes are in shade during the day. The vertical walls should be solid so that ventilation is in a vertical direction; sometimes a bought-in plastic sheet is used to ensure the walls are sealed. The simplest system has a ventilated floor 0.6 m off the ground so that air can pass through the tubers all the time and animals or rodents cannot reach the tubers at the base of the store. To ensure ventilation is primarily during the night, the store has flaps (Fig. 5.3) that can be opened in the evening and shut in the morning. This can be particularly useful where there is a large diurnal variation such as at altitude. These systems can work well and often potatoes can be stored for up to 3 or even 4 months. If two harvests are grown per year, this can be enough to provide a year-round supply of potatoes.

Bishop and Stenning (1997) describe a further development of the naturally ventilated store using night ventilation where a solar collector is included in the design. This consists of stones laid out on a sloping tray at store roof height, covered with a clear polythene sheet. The solar-heated tray and the store below form a continuous sealed enclosure, with flaps at top and base to prevent ventilation

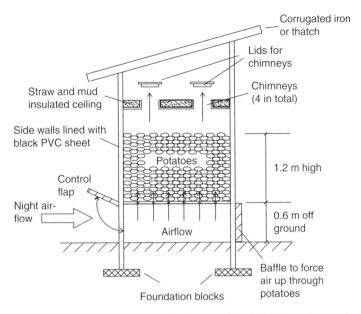

Fig. 5.3. Naturally ventilated rustic tropical store with night-air cooling system.

during the day. In the cool evening, the flaps are opened, and the stored heat from the rocks together with the metabolic heat from the stored crop results in natural convection and cooling of the crop. The initial trials carried out in Kenya at sea level near Mombasa gave some temperatures slightly lower than with night ventilation without the solar collector but additional development work is required.

There are examples, such as in the Mexican highlands, of tropical ambient-air ventilated stores of 3000 t. In most cases, such large stores are refrigerated and used for seed.

5.1.3 Evaporatively cooled tropical stores

When air below 100% RH passes through a wetted pad or fine mist, it will absorb water, increasing its RH. The latent heat absorbed in this process also cools the air. Evaporative cooling would appear to have considerable potential to cool potatoes while minimizing weight loss in warm dry climates.

One design of an evaporatively cooled store is a mud or blockwork walled hut on top of a floor that is kept constantly wet (Fuglie *et al.*, 2000). The floor is made up of a series of low walls, with wet sand in between and covered with a bamboo mat. Wire mesh-covered openings at the bottom of the hut walls, on the two opposite sides, allow air to enter the base of the store to absorb moisture from the wet sand before rising up through the crop. A wind-operated exhaust turbine, located in the roof, encourages this air movement.

While the system did show considerable potential to reduce crop weight loss and improve quality, the increase in value of the crop was negated by the cost of the stores. In addition, the potential was not as high as it might have been due to the owners of the stores not keeping the floors wet at all times.

Another design used moistened permeable pads, fitted into the sides of the stores, so that the air entering the stores was humidified (Hunt, 1990; Bishop and Stenning, 1997). However, the additional resistance to airflow caused by the pads negated their cooling potential in stores relying on natural ventilation. Although the technology is simple these systems are susceptible to problems such as keeping the walls uniformly wet, preventing algae from growing on the wet pads, holes developing in the pads, or saline build-up in recirculated water. Good management is therefore essential. For the pads to work, two or three times as much water has to trickle down the pad as is evaporated and pad height has to be kept to below 1.5 m to ensure the whole pad is kept wet. For effective cooling of side-ventilated stores forced ventilation is required. Evaporatively cooled stores in practice have so far promised more than they have delivered.

5.2 Building Structures

5.2.1 Portal frame buildings

Portal frame buildings in the UK are now the industry standard for potato storage and processing. Their clear span allows operation of potato elevators and forklifts free from obstructions (Figs 5.4 and 5.5). Their extensive use for industrial buildings means that they are competitively priced.

Fig. 5.4. Portal frame seed store with adjoining grading area. (Ordens, Portsoy, Aberdeenshire, UK.)

Fig. 5.5. Internal view of a clear-span, portal frame, pre-pack store. (Courtesy of W. Leslie, Farm Electronics, Lincoln, UK.)

When erecting portal frame buildings, certain aspects should be considered:

- The portal frames require to be strengthened to resist the pressure of the bulk stored potatoes on the load-retaining walls.
- Their lightweight insulation compared with underground stores makes these stores vulnerable to solar radiation in the daytime and rapid cooling on cloud-free nights.
- Wind pressure, especially over high, low-pitched roofs, causes high-pressure differences between windward and leeward sections of the building. This increases air leakage through any gaps in the building fabric, louvres or doors.
- Normal portal frame construction results in the use of sheeting rails and purlins, which can collect dust and disease spores over time.

5.2.2 Sloped wall buildings

To reduce the high cost of vertical retaining walls, bulk stores, particularly in North America, are often designed with sloping walls (Figs 5.6 and 5.7). The pressure exerted by the crop on a 60° sloping wall is between 25 and 33% that on a vertical wall (Waelti, 1989). The building is designed as a single structural entity, using timber or steel frames, with the outward force of the crop being countered by a reinforced concrete wall set into the ground. The floor may be made of either tamped loam soil or concrete.

The stores have a central ventilation duct, with crop stored in bulk piles between 5.5 and 7.3 m high on either side. The ventilation fans, media humidifiers

Fig. 5.6. Bulk stores with sloping retaining walls. (Courtesy of M.J. Frazier, Idaho, USA.)

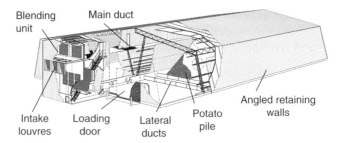

Blending unit   Main duct

Intake louvres   Loading door   Lateral ducts   Potato pile   Angled retaining walls

Fig. 5.7. Cut-away of a bulk store showing ventilation system. (Courtesy of Suberizer Inc., Washington, USA.)

and air-blending chamber are all located within the extension at the front of the building. Intake louvres are fitted in the front of the extension. The doors on either side of the extension allow access to either side of the store for loading the crop.

Potatoes arrive at the store in bulker truck or trailer, fitted with a horizontal conveyor discharge. This discharges on to a cleaner/sizer, feeding an elevator or piler, which loads the store. If above-ground lateral ducts are used, these are connected into the main duct and put in place, starting at the back of the store, as the piler retreats.

For unloading, all the machinery is turned around. An elevator ('hog' or 'scooper') removes potatoes from the pile and elevates them to the truck directly or via a cleaner.

5.2.3   Heated wall buildings

Sudden reductions in external temperature due to a change in the weather can result in condensation on the roof and walls of a store, or even condensation forming on the potatoes themselves. This is particularly a problem with stores fitted with continuously humidified air ventilation systems. One solution to this problem is to build stores with a double skin and maintain the airspace between the skins at 2–4°C above the crop set-point temperature. In this way the inside wall of the store is always warmer than the crop and the store air so that condensation is prevented. The popularity of these stores in the USA has been increasing in recent years (T. Forbush, Techmark, Michigan, USA, 2007, personal communication).

5.2.4   Underground buildings

In continental climate areas of the world, like the mid-west USA, Turkey and Eastern Europe, low temperatures in winter and high temperatures in summer are the norm. By burying the building underground, the thermal mass of the soil above not only insulates the stored crop from extremes of hot and cold weather but also greatly reduces the effect of solar heat gain.

While the thermal concept of these buildings is sound, it does mean that the building structure has to carry the soil that covers the building. This adds to the initial capital cost. This, together with the increased availability of refrigeration and the relatively low cost of electricity, has meant that these types of buildings are no longer built. In some areas of the world, caves or old mine workings are used to achieve the same effect but without the high initial cost.

5.2.5   Underground cellars

Modelled after storage units used by Idaho farmers in the 1930s, cellars to store 25 t of potatoes in places like Afghanistan are being promoted by the Citizens' Network for Foreign Affairs (Rowe, 2005). The stores consist of a hole dug in preferably sloping ground, 3.5 m deep, 5 m wide and 8 m long, with a long sloping passage to allow access (Fig. 5.8). Foundation walls are built around the sides of the excavated hole, 2.5 m high on the high side and 2.0 m high on the low side. Beams 5 m long (Fig. 5.9) sit on the foundation walls, to support the roof. One metre long, 200-mm-wide boards, with gaps left between them for air movement, are nailed on to the beams. Retaining walls, built 1 m high round the perimeter of the hole, form the edge of the roof. Straw, 1 m deep, is placed on top of the roof boards and is filled level with the top of the retaining walls. The retaining wall on the lower side has drains in it to allow any rainwater leaking into the straw to escape. Lastly, the straw and retaining walls are covered with 100 mm of loose soil.

An 'A'-shaped perforated duct stretches the length of the store, and is fitted to an air intake box at one end. This allows air within the store to circulate naturally, down the intake box, along the duct and up through the pile. Two ventilation

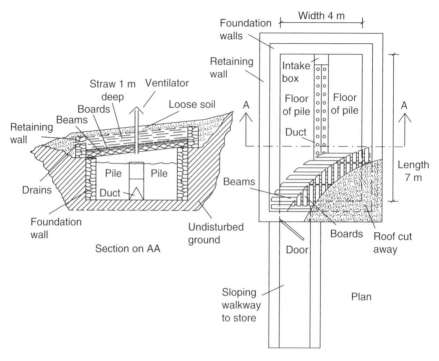

Fig. 5.8. Section and plan of an underground 25-t store.

Fig. 5.9. Underground store under construction. (Courtesy of Citizens' Network for Foreign Affairs, Washington, DC, USA.)

Fig. 5.10. Indian multi-storey sack cold store. (Courtesy of W. Leslie, Farm Electronics, Lincoln, UK.)

pipes, located in the roof, allow ambient air to circulate through the store when cool night air is available. Ventilation is controlled by the store owner, using a thermometer lowered into the store to check when cooling is required.

5.2.6   Sack stores

In parts of the world such as India where labour is relatively low in cost, potatoes are often stored in manually loaded multi-storey refrigerated cold stores (Figs 5.10 and 5.11). Loading is either manual using ladders or by a form of 'dumb waiter'-type lift arrangement. The store illustrated has a steel framework five floors high fitted with porous floors. The sacks, usually 50 kg, are stacked flat, in a series of piles on the floor (Figs 6.29 and 6.30). Refrigeration cooling coils, suspended from the roof, emit cooled air, which passes over the piles of sacks on the upper floor. This forms a circulatory air movement within the store, with air being drawn up through the gaps between the piles of sacks. Many stores rely only on the refrigeration cooling coil fans to create the circulation of air. In the sketch shown (Fig. 5.11), based on a design by Tolsma Techniek, The Netherlands, additional air recirculation fans are fitted to improve the air movement through the sacks of potatoes.

5.3   Building Complexes

In building complexes, where a number of stores include a grading or dispatch area, stores should be accessed via this common area. Access to individual stores should be restricted to a single, high-quality, well-sealed loading/unloading door

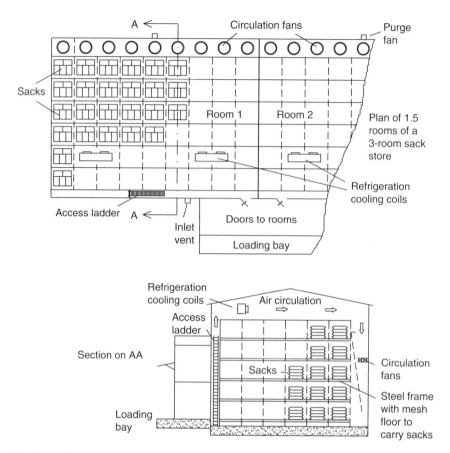

Fig. 5.11. Refrigerated multi-storey sack store. (Redrawn from Tolsma Techniek, Netherlands plan.)

with a second personnel door fitted to ease staff access. Doors are potentially a source of air leakage so the fewer the better. The enclosed grading area that adjoins the stores acts as an air lock. If the store door is open, the grading area door should be kept shut. This reduces the risk of freezing ambient air damaging the stored crop or warm moist ambient air condensing on cold potatoes held in the stores.

5.3.1 Bulk bin complex

One layout popular in The Netherlands, used to separate potatoes into 200–500-t batches, uses a series of bins opening out into a central corridor (Fig. 5.12). The design shown contains 530 t of potatoes in each of six stores, or 3200 t in total. A mobile elevator, fed from trailers or trucks, loads the bins. As duct lengths are 26 m long, above-ground tapered ducts are used to ensure uniform airflow through the crop. When the bin is nearly full, removable timber battens, or half-height doors, are closed in the front of the bin to act as a front retaining wall. Loading of

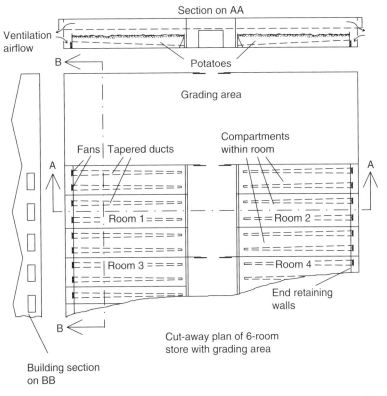

Fig. 5.12. Building complex of bulk stores holding a total of 3200 t, with grading area attached. (Redrawn from Tolsma Techniek, Netherlands plan.)

potatoes continues over the wall or doors. This prevents the potatoes rolling forward and provides a vertical front retaining wall. The sealed door in front of this retaining wall is then shut.

To unload the bins, the bottom timber battens are removed, or the bottom half-door opened, allowing potatoes to flow out the bottom of the bin on to an elevator conveyor feeding a central conveyor running the length of the central corridor. The conveying system takes the potatoes from the bins to a grading system at the end of the storage building complex.

5.3.2 Box store complex

The layout for a group of box stores for ware potatoes is similar to that of the bulk bins, with each store holding 500–1000 t of potatoes. Each store should be sized so that it can be filled in 4–6 days, to minimize the exposure to changes in weather. Once the doors are closed, the internal climate can be controlled.

Air movement and cooling can be provided by a blending box/refrigeration system fitted with vertical ducts which throw air over the top of a stack of boxes. Boxes are stacked to maximize airflow through the pallet apertures.

If a letterbox system for seed potatoes is required, the end retaining wall shown in Fig. 5.12 is replaced by a letterbox duct (Ch2.8), with the duct being pressurized by fans jetting air downwards into the duct. Boxes are then stacked against the duct for ventilation.

Loading and unloading is by forklift truck.

5.3.3   Grading area

Where possible, the grading and packing area should be under the same roof as the stores it is drawing potatoes from. This allows the transfer of potatoes from store to grading area to be unaffected by frost, snow or wet weather. It also allows stores to be opened without wind adding to the likelihood of outside air entering the store. If possible the grading area should be equidistant from all stores.

5.4   Store Aspect and Exposure

If loading of lorries to dispatch crop has to be done outside the grading area or individual stores, the loading area should be designed to give shelter from the prevailing winds. In exposed areas, buildings can either be surrounded by trees or surrounded by an earth bund.

Building layout and colour should be selected to minimize solar heat gain. For the northern hemisphere, the relative solar radiation on roof and walls can be found from Table 5.1. The figures give the temperature difference that should be added to roof and wall heat gain calculations to take account of solar heat gain (Box 5.1).

Table 5.1. Allowance for solar radiation in °C. (From Searle, 1975.)

Type of surface	East wall	South wall	West wall	Flat roof
Dark	4.4	2.8	4.4	11.0
Medium	3.3	2.2	3.3	8.3
Light	2.2	1.0	2.2	5.0

Box 5.1. Solar heat gain expressed as a temperature differential

To calculate heat gain through a roof or wall, the ambient dry-bulb temperature for that part of the world and a store set-point temperature are selected. These may be 18°C and 3°C, respectively. The temperature differential across the structure will therefore be $18 - 3 = 15°C$. To take account of solar heat gain, the figures in Table 5.1 provide an additional temperature differential that should be added to this figure. For a dark roof, a further 11°C should be added. This increases the temperature differential to $15 + 11 = 26°C$.

The table shows that dark roofs in sunny climates should be avoided and west- and east-facing walls receive more solar gain than south-facing walls. There is zero solar gain in north-facing walls. External temperature sensors should be on the north-facing wall to avoid being affected by the sun. Where potatoes are stored in boxes, empty boxes can be stacked on the sunniest side of the store to provide shading for the wall. In tropical countries, a twin-skinned roof is often used so that natural air movement carries away the radiant solar heat.

5.5 Designing for Store Hygiene

Stores, grading areas and grading equipment should be designed to minimize the risk of crop being infected with spores or contaminated with exudates from rotting tubers. Spores are present in the soil, which when dry may turn to dust, become airborne and settle on floors, ledges and potatoes. Contamination by exudates occurs on conveying, washing, grading, brushing and packaging machinery.

5.5.1 Exclusion of soil from store

The quantity of dust brought into store can be minimized by effective cleaning of crops by the harvester and into-store cleaning equipment (Ch2.2). The greatest amount of dust is generated once crops have dried out and are moved for the first time after becoming dry. By enclosing box-tipping equipment or drops in conveying equipment within a dust hood, a proportion of the dust can be removed by dust extractors. Boxes conveyed by forklift spill considerable quantities of dust on to the floor. This is dispersed into the air as the forklift passes to collect another box. To minimize dust dispersal, forklifts should be driven slowly, transport distance to the tipping point should be kept as short as possible and floors should be regularly vacuumed.

5.5.2 Store structure

The structure of the store can help minimize the amount of dust in store. Normal store construction has exposed horizontal sheeting rails and roof purlins (Fig. 5.13), where dust can settle. Any air movement can cause the dust to become airborne and potentially infect stored crop. Either sheeting rails and purlins should be routinely vacuumed, or, where hygiene is paramount, the inside of the store can be lined, for example with an extruded polystyrene board laminated to a white, rigid PVC sheeting, 1.5 mm thick, such as that made by Owens Corning Building Products (UK) Ltd (St Helens, UK). This is fitted on the inside of the sheeting rails and purlins to give a smooth, white, washable, dust-free surface. Where the cladding and insulation are to be provided by composite panels, rather than using one thick composite panel, two composite panels half the thickness can be fixed on both outside and inside of the sheeting rails. This will significantly add to the price of the building.

Fig. 5.13. Sheeting rails and purlins where dust can settle. (Courtesy of W. Leslie, Farm Electronics, Lincoln, UK.)

5.5.3 Ease of cleaning

Store design should take into account ease of cleaning. This will mainly involve suction cleaning (vacuuming), which may require long extensions to cleaning equipment to reach the top sheeting rails or the use of a travelling gantry. A slight slope on the floor allows water to drain when washing floors.

5.6 Insulation Materials

The RH in a potato store ranges between 85 and 100% (Pringle *et al.*, 1997). In cold weather internal store surfaces may become wet with condensation. Only insulation with a high resistance to water vapour transfer should be used. If vapour penetrates the insulation, it will reduce its insulation value. Values for thermal conductivity and resistance to vapour transfer for various insulation materials are shown in Table 5.2. These are average figures taken from trade literature and mostly relate to published standards for thermal conductivity (BS EN 12667:2002; BSI, 2001) and water vapour resistance (BS EN 12086:1997; BSI, 1997a). The water vapour resistance is the reciprocal of the rate of water vapour transmission, which is a measure of the weight of vapour (kilograms) flowing through a material per second with a pressure difference measured in giganewtons. The resistance is therefore giganewton seconds per kilogram of moisture (GN s/kg). A high value indicates a good resistance to water vapour penetration. Box 5.2 describes the terminology used to define the thermal properties of buildings.

Table 5.2. Insulation material characteristics.

Insulation material	Normal form	Thermal resistivity (m °C/W)	Thermal conductivity (W/m °C)	Water vapour resistance (GN s/kg)[a]	Trade names	Comment
Polystyrene – closed cell, extruded	Board	36.0 36.0 33.3	0.028 0.028 0.030	625 NA 456	Styrofoam Styrodur Polyfoam	Insulation of choice before composite panels became popular
Polyurethane foam	Cavity/spray-on foam	43.5	0.023	600–1000	Urethane spray foams	BUFCA Ltd[b] controls foam quality
Polystyrene – expanded (EPS)	Board	29.4–27.0	0.034–0.037	–	EPS Hemsec polystyrene panels	Needs vapour barrier each side to resist moisture
Polyisocyanurate (PIR)	Composite panel	50.0	0.020	500 taped 100 untaped	Baxendale Chemicals Ltd	U-value of 80 mm panel = 0.25W/m^2 °C
Urethane foam	Composite panel	47.6	0.021	269	Kingspan Aeroliner	Good seal between panels is vital

NA, not available.
[a] GN s/kg = MN s/g.
[b] BUFCA Ltd = British Urethane Foam Contractors Association.

Box 5.2. Terminology and values used to define building thermal properties. (Based on CIBSE, 2006)

The *thermal conductivity*, λ, is the rate of heat transfer (W/m °C) through insulation material 1 m thick with a 1°C temperature difference between one side and the other. It is used to compare the conductivity of different insulation materials. Sometimes values of its reciprocal, *thermal resistivity*, $\frac{1}{\lambda}$ (m °C/W), are quoted by insulation manufacturers instead. *Thermal conductance* (W/m²°C) is the rate of heat transfer per 1°C temperature difference through a material of thickness *L*. The reciprocal is the *thermal resistance* (m²°C/W).

The *surface conductance* is the rate of heat transfer between a surface and air per 1°C temperature difference between the two. Its reciprocal is *surface resistance* (m²°C/W).

The thermal and surface resistances of all the components of a wall, floor or roof are added together to find a *total resistance* for the structure. The reciprocal of the total resistance is the *thermal transmittance* for the structure, called the *U-value* (W/m²°C).

In the UK porous insulation materials such as glass fibre or rock wool are usually avoided in new potato stores, as they are liable to become soaked with condensation and lose their insulation value if the vapour barrier used (e.g. polythene sheeting) becomes punctured or the installation workmanship is substandard.

Board insulation materials used for stores should be rigid extruded polystyrene board (e.g. Styrofoam, Styrodur, Polyfoam, etc.) with a closed cellular structure, which resists vapour movement. Alternatively they can be polyurethane foams made into boards by injecting foam between aluminium foil, PVC sheets or plastic-coated steel sheet vapour barriers. With the polyurethane foams, vapour can enter the insulation via the exposed ends of the sandwich, so a sealant should be applied on exposed or newly cut edges.

Polyurethane foam can also be sprayed directly on to the store sheeting. This has the advantage of sealing the store from air leakage and is an excellent way of renovating buildings that are nearing the end of their life.

Extruded polystyrene sheeting, similar to the material used in disposable coffee cups, is sometimes used as insulation for its cheapness, but unless it is sandwiched between vapour barriers it will lose a proportion of its insulation value over time due to surface moisture absorption.

Recommended thermal transmittance values (U-values) for various storage systems and store locations are shown in Table 5.3. Since ambient-air cooled ventilated stores usually only store potatoes for 4–6 months into early spring, the U-value specified is less than for refrigerated stores, where storage may last for 8 months into early summer. More insulation is therefore required to minimize heat ingress due to the warmer summer temperatures. However, higher levels of insulation for ambient stores are an advantage as 2–3 weeks can pass when no cool ambient air for cooling is available. The better the insulation and store sealing, the slower the crop will warm during periods of warm weather.

Table 5.3. U-values for potato stores and thickness of foam and rigid board. (From Cargill, 1976; Cargill *et al.*, 1989.)

	Walls			Roof		
		Thickness (mm)			Thickness (mm)	
Store type	U-value (W/m²°C)	Foam/ PIR	Board	U-value (W/m²°C)	Foam/ PIR	Board
UK						
Ambient	0.42	50	60	0.36	60	80[b]
Refrigerated	0.36	60	70	0.26	80	100
Continental						
Cold	0.27	80	100	0.20	110	140
Very cold[a]	0.20	120	140	0.14	160	200

PIR, polyisocyanurate.
[a]Insulation higher than necessary for temperature control to prevent condensation on the internal structure.
[b]70 mm would be sufficient but sheet does not come in this size.

Basic U-values for different wall, roof and floor configurations can be calculated (Box 5.3). For more complex configurations the CIBSE guide (CIBSE, 2006) should be consulted.

5.7 Building Fabric

5.7.1 Corrugated sheeting and board insulation

The traditional structure for potato stores in the UK, until the mid-1990s, was to externally clad a portal frame building with insulation board, and then protect that from the weather with corrugated plastic-coated steel sheeting (Fig. 5.14). This construction avoids any cold bridges, as stanchions are within the outer insulated shell. Air leakage between the boards (Box 5.3) is minimized by:

- Using tongue-and-groove insulated boards.
- Taping the joins between the boards.
- Using two layers of boards staggered to ensure joins in the boards do not line up.

The tongue and groove is of limited benefit, as often the fit is not tight, the boards can shrink over time and only the sides, not the ends, are grooved.

Sealing of every join with mastic or silicon sealant, while effective, is very labour-intensive and needs constant supervision of labour.

Taping of joins is successful only if applied externally, prior to the boards being fixed. If done on the inner face of the boards, the tape loses its adhesion over time and hangs in arcs from the roof. Staggering the boards is successful but more expensive to install compared with fitting a single thicker board.

Box 5.3. Calculating U-values of walls, roofs and floor

The U-value, or thermal transmittance, of a wall, roof or floor can be calculated if the individual thermal resistances of the materials that make up the wall, roof or floor are added together and the reciprocal taken. Since there is resistance to heat transfer between a material surface and air, and through airspace (Tables B.5.1 and B.5.2), these too should be added.

Table B.5.1. Resistances to heat transfer to and from surfaces. (From CIBSE, 2006.)

Location	Structure	Resistance ($m^2 °C/W$)
External	Walls	0.055
	Roof	0.045
Internal	Walls	0.123
	Roof	0.106
	Floor	0.150

Table B.5.2. Resistance to heat transfer of airspaces. (From CIBSE, 2006.)

Size of airspace (mm)	Resistance ($m^2 °C/W$)
5	0.11
≥20	0.18

The insulation value of materials is often quoted by their conductivity per metre of material rather than their resistance to heat flow. The resistance R of a component of the structure is calculated as:

$$R = (1 / \text{conductivity}) \times \text{thickness (m)}.$$

For example, for Polyfoam:

$$R = 1 / 0.030 \times 0.080 = 2.667 \, m^2 °C/W.$$

If a wall construction is as in Fig. B.5.1 and the resistance for 150-mm concrete block is 0.104 $m^2 °C/W$ and for Polyfoam is 2.667 $m^2 °C/W$, the total resistance for the walls is calculated as follows.

Component	Resistance ($m^2 °C/W$)
External surface	0.055
Concrete blockwork	0.104
Airspace	0.180
Extruded polystyrene	2.667
Internal surface	0.123
Total resistance (R)	3.129

Continued

Box 5.3. Continued

Thus:

   U-value = 1 / R = 1 / 3.129 = 0.32 W/m²°C.

For more precise calculations of U-values, see the CIBSE guide (CIBSE, 2006). Values for thermal conductivities of a number of common insulation materials are given in Appendix 3.

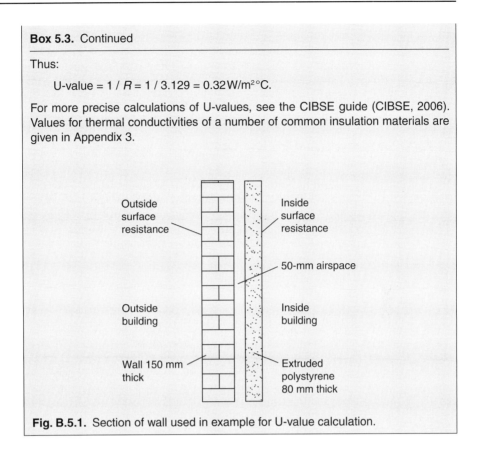

Fig. B.5.1. Section of wall used in example for U-value calculation.

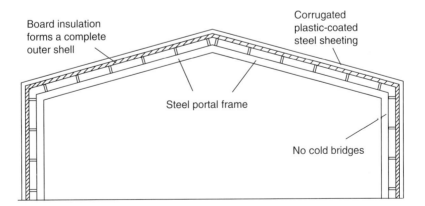

Fig. 5.14. Insulation external to steelwork avoids cold bridges.

Board insulation is still used to insulate existing non-insulated buildings, but in that situation it is fixed to a timber framing fixed to the sheeting rails and purlins and forms an internal shell to the building. The portal frames have to be boxed in to prevent cold bridges. So long as sealing between boards is achieved, the use of board insulation makes a very effective insulation system.

5.7.2   Spray foam applied to store cladding

The great benefit from spray-on polyurethane foam is that it seals every crack or gap in the store, so that leakage through the building fabric is prevented altogether. So long as doors and louvres are well sealed, the building will be airtight. The foam should be applied at ambient temperatures above freezing, using CFC-free foam applied by a reliable contractor, who applies the correct thickness (usually 60–80 mm in the UK) of material throughout. The final result will be a well-insulated, well-sealed store.

Care has to be taken regarding the fire rating and insurance cover. If the foam itself does not comply with the fire rating ISO 9705:1993 (ISO, 1993) or BS 476: Part 7:1997 (BSI, 1997b), it will have to be spray-coated with an intumescent paint which expands when a flame is applied to it. Only foams which are acceptable to the client's insurance company should be used.

The foam forms an insulated shell on the inside of the building, rather than on the outside when board insulation is used. All stanchions, purlins and sheeting rails must therefore be sprayed with foam to avoid cold bridges and prevent condensation forming on the portal frames.

The foam surface is rough, with the result that dust can collect on the vertical surfaces. The walls may therefore require more frequent cleaning compared with situations where the insulation has a smooth surface. Its colour can darken in time to a dull yellow and the surface can be damaged though collisions with machinery or boxes on forklifts. If power washers are used to wash the foam, they should be limited to 3.4 MPa (34.5 bar) pressure if the foam is not to be damaged (W. Elder, CPS, Fife, UK, 2005, personal communication).

5.7.3   Composite panels

Composite panels consist of 600–1000-mm-wide sheets of polyisocyanurate or urethane foams sandwiched between corrugated plasticized steel sheets (Fig. 5.15). They are manufactured in lengths to match the full wall height and roof span of the building. The sides of each panel are designed to locate with its neighbour to provide an airtight seal. The sandwich construction makes the composite sheets stiff, so that purlin spacing can be wider than with corrugated steel sheeting alone.

The interior of the panels can be white. This produces a bright, clean-looking effect, liked by supermarket buyers. The surface is easily cleaned, is less prone than applied spray foam to hold dust, and maintains its fresh appearance indefinitely. These features, together with their cost becoming more competitive with other

Fig. 5.15. Section of composite panel.

forms of insulated cladding, have made this the preferred construction method for new stores in the UK.

In the past a major weakness of this form of construction was the seal between neighbouring panels and where the panels meet the floor, the roof sheeting, the gables and the ridge. This has resulted in a number of buildings incapable of producing a quality product or with excessive refrigeration operating costs (Box 5.4).

Recent composite sheeting now uses three beads (i.e. extruded lines) of mastic to seal the joins between the panels. If there is any doubt as to the sealing method, the contractor can be left to select how he wants to seal the building, with the building contract specifying that the building be subject to an air pressure test when complete. Air leakage of $10\,m^3/m^2/h$ at $50\,Pa$ pressure for composite sheeting is quoted as the worst case that these panels will achieve in practice (Kingspan Insulation Ltd, Leominster, UK; http://www.insulation.kingspan.com/).

As with spray foam buildings, the client's insurance company should be approached to verify that the foam and type of construction of composite panel has a fire rating acceptable to them.

5.7.4   Load-bearing panels

Load-bearing walls taking the lateral force of potatoes in a pile should be within the insulated shell of the store. This will minimize the possibility of potatoes being affected by frost, low-temperature sweetening, or condensation forming on the panels and the potatoes nearest the walls.

Box 5.4. Consequences of air leaks in potato stores

Stores with leaks in their building fabric or which have no way of keeping doors closed during cleaning, grading or packing are liable to:

- Condensation on the crop when warm humid air enters the store and meets cool potatoes.
- Freezing injury to the crop in freezing weather conditions.
- Lack of crop temperature control.
- Require excessive cooling ventilation to compensate for warm air entering the building.

The building shown in Fig. B.5.2 was manufactured and built by an enterprising farmer, but the crops he stored were severely infected with silver scurf. Air was found to be leaking into store between the joins in the composite sheeting. The leaks have subsequently been sealed with mastic, a major undertaking. It is too early to know whether the leaks have been adequately sealed and the disease problems solved.

Fig. B.5.2. Composite panel building that experienced air leakage between vertical sheeting.

5.8 Floors

5.8.1 Uninsulated floor construction

In the UK, floors for stores are rarely insulated. The primary considerations are to ensure that the floor is strong enough to take the weight of forklift trucks with industrial-type, solid rubber tyres, carrying or raising potato boxes. Often two boxes, holding 1 t of potatoes each, are put on the top of the stack of boxes, as the forklift has insufficient reach to put the top box up on its own. To cope with these loads, floors should be reinforced with steel mesh or steel fibre to ensure that they do not crack over time. A detail of a typical floor for forklift trucks is shown in Fig. 5.16a.

In bulk stores, the pneumatic tyres on front loaders are likely to exert less force on the floor than forklifts so floor strength can be reduced.

The floor should have sufficient fall to ensure that wash water will move to drains, but should be sufficiently level to ensure stacks of boxes remain near vertical.

5.8.2 Floor insulation

In continental climates, where ambient temperatures can be very low, floors should be insulated. This is usually done by installing an extruded polystyrene insulated board, capable of taking a distributed load, on top of a 150-mm-thick concrete slab (Fig. 5.16b). A further 200-mm reinforced concrete slab is laid on top of the insulation. The important detail is to ensure that heat is not lost through the edges of the store. Thermal insulation blockwork or insulation board placed vertically will prevent this happening.

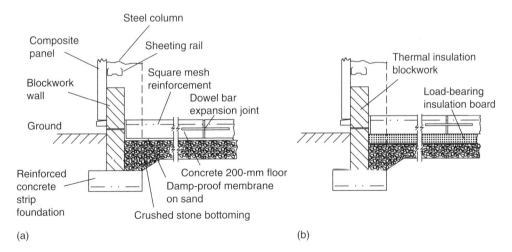

Fig. 5.16. Detail of (a) an uninsulated and (b) an insulated floor designed to carry a forklift for stacking boxes. (Courtesy of C.A. Johnston, CEng MICE, Aberdeenshire, UK.)

5.8.3 Ventilated floors

In bulk stores the ducts are often under the floor to ease the removal of potatoes from the store. Most underfloor systems are specified and installed by the suppliers of the ventilation equipment (Ch6.4). Their involvement should be sought in the early stages of building design.

5.9 Doors

Doors are commonly the major source of air leakage in potato stores. The best arrangement is for there to be only one crop loading door into the store, and it should open into a second building such as a grading area so that it acts as an airlock. The door should be well-sealed and easy to open and close. In the UK it is preferable, but not essential, for the door to be insulated. In continental climates insulation of the door panels, or a secondary insulated curtain, is required. A personnel door should be fitted to avoid having to open the main door for staff to enter.

5.9.1 Up-and-over doors

Up-and-over doors can be insulated or uninsulated, and manually or electrically operated (Fig. 5.17). They can be fitted with sealing strips which, so long as they are adjusted correctly, can be fairly airtight. An 'I' beam located in the concrete floor where the door meets it minimizes wear and maintains the seal over time. If any light leaks into the store through the sides or bottom of the door, then air will leak too.

Fig. 5.17. Up-and-over insulated roller shutter door with an 'I' beam in the floor to maintain a seal at its base. (Yorkies Hill, Mintlaw, Aberdeenshire, UK.)

5.9.2 Sliding fridge-type doors

Sliding doors designed for food chillers also make good doors for potato stores (Fig. 5.18). The track is so designed that when the door is nearly shut, it moves laterally, so that it seals against the frame of the doorway on all four sides.

5.9.3 Speed of door opening

The speed of electrically operated doors can be a significant issue if stores open to the outside. Slow-opening doors will tend to be left open by forklift drivers, so that they can maintain a rapid unloading or loading rate. In some store complexes, the doors of a number of stores are opened in the morning and stay open until late afternoon. Environmental control is therefore lost during daytime. Warm humid air can enter the stores and condense on the crop, and the cooling has to work overtime during the night to cool the crop back down to temperature.

If a forklift travels at 5 m/s and the door takes 10 s to open, the door has to start opening 50 m before the forklift arrives if its entry is not to be impeded. If the door takes only 5 s to open, only 25 m is required. Door opening by radio control operated from the forklift cab will greatly improve the chance of doors being opened only when access is required.

Fig. 5.18. Fridge-type sliding door. (Wood Farm, Norwich, Norfolk, UK.)

5.9.4 Conventional sliding doors

Most old stores in the UK are fitted with sliding doors. Even when fitted with brush strip seals these tend to allow considerable ingress of ambient air. They are particularly difficult to seal at ground level. They should not be fitted to new stores.

Plastic strip curtains can be fitted inside sliding doors to both improve the sealing of the door and act to reduce air ingress when the door is open, but their effectiveness is limited (Box 10.1).

5.10 Health and Safety

Stores and grading areas can be dangerous workplaces, so a full written assessment should be made of the potential risks to staff (BPC, 2005a). These should be updated regularly and staff training undertaken to ensure that the recommendations are part of routine store operation.

5.10.1 Operation of rotating machinery

Harvesting, crop cleaning, crop sizing and the elevating of the crop into store all involve the use of rotating machinery. These should all be guarded to ensure fingers, hands or pieces of clothing cannot be drawn into moving machinery. Guards that have to be removed for machine adjustment must be replaced prior to restarting machines. Staff should receive training on the safe operation of the plant.

5.10.2 Mobile machinery

In box stores, forklifts often pose the greatest danger to staff in the store. They can appear suddenly through a plastic curtain doorway at frightening speed. Forklift movement areas should be clearly marked on the store floor, only essential staff should have access to vehicle movement areas and high-visibility clothing should be worn at all times. Forklift drivers should be trained to keep speeds down, to use lights and horns in areas of poor visibility and to use mirrors when reversing.

5.10.3 Working at heights

Boxes may be stacked to heights of 7–8 m. A fall from this height on to concrete is often fatal. Staff should only access the top boxes when this is necessary to install sensors or to sample crop. Sampling of top boxes should follow guidelines described in the work at height regulations (HSE, 2005b). Access should be via a fixed platform, fixed ladders, staircases, access towers, scissor lifts, or a purpose-designed

Fig. 5.19. Short ladders and use of a head torch make for safe sampling of potatoes in boxes. (Courtesy of G. Stroud, Potato Council, Oxford, UK.)

cage which allows staff to be elevated safely by forklift. A simple solution is multiple-step short ladders (Fig. 5.19).

Boxes should be stacked so that any gaps do not exceed 200 mm, to prevent staff falling between boxes. Staff should be protected from falling over the edge of a stack of boxes by erecting high-visibility tape fixed to canes inserted into the corner of the box one-in from the edge of the stack. An alternative measure is to use a work restraint system, consisting of a lanyard fixed at one end to the belt of the staff member and at the other to a steel rope. Stops on the rope are adjusted to prevent the staff member reaching the edge of the boxes. Stops should be altered each time boxes are removed from store. A fall arrestor system could be used instead, but this adds to the cost and complexity of the safety system.

Store lighting should be 50 lux or more to see dangers and obstructions when carrying out sampling. The use of a head torch can be extremely useful for examining samples more closely (Fig. 5.19).

5.10.4 Sample collection

Samples of potatoes should be collected into bags or containers not exceeding 8 kg, so that they can be safely passed down to a colleague below.

5.10.5   Box construction and condition

Boxes are usually constructed of wood, which degrades and can be damaged over time. In the UK, 1-t boxes should be constructed to BS 7611:1992 (BSI, 1992). Failure of pallet bases or corner posts can lead to a column of boxes toppling. Staircase- or short ladder-fixing systems should hook over the centre or corner posts as damaged top boards can be pulled off if a side load is applied. A better approach is to provide permanent access at the design stage of the store, to a top gantry supported from the roof or to the top of a ventilation duct (Fig. 5.20).

5.10.6   Ventilation equipment

Access to ventilation tunnels and fans is required for opening and shutting laterals and for maintenance. Prior to personnel access, fans and equipment should be shut down and an isolator switched to the off position. Duct doors should open outwards, so that fan pressure aids rather than prevents escape in the event of a fan being switched on inadvertently when a member of staff is in the duct. Electrical services and lighting in the duct should be completely waterproof, to avoid the formation of condensation within luminaries and the associated risk of the bulbs fusing and staff being put in danger.

5.10.7   Store atmosphere

Potatoes consume oxygen and produce carbon dioxide. In extremely well-sealed stores, oxygen levels can fall and carbon dioxide levels can rise to levels that can affect staff. It can also affect the crop, resulting in blackheart in the crop (Ch3.1). Carbon dioxide levels can be increased during rapid store loading when crop respiration is at its highest, following CIPC treatment, or where forklifts are working within a closed store. Carbon dioxide should be monitored routinely, particularly during such operations, to alert staff to dangerously high levels of the gas. Flushing of stores routinely for a few minutes is practised as a safety measure, but monitoring equipment will indicate if this is actually required.

5.10.8   Work environment

Dust, especially when working on grading lines (Ch10.4), can be a major hazard. Staff may become allergic to the dust and be unable to continue to work in environments where dust is present. Employers may subsequently become liable for costly compensation. It is the responsibility of the employer to ensure that dust levels are kept below acceptable levels. In the storage complex, dust is usually worst when potatoes have been stored and dried and are subsequently moved. Dust falls through boxes on the forklift or is generated on cleaners or conveying machinery. Good design of the store can minimize forklift travel distances and separate off dusty areas from where staff are working. Where dust cannot be controlled, staff should be supplied with masks

Fig. 5.20. Access to top of boxes built in at the store design stage. (Wood Farm, Norwich, Norfolk, UK.)

(respiratory protective devices) which conform to BS EN 12942:1999 (BSI, 1999) or similar. These should be stored in a clean, dry place when not in use.

Noise from plant and equipment can restrict proper communication between staff and could mask calls for help. When engaged in any work beside noisy equipment, plant should be switched off for the period of the job. The temporary switching off of ventilation or refrigeration will have little effect on crop condition so long as members of staff remember to switch them back on!

Potato stores can operate near freezing, so staff should be dressed in warm clothing to keep them alert and to prevent numbness and muscle strain due to the cold environment.

5.10.9 Fire escape routes

Stores should be designed so that there is a 900-mm-wide walkway to a personnel escape door. The maximum distance should be 18 m if only one walkway is available or 45 m if there is more than one. A safety light should mark the escape door (Building Regulations, 2007).

5.11 Summary

This chapter has focused on the different types of stores used worldwide and what criteria should be considered when selecting and designing a store. The main aspects to consider are the following.

- Low-cost in-field storage is feasible but is labour-intensive and vulnerable to high crop loss through rotting and disease.
- Stores for the higher latitudes and cold continental areas need to be well-insulated, well-sealed, should be able to store crop for 6–9 months and be environmentally controlled.
- Stores for tropical areas can be lightweight structures, storing 1–2 t of potatoes for 2–3 months and use manually operated flaps to make use of night-air cooling.
- Evaporatively cooled small-scale tropical stores have yet to live up to their potential.
- Different evolutions of store technology in different parts of the world have produced different types of structure.
- While the USA and Scandinavia tend to use continuous humidified-air ventilation systems with bulk storage developed in the 1970s and 1980s, The Netherlands and the UK have kept to the non-humidified ventilation systems used previously.
- In India and southern Asia, high temperatures the year round combined with low labour costs have favoured the general-purpose refrigerated cold store for keeping potatoes and other food produce.
- Store complexes with a grading building attached allow potato unloading, grading and dispatch to be carried out without restrictions from freezing, snow or wet weather.

- Building location, aspect, colour and shelter influence store air leakage and solar heat gain. Sealed stores, light in colour and protected by trees all ease store environmental control.
- Building designs which minimize ledges for dust to settle, ease washing of stores and minimize forklift travel distance help minimize dust dispersal, worker discomfort and potential spore contamination of crops.
- To cope with the high store humidity and prevent condensation within porous insulation materials, new stores tend to be insulated with foam-based insulation materials.
- Composite-sheeted buildings are becoming the norm in the UK for new stores but must be sealed between sheets, at eaves, apex and floor.
- Spray-on foam is popular for upgrading existing buildings and for sealing leaky buildings.
- Floors must be designed to take the high point loads exerted by solid-wheeled forklift trucks when putting two boxes up together.
- Well-sealed, good-quality doors are essential if stores are to be environmentally controlled. They should be able to be open and shut rapidly where repeated access is required.
- Risk to staff from rotating machinery, forklift movement and falling from the top of stacks of boxes should be assessed and safe operating procedures put in place.
- Staff exposure to dust can severely impair health, so dust extraction equipment should be installed wherever dry potatoes are moved.
- Carbon dioxide concentration in stores should be monitored routinely to ensure staff safety and to avoid blackheart in the crop.
- Fire escapes should be provided with lights to indicate exits.

6 Store Ventilation

Topics discussed in this chapter:

- Determining airflow required.
- Causes of backpressure.
- Determining backpressure on fans.
- Bulk storage ventilation systems.
- Airspace ventilation systems for boxes.
- Positive ventilation systems for boxes.
- Box designs and bag containers.
- Warming of potatoes prior to dispatch.

6.1 Introduction

A ventilation system is installed into a potato storage facility to dry the crop, remove crop respiration heat, prevent condensation that often occurs just under the surface of newly stored potatoes (subsurface condensation) and allow cooling of the crop when the weather permits. This chapter describes the procedures that should be followed in designing a ventilation system. This involves calculations but these are kept as basic as possible. More detailed supporting mathematics can be found in Appendix 4.

6.2 Components of a Ventilation System

In any ventilation system, air has to be brought into the building, blended with air in the store to reduce the crop/ventilation-air temperature difference (Ch3.3) and then distributed uniformly through the potatoes to remove the heat from the stored crop. The resulting warm air is then expelled from the store, taking the heat removed from the crop with it (Fig. 6.1).

©CAB International 2009. *Potatoes Postharvest* (R. Pringle, C. Bishop and R. Clayton)

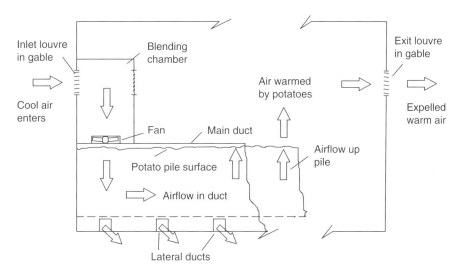

Fig. 6.1. Side view of a bulk store ventilation system.

6.2.1 Intake louvres

In the mild climate of the UK, inlet louvres tend to be fitted vertically into the external cladding of the store (Fig. 6.2). If there is risk of rain being carried into store by the incoming air and landing on potatoes, then a rain hood is put over the inlet. If the air first goes into a duct, the rain will drop harmlessly on to the floor of the duct, so no hood is required.

In countries within a large continental mass, louvres are often fitted horizontally under an overhanging entry duct, so that snow and ice cannot form on them and prevent them opening or closing (Fig. 6.3). In North America, ram-operated vertical insulated panels are used instead of blade-type louvres, as panels tend to be less likely to freeze and jam. In periods of very cold weather, additional insulated

Fig. 6.2. Blade-type louvre inlet.

Fig. 6.3. Horizontal louvres under the building eaves protect blades from icing up. (Courtesy of A. Johansson, Norrköping, Sweden.)

panels are sometimes fitted manually on the outside of the louvres to prevent them freezing up, with the possibility of cold air leaking into store.

The louvres should seal tightly when closed, and no light should be seen when the store doors and louvres are shut. Air speed through louvres is normally restricted to between 3 and 5 m/s to minimize backpressure on the fans (Box 6.1). The higher the air speed through the louvre, the greater the backpressure on the fan (Box 6.2); the lower the air speed, the less the backpressure, but the more expensive the louvre. A compromise between the amount of backpressure and the cost of the louvre has to be found.

6.2.2   Blending chamber

The usual arrangement is for the inlet to feed the incoming air into a blending chamber. The inlet louvre is usually placed opposite to a recirculation louvre

---

Box 6.1. Terminology for resistance to airflow

When a fan is used to move air through a ventilation system it has to overcome the *resistance* caused by the friction between the flowing air and the items it passes through such as louvres, safety guards, main duct and lateral ducts. If the airflow has to suddenly contract or expand, or change direction, these too cause a resistance. The sum of these resistances, or *pressure losses*, measured in N/m^2 or Pascal (Pa), is referred to as the *system resistance*. This system resistance causes a *backpressure* on the fan. For any fan, the higher the system resistance, the greater the backpressure and the less air will flow from the fan. A fan has to be selected with sufficient *backpressure capability* to overcome the system resistance of the ventilation system and provide the desired quantity of air.

Box 6.2. Sizing louvres

Inlet and outlet louvres are usually sized so that the air speed through them is kept to below 3–5 m/s. This keeps the backpressure through the louvre between 8 and 24 Pa (Table B.6.1), and so enables low backpressure propeller or axial fans to be selected. The backpressures were calculated as shown in Ch6.3.

Table B.6.1. Resistance of an inlet or outlet bladed louvre.

Air speed (m/s)	Backpressure (Pa)
3	8
4	15
5	24

(Fig. 6.4). Alternatively a single flap or damper is sometimes used, with the inlet opening as the recirculation inlet closes. Two alternative arrangements for feeding the air into the distribution system of a bulk store can be used. In Fig. 6.4a, a fan, with its impeller blades horizontal, draws air in from the blending chamber and discharges it into the lateral ducts that branch off the main duct. In Fig. 6.4b, each lateral duct has its own fan which sources its air from the blending chamber.

In airspace ventilation systems for boxes, fans on the outlet of the blending chamber discharge the air through overhead ducts into the store. In positively ventilated stores, the fan(s) discharge air into the letterbox duct, which in turn distributes the air into the stack of boxes placed against it (Ch6.8).

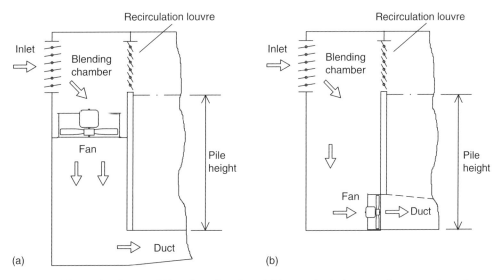

Fig. 6.4. Alternative forms of blending chamber: (a) large single fan supplies a number of (underfloor) ducts; (b) one fan per (above-floor) duct.

6.2.3 Outlet louvres

The exit louvres (Fig. 6.1) are usually located as high as possible in the building, as the exhaust air, having been warmed during its passage through the crop being cooled, rises to the apex of the store. In continuous humid ventilation systems, the outlets act like weirs in a river, allowing the warm exhaust air to spill out of the store.

So long as the air always exits the louvres at more than 3 m/s it is unlikely that rain will enter through outlet louvres. Exit louvres should preferably not be in both gables, as when the wind blows, the louvre in the windward gable can become an inlet while the louvre in the leeward gable becomes an outlet. Ventilating air should only be allowed to enter the store via the inlet duct and blending chamber so that its temperature can be controlled (Ch3.5 and Ch8.3).

If there is a risk of rain wetting potatoes near an outlet louvre, rain hoods should be fitted on these.

6.3 Calculation of Airflow Rate and Backpressure in Ducted Systems

Once the overall design of the ventilation system has been sketched out and the airflow capacity of the fan or fans and their backpressure calculated, changes in the ventilation system design may be made to reduce the backpressure to a minimum. The greater the backpressure, the more powerful a motor the fan requires. As the heat given off by the fan motor usually warms the ventilating airstream, the fan with the lowest possible power rating should be used if drying of the air and associated desiccation of the stored crop are to be avoided.

6.3.1 Calculating airflow required

The quantity of air required to ventilate potatoes is calculated by multiplying the desired ventilation rate per tonne by the number of tonnes to be stored. Specifying the airflow the fan has to supply is therefore straightforward.

The airflow rates are as in Table 6.1.

Table 6.1. Airflow rates used in the three common bulk potato storage systems.

System	Airflow rate (m³/s)
USA and Scandinavia	0.0070/0.0035[a]
UK	0.02
Netherlands	0.04; in the past 0.02

[a]The first value is for drying and wound healing, the second is for long-term storage.

Example

If a store is to hold 1000 t of potatoes, and the UK system of ventilation and a single fan is to be used, then:

Fan airflow required $= 1000 \times 0.02 = 20.0\,\mathrm{m^3/s}$.

Airflow rates for positive ventilation of boxes are often higher to compensate for leakage between the boxes.

6.3.2   Calculating fan backpressure

To fully specify the fan, the backpressure it has to work against should be calculated. The total backpressure, or system resistance, is the sum of all the pressure drops that occur as the air flows through the ventilation system. In the diagram shown (Fig. 6.5), this includes the sudden contraction (1) as the air enters the inlet louvre, the resistance of the louvre itself (2), the sudden expansion (3) as it leaves the louvre, the loss due to a change in direction (4), the sudden contraction (5) as the air enters the fan, the resistance created by the fan guard (6), the sudden expansion (7) as it leave the fan, the loss due to a change in direction from vertical downwards flow to horizontal flow (8) along the main duct, the loss due to the air resistance (9) due to air flowing along the duct, the sudden contraction of air (10) as it enters the lateral, the resistance to air (11) as it travels through the lateral, the sudden expansion (12) as the air leaves the lateral, the resistance to airflow caused by the pile of potatoes (13), the sudden contraction as the air enters the exhaust louvres (14), the resistance of air as it flows through the louvres (15) and finally the sudden expansion (16) as the air leaves the exhaust louvre. These losses exclude items that could be present such as mesh grids over the louvres to prevent birds entering the store, and rain louvres on either inlet or outlet louvres or both.

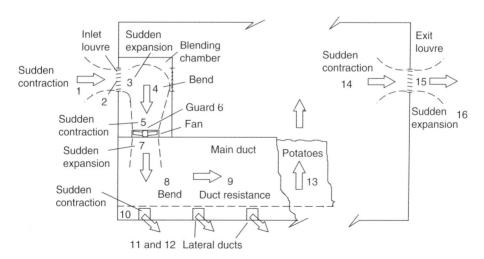

Fig. 6.5. Components that make up the system resistance of a bulk ventilation system.

The static backpressure caused by each of these constrictions is calculated by multiplying the velocity pressure (Box 6.3) by a constant, k, to give:

$$P_s = k \times \frac{1}{2}\rho V^2, \tag{6.1}$$

where

P_s = static pressure caused by the constriction
k = constant for the constriction
ρ = air density (1.23 kg/m^3)
V = air velocity through the constriction.

Values of k for each of these constrictions are available (FläktWoods Ltd, Colchester, UK; http://www.flaktwoods.com/). The selection of the precise value for k depends on the tightness of bend, the size of the mesh guards, the ratio between the main duct and the lateral, and the roughness of the duct. In the example shown (Fig. 6.5 and Table 6.2), all the losses have been included to show what goes to make up the pressure loss of the whole system and how the removal of a guard, the fitting of a bell mouth to the fan or the elimination of a bend can help reduce the backpressure on the fan. To ease routine calculation of total backpressure, some of the losses can be combined. If the fan being selected has been tested with a sudden contraction at entry, the first sudden entry pressure drop should be omitted.

What is evident from Eqn (6.1) is that backpressure is proportional to the square of the air velocity. To minimize backpressure at any point, the air velocity should be kept as low as possible. Air speeds of greater than 6 m/s cause a disproportionate increase in backpressure; so for potato storage, air speeds in ducts should be kept at or below this value (Fig. 6.6).

High backpressures can be avoided by:

- Specifying large-diameter fans.
- Restricting air speeds in ducts to 6 m/s.
- Avoiding restrictions such as bird mesh and guards in high-velocity areas.
- Minimizing changes in air direction, sudden expansions and sudden contractions.

Large-diameter, slow-running fans cost more than small-diameter, fast-running fans, but they will have lower operating costs and reduce crop weight loss as they produce less heat.

6.3.3 Continuity of air in ducted systems

If a fan causes air to flow through a series of ducts of different cross-sections, the flow Q will be constant, but the velocity V of air will change depending upon whether the duct has a small (A_1) or a wide cross-section (A_2) (Fig. 6.7). This gives rise to the continuity equation below:

$$Q = A_1 V_1 = A_2 V_2, \tag{6.2}$$

Box 6.3. Total pressure, static pressure and velocity pressure

Air being forced down a duct by a fan has kinetic energy due to its movement and potential energy if the exit to the duct is constricted. The total energy of the air in the duct is the sum of the kinetic energy and the potential energy. This energy is normally measured as pressure (N/m² or Pa), potential energy being referred to as 'static pressure' and kinetic energy as 'velocity pressure'. The sum of the two at any point in the duct system is the total pressure at that point. The equation for velocity pressure and static pressure is:

$$\text{Total pressure (Pa)} = \text{Velocity pressure} + \text{Static pressure} = \frac{1}{2}\rho V^2 + P_s,$$

where

ρ = density of air (kg/m³)
V = velocity of air (m/s)
P_s = static pressure in the duct (Pa).

The backpressure against which the fan is working is the static pressure.

The static pressure is measured using a manometer, one end of which is connected to the duct, the other to atmosphere (Fig. B.6.1a). The weight of fluid supported, divided by the cross-sectional area of the manometer tube, gives the static pressure in Pa (N/m²). Alternatively the static pressure can be quoted in mm of water, which assumes that the fluid in the manometer is water, or that the manometer scale is calibrated to read in mm of water even though the liquid in the manometer is dyed paraffin.

The manometer can also measure total pressure (Fig. B.6.1b) by pointing the intake tube of the manometer into the airstream. This measures the sum of the velocity pressure and the static pressure.

To measure the velocity pressure using the manometer directly, the tube facing into the airflow is connected to one side of the manometer and the tube measuring the static pressure is connected to the other side (Fig. B.6.1c). The manometer reads the difference between the two measurements, the total pressure minus the static pressure, which gives the velocity pressure.

In practice a Pitot tube (Fig. B.6.1d) is connected to the manometer as this is designed to minimize turbulence, which could affect the readings.

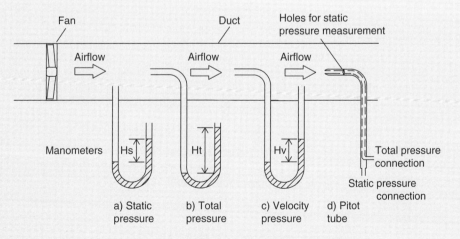

Fig. B.6.1. Measurement of static, velocity and total pressure in a duct.

Table 6.2. System resistance of a bulk storage ventilation system (Fig. 6.5).

Location on Fig 6.5	System element	Free area, A (m²)	Flow, Q (m³/s)	Velocity, V (m/s)	k	Velocity pressure, P_v (Pa)	Duct length (m)	Pressure drop, P_s (Pa)	% of system resistance
1	Inlet louvre	5.00	20.0	4.0	1.525	9.8		15.0	7
4	90-degree turn	5.80	20.0	3.4	0.500	7.3		3.7	2
5	Sudden contraction	1.97	20.0	10.2	0.200	63.4		12.7	6
6	Guard (50-mm mesh)	1.97	20.0	10.2	0.130	63.4		8.2	4
7	Expansion after fan	1.97	20.0	10.2	1.000	63.4		63.4	32
8	90-degree turn	5.80	20.0	3.4	0.500	7.3		3.7	2
9[a]	Duct resistance	5.40	20.0	3.7		0.5		14.0	7
10[b]	Resistance through outlet to lateral			6.0	1.000	22.1	28.0	22.0	11
11[a]	Lateral duct resistance					0.5	10.0	5.0	2
12[b]	Sudden expansion from lateral			4.0	1.000	9.8		9.8	5
13[c]	Resistance through potatoes							28.7	14
15	Exit louvre	5.00	20.0	4.0	1.525	9.8		15.0	7
								201[d]	100

[a]For locations 9 and 11, duct resistance is estimated to be about 0.5 Pa/m. This value is a gross simplification but allows P_s to be calculated by multiplying this figure by the length of the duct.

[b]For locations 10 and 12, the resistance due to 'sudden expansion from the duct outlet to laterals' and from the 'laterals to the potatoes', the outlet area A is chosen to give air velocity of 6 and 4 m/s, respectively. These values are used to calculate P_s for the two locations.

[c]For location 13, the resistance of the crop $P_s = cv^n$, where $c = 2141$, $n = 1.34$ and $v =$ approach velocity for the crop (m/s) (see Table 6.6).

[d]As explained below, with a safety margin, = 225 Pa.

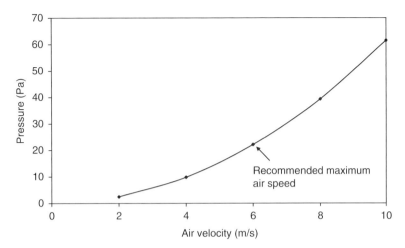

Fig. 6.6. Backpressure of an outlet ($k = 1$) increases as air velocity through it increases.

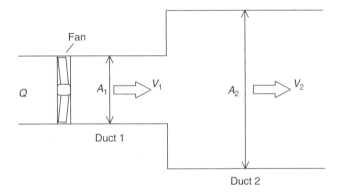

Fig. 6.7. Velocity of air decreases as duct size increases.

where

$Q =$ flow of air (m³/s)
A_1, A_2 = cross-sectional areas of duct sections 1 and 2 (m²)
V_1, V_2 = respective air speeds in the two sections of duct (m/s).

Eqn (6.2) is used to determine the speed of air at any point in the ventilation system, so that the pressure drop for each item that causes a pressure drop can be calculated using Eqn (6.1).

Example: Calculation of backpressure at the inlet louvre
If the fan is to produce an airflow of 20.0 m³/s as calculated in the example in Ch6.3.1 above, and an inlet louvre having a free area of 5 m² is chosen, the velocity of air flowing through the louvres is calculated using a rearranged form of Eqn (6.2):

$$V_1 = \frac{Q}{A_1} = \frac{20.0}{5.0} = 4.0 \, \text{m/s}.$$

The combined k value for a sudden contraction ($k = 0.5$), the resistance through a bladed louvre ($k = 0.025$) and a sudden expansion ($k = 1.0$) is 1.525.

Using Eqn (6.1) with $\rho = 1.23 \, \text{kg/m}^3$ gives:

$$P_s = k \times \frac{1}{2}\rho V^2$$

$$= 1.525 \times \frac{1}{2} \times 1.23 \times (4.0)^2$$

$$= 15.0 \, \text{Pa}.$$

6.3.4 Calculation of system resistance for the whole bulk storage ventilation system

This procedure is followed for all locations along the airflow path (Fig. 6.5), points 1, 4–13 and 15. The calculations are shown in Table 6.2.

The total system resistance, or backpressure, that the two fans have to work against is 201 Pa (Table 6.2). It is usually wise to add some extra backpressure capability to take into account any unforeseen resistances such as bird screens or guards. For this example we shall increase the design backpressure to 225 Pa.

6.3.5 Sizing axial fan to suit ducted system

Axial flow fans (Fig. 6.8) predominate in potato storage. They provide the large volumes of air required, with a minimum of heat output from the fan. Old installations may use centrifugal fans, but these were often used both for grain drying and potato

Fig. 6.8. Axial fan with guard to prevent fingers touching blades. (Courtesy of W. Leslie, Farm Electronics, Lincoln, UK.)

storage, so were fitted with motors about three times the size of those used in axial flow fans for potatoes to cope with the 1000 Pa of backpressure required for grain drying. While they will not use their whole power capability at the 150–250-Pa backpressures found in bulk potato stores, their power use will be higher than that of axial flow fans selected purely for potato storage. The use of a high backpressure centrifugal fan not only increases the running costs of ventilation but the additional heat produced by the electric motor will dry the air more than if an axial flow fan is used instead. If a centrifugal fan must be used, the heat from the motor should be kept out of the airstream by separating the motor from the ventilating air.

To select a fan, find one with a performance curve with the required volume flow near the centre of the *x*-axis and a static pressure near the centre of the *y*-axis (Fig. 6.9). Draw a line vertically upwards from the desired volume flow until it meets the required static pressure, shown by the upward curving dotted line and marked PsF. Mark this meeting point on the chart and determine the blade pitch angle from the series of curving lines marked from left to right in degrees. The fan selection is now almost complete. Draw the vertical line down until it meets the appropriate pitch angle on the absorbed power plot. Read off the power on the *y*-axis.

Additional information may show the efficiency of the fan at this operating point and its noise output in decibels. Repeat this for a number of different diameters of fans and see which provides the flow and static pressure you want for reasonable cost. Manufacturers produce selection programs on the World Wide Web or on a CD to make this selection easier.

If variable speed control is being considered, this should be stated in the specification as the coils of the fan motor need to be wound to suit.

Example

On Fig. 6.9, for the fan duty we selected in the section above, select the flow as 10.0 m^3/s because there are two fans providing the combined flow of 20 m^3/s. Draw the line up to point A, where the static pressure is 225 Pa (Table 6.2). Note the pitch of the blades to the right of this point is 28°. Extend the line from A down into the absorbed power plot, meeting the 28° blade pitch curve at point B. The absorbed power taken of the *y*-axis to the left of point B is 4.6 kW.

6.3.6 Fan noise

Fan noise is of increasing importance as farming areas become ever more populated with rural dwellers. The whine from axial fans, similar to that from an aircraft engine, can be very annoying to neighbours. When selecting fans, their noise rating should be checked and the planning authority consulted to see if the noise levels are acceptable. The slower the fan runs the lower its noise output (Table 6.3). By choosing large-diameter fans to give low air speeds in ducts, which favours uniform airflow distribution, the additional benefit from low noise will be achieved.

The noise level to neighbours can be further reduced by fitting a noise baffle on the fan inlet (Fig. 6.10) or by surrounding the building with a soil bund. Trees do not make a very good sound barrier.

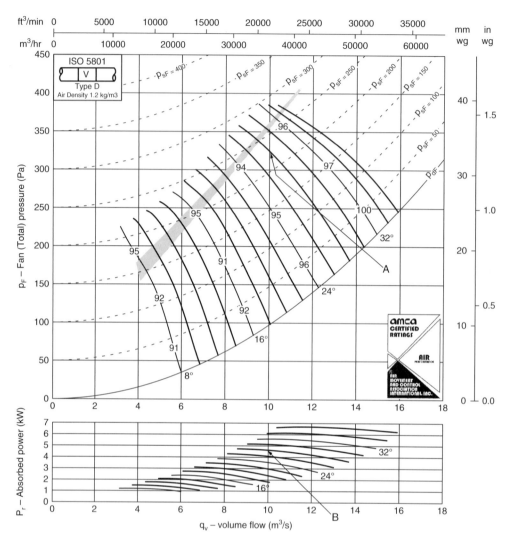

Fig. 6.9. Manufacturer's performance curves for an axial fan. (Courtesy of FläktWoods Ltd, Colchester, UK.)

Table 6.3. Sound level at 3-m distance from 0.97–1.22-m diameter aerofoil fans running at different speeds. (From FläktWoods Ltd, Colchester, UK.)

Fan speed (rev/min)	Sound rating (dB)
715	59–63
960	68–72
1475	78–84

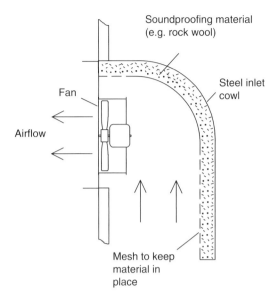

Soundproofing material
(e.g. rock wool)

Steel inlet
cowl

Fan

Airflow

Mesh to keep
material in
place

Fig. 6.10. Soundproofing a fan inlet.

6.3.7 Fan guards and safety features

Many fan guards are bolted directly on to the fan casing (Fig. 6.8) and are designed to prevent fingers touching fan blades. These guards can impose large and unnecessary backpressures on the fan. If the guard can be offset from the fan by the length of a finger then the mesh can be sized to prevent a hand passing through the mesh. If it can be offset by an arm's length, then the mesh can be sized to prevent a body passing through it. This will reduce the backpressure on the fan significantly and allow the lowest backpressure fan possible to be used, though a further mesh in a low air speed area of the ducting may be necessary to keep birds out.

6.4 Air Distribution in Bulk Stores

The air distribution system can consist of:

- A main duct with smaller ducts or laterals off it.
- A series of tapered ducts, with one fan per duct.
- Pedestal duct fans.

6.4.1 Main duct with laterals

The traditional system for ventilated bulk storage is to have a fan blowing into a main duct with porous lateral ducts coming off one or both sides of the main duct (Figs 6.11 and 6.12). The laterals may be above or below ground.

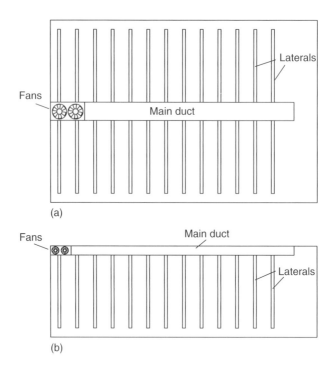

(a)

(b)

Fig. 6.11. Plan of alternative main and lateral duct layouts for bulk stores: (a) central main duct; (b) side main duct.

Fig. 6.12. Main duct with bulk potatoes on either side. (Courtesy of W. Leslie, Farm Electronics, Lincoln, UK.)

Main duct

The main duct should be sized to allow the store manager to walk through without bending. Cramped ducts deter managers from opening and closing laterals, vital to ensure that the parts of the crop that need extra ventilation get it (Holloway, 1990). A light switch and fan stop button should be installed inside the duct, to allow staff inadvertently trapped within the duct to stop the fan should it start on automatic. Alternatively, doors to ducts should open outwards so they can be opened with the fan running. The duct lights should be waterproof so that they are not affected by condensation.

Above-ground laterals

The use of above-ground ducts saves the expense of installing a ventilated floor. They do not become clogged with soil and they can be tapered for uniform air distribution. Care has to be taken during store loading to prevent above-ground ducts being pushed out of position due to the pressure of potatoes.

   Above-ground ducts are less suitable for stores that are unloaded by front-end loader, as the loader wheels can damage the ducts. If they are to be used with a front-end loader unloading system, they should be spaced so that the loader can work between the laterals.

   Above-ground ducts are normally of an 'A' or semicircular cross-section (Fig. 6.13a and b), although plastic drainage piping can be used as a temporary measure (Bishop, 1994).

Below-ground laterals

Below-ground ducts allow the store to be filled and emptied without the risk of damaging the ducts. These range from:

* Simple channels covered by 50–75-mm-thick wooden slats (Fig. 6.13c).
* Concrete floors with vents cast in them.
* Ventilated hardwood floors that can be placed on top of existing concrete floors (Fig. 6.14).

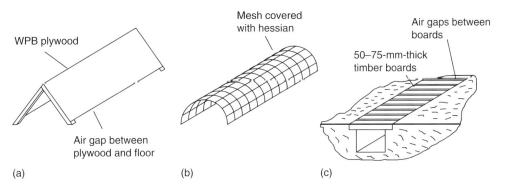

Fig. 6.13. Alternative designs of lateral ducts: (a) timber A-frame ducts: (b) weld mesh hoop ducts; (c) underfloor ducts with base rising to form a taper.

Fig. 6.14. Hardwood ventilated floors laid on a plain concrete base. (Courtesy of Flach and Le-Roy Ltd, Huntingdon, Cambridgeshire, UK.)

Dual-purpose grain/potato floors need a fine mesh to prevent grain from falling through the porous area. This mesh can get plugged up with soil, so needs cleaning annually. The mesh is recessed in troughs about 10 mm below the surface of the floor to protect it from damage by wheel lugs and the edge of unloading buckets. If buckets are used to empty the store, potatoes sitting in the recessed troughs can be sliced by the edge of the bucket, resulting in a small but significant proportion of the crop being unmarketable.

Lateral spacings

The maximum lateral spacing for above-ground ducts should not exceed the minimum pile height, if an acceptable uniformity of ventilation through the potatoes is to be achieved (Fig. 6.15a). This should take into consideration that in some years the crop being stored will not fully fill the store, so that the depth of storage may be lower than the design specification. Narrower spacings between laterals will improve the uniformity of air through the crop. With below-ground ducts the recommended spacing is 0.8 times the pile height (Rastovski and van Es, 1981), as they do not spread the air sideways as effectively as above-ground ducts.

Laterals are spaced as shown in Fig. 6.15b. The ends of the laterals should terminate at half the spacing between laterals from the sidewalls. The lateral nearest the end wall should be no closer than half the spacing between laterals. These

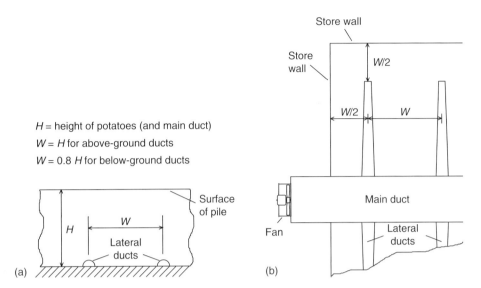

Fig. 6.15. Distance of laterals from load-bearing retaining walls: (a) elevation of main duct and laterals; (b) plan of main duct and laterals.

recommendations prevent air short-circuiting up the lower-resistance air path at the junction of where the crop meets the load-bearing walls.

Achieving uniform airflow in ducted ventilation systems

When a fan blows air into a porous duct with fixed cross-section, such as a main duct with laterals, the total pressure in the duct stays almost constant for the length of the duct. The small loss that does occur is due to friction between the moving air and the duct walls and framing.

At the start of the duct, the air from the fan is moving at maximum velocity. At the end of the duct, the air velocity is zero, as the duct has a blank end. Between the beginning and the end, all of the air has been discharged through the laterals. Between the start and end of the porous duct, the velocity pressure therefore declines from a maximum at the start to zero at the far end (Fig. 6.16). In contrast, the static pressure increases from a low value to a maximum at the far end from the fan. This increase is termed static pressure regain.

The amount of air that flows from each lateral depends on the potential energy (i.e. static pressure) in the duct at the entry to the lateral. As shown in Fig. 6.16b, as the static pressure is highest nearest the end of the main duct, more air will come out of the laterals at the far end of the main duct than the end nearest the fan. This is what happens in practice and is a major problem when trying to achieve uniform airflow in laterals of potato stores.

In grain storage, the backpressure created by the grain serves to make the airflow from the ducts more uniform. With potatoes, in contrast, the backpressure is so low that this effect is much reduced. Air distribution systems should therefore be designed for uniform distribution regardless of whether potatoes are present or not.

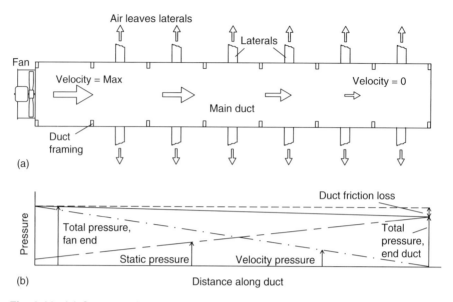

Fig. 6.16. (a) Cross-sectional plan of main bulk ventilation duct showing how air velocity falls due to air leaving through laterals; (b) plot showing how total pressure, static pressure and velocity pressure change along the main duct.

Designing for uniform airflow to laterals from constant cross-section main ducts

The analysis of air distribution from a main duct of constant cross-section into side laterals is described in Appendix 4. A number of conclusions come from this analysis:

- The ratio of airflow between the last and first lateral for different ratios of total lateral outlet area to main duct area is shown in Table 6.4.
- The smaller the discharge apertures to the laterals, the higher the static pressure will become in the main duct and the more uniform will be the airflow to the laterals.
- Uniformity of airflow distribution from laterals improves as air speed in the main duct reduces, with maximum duct air speed of 6 m/s chosen as a reasonable compromise.

Table 6.4. Uniformity of airflow to laterals and backpressure loss at outlets to laterals for various ratios of total lateral cross-sectional area to main duct cross-sectional area. (From Rastovski and van Es, 1981.)

Total lateral cross-sectional area/main duct cross-sectional area	Airflow in last lateral/ airflow in first lateral	Backpressure at lateral outlets assuming 6 m/s at entry to main duct (Pa)
0.5	1.13	89
1.0	1.35	22
2.0	2.00	6

Table 6.4 shows that where the flow of air from the main duct into laterals is more uniform, the backpressure on the fan is increased.

Even the best designed main duct and lateral systems tend to have higher airflows in the ducts furthest from the fan. Slides are usually put in the entrance to each lateral, partly to allow selective ventilation of new crop being loaded into store and partly to allow the exit airflow to each lateral to be adjusted. Slides can be adjusted to make the airflow from each lateral more uniform. This is a labour-intensive task as an adjustment in one lateral affects all the others. If it is done once, and the slide settings marked, then adjustment thereafter is simplified.

Avoidance of Venturi effect
An additional requirement in duct design is to avoid the formation of jets through static air masses, especially near the outlets to the first side laterals in the main duct. This is to avoid the Venturi effect (Fig. 6.17), which can cause air to be sucked via these first laterals into the main duct. The Venturi effect is

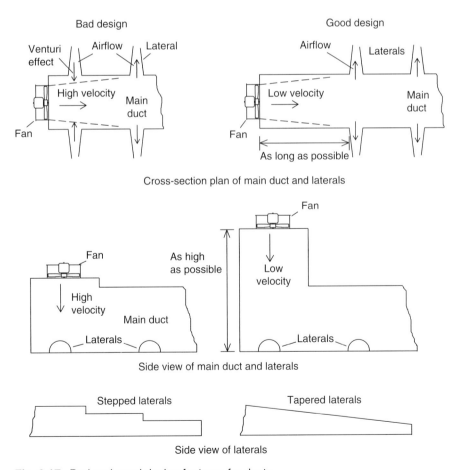

Fig. 6.17. Bad and good design features for ducts.

minimized if air jets through stationary air masses are avoided. Air jets are minimized by:

- Selecting the largest-diameter fan that can be afforded.
- Expanding the air from a fan in a tapered transformation piece with an angle of 11° to the direction of the airflow.
- Placing the fan as far from the first lateral as possible.

Designing for uniform airflow from laterals

As constant cross-section lateral ducts are porous in the same way as main ducts, they experience the same problem of more air discharging from the far end compared with the end that connects to the main duct. The easiest way of improving uniformity of air distribution is to have tapered laterals. The static pressure is kept almost constant by the taper in the duct. In above-ground ducts this requires the ducts to be specially manufactured by specialist ventilation companies (Fig 6.18). Underground ducts can be tapered simply by making them deep to start with and shallower near their end. Laterals tapered in steps, to simplify construction, provide a more uniform air distribution than fixed cross-section ducts but less uniform airflow than fully tapered ducts. The amount of taper to achieve a uniform air distribution is given in Table 6.5.

The beauty of using tapered ducts is that they provide a uniform air output without having to constrict the openings to create backpressure. This allows a fan to be selected with a lower static pressure than would be required for a constant cross-section duct system.

Fig. 6.18. Tapered duct. (Courtesy of Tolsma Techniek, The Netherlands.)

Table 6.5. Ratio of duct inlet to duct end cross-section for different lengths of duct. (From Rastovski and van Es, 1981.)

Duct length (m)	Duct inlet cross-section/ duct end cross-section
5	4:1
10	7:1
15	10:1
20	13:1

Resistance due to crop

The backpressure added by the resistance of the potatoes is shown in Table 6.6. The equation used for these calculations is (Burton, 1966):

$$P_c = cv^n, \tag{6.3}$$

where

P_c = pressure drop per metre of crop height (Pa/m)
c = constant for the crop (2141 for clean potatoes)
v = approach velocity (m/s)
n = constant for the crop (1.34 for potatoes).

The resistance of the potatoes (Table 6.6) increases as air velocity increases, so the Dutch system has considerably higher values than the US/Scandinavian system. The potential for the Dutch system to desiccate the crop is limited as the high airflow means that fans need only be run for short periods to dry or cool the crop (Ch3.7).

Example: Procedure for sizing of ducts

For our example we shall consider the bulk store for 1000 t of potatoes discussed above, with a 23-m-long main duct, 1.22 m wide and 3.6 m high inside, with tapered underground ducts and potatoes stacked 3 m high.

Table 6.6. Pressure drop due to resistance of potatoes (P_c) for different air speeds and for different heights of pile (specific volume of potatoes = 1.5 m³/t).

Country standard	Airflow rate (m³/s/t)	Pile height (m)	Approach velocity[a] (m/s)	P_c (Pa)	Pile height (m)	Approach velocity (m/s)	P_c (Pa)
USA/ Scandinavia	0.007/0.0035[b]	3	0.014	7.0	5	0.023	13.7
UK	0.02		0.04	28.7		0.067	57.2
Netherlands	0.04/0.02[b]		0.08	72.6		0.133	143.5

[a]Approach velocity = (volume of potatoes/m² floor area/specific volume of potatoes) × airflow rate per tonne (see Box 3.10).
[b]Highest of the two rates was used for calculations.

STEP **1**: SIZING OF MAIN DUCT   The main duct is sized mainly by the height the potatoes are to be stored, wide enough to allow a person to walk along it, and if possible sized to suit the 1.22- or 2.44-m size of plywood sheet if this or similar sheet material is to be used. Eqn (6.2) should be used to check that the air speed is no higher than 6 m/s.

Airflow (Q) is 20 m³/s. Required duct area is $Q/6 = 3.333 \, \text{m}^2$. Actual duct area is $1.22 \times 3.6 = 4.39 \, \text{m}^2$; this is above the required size so is satisfactory. Initial air speed is $20/4.39 = 4.55 \, \text{m/s}$.

STEP **2**: NUMBER OF LATERALS   As the depth of potatoes is to be 3 m, the duct spacing is $0.8 \times 3.0 = 2.4 \, \text{m}$. The 23-m-long duct will have nine laterals each side. The width of storage area between the duct and the load-bearing wall is 10 m, so the duct will be $10 - 1.2 = 8.8 \, \text{m}$ long (Fig. 6.11).

STEP **3**: SIZE OF LATERALS   The inlet of the laterals should be sized to have an air entry speed V_{max} of 6 m/s or below. Since there are 18 laterals in total, the amount of air in each lateral (Q_o) is:

$$Q_o = 20.0 / 18 = 1.11 \, \text{m}^3/\text{s}.$$

The area of laterals at the start is:

$$A_o = Q_o / V_{max} = 1.11 / 6.0 = 0.185 \, \text{m}^2.$$

A decision as to the width of the laterals has to be taken. If they are assumed to be 0.4 m wide internally, their initial depth can be calculated:

$$\text{Depth (initial)} = A_o / \text{width} = 0.185 / 0.4 = 0.46 \, \text{m}.$$

The area at the closed end of the lateral is taken from Table 6.5. The ratio of the duct inlet/duct end cross-sectional area for a 8.4 m duct is about 6:1. Therefore the area at the end is:

$$\text{Initial area/ratio} = 0.185 / 6 = 0.0308 \, \text{m}^2.$$

If the width is kept constant at 0.4 m wide, the depth at the far end of the duct will be $0.0308/0.4 = 0.077 \, \text{m}$, or about 80 mm below the bottom of the duct covers.

STEP **4**: SIZE OF LATERAL DUCT COVERS   For tapered ducts the static pressure in the duct should be even. An outlet velocity of 4 m/s is recommended (Rastovski and van Es, 1981) to ensure the flow is even. As potatoes block off between 65 and 75% of the air outlets, the outlet area has to be 3 to 4 times that needed to give 4 m/s outlet air speed with no potatoes in place.

a) Free area per lateral

For the lateral ducts sized in step 3, the free area is calculated as follows:

$$\text{Airflow per lateral } (Q_o) = 1.11 \, \text{m}^3/\text{s}.$$

Free area required for the whole duct, if no potatoes in place, is:

$$Q_o / V_{min} = 1.11 / 4 = 0.2775 \, \text{m}^2.$$

Free area required to take account of the potatoes is $0.2775 \times 3 = 0.8325 \, \text{m}^2$.

As laterals have an internal width of 0.4 m, the total amount of gaps between wooden slats should be:

Total gap = 0.8325 / 0.4 = 2.08 m.

b) Number, width and spacing of boards per metre
Gap per metre is calculated as:

Total gap / length of lateral = 2.08 / 8.8 = 0.236 m per metre of lateral.

If boards are 60 mm wide, and the space between them is X (m), then

$$(0.060 + X) \times n = 1.00 \, \text{m} \tag{i}$$

where n is the number of boards per metre. As $n \times X = 0.236$ m, then

$$n = 0.236 / X \tag{ii}$$

Thus, substituting n from (ii) into (i):

$$(0.060 + X) \times 0.236 / X = 1.00 \, \text{m},$$

or

$$
\begin{aligned}
X &= [(0.060 + X) \times 0.236] \\
&= (0.060 \times 0.236) + 0.236X \\
&= 0.01416 / 0.764 = 0.0185\text{-m or } 18.5\text{-mm gaps.}
\end{aligned}
$$

If the boards are 60 mm wide, the gaps should be 18 mm wide.

Gaps should not be wider than 25 mm to prevent small tubers being pushed into the gaps. They should preferably be greater than 15 mm to prevent them being plugged up with soil.

STEP 5: SIZE OF OPENINGS IN MAIN DUCT TO GIVE REASONABLY UNIFORM AIRFLOW   As the main duct is a constant cross-section, the only way to obtain a relatively uniform airflow into the laterals is to raise the static pressure in the duct by constricting the outlets to the laterals. To minimize the variation to 1:1.35 (Table 6.4), the total area of the lateral outlets should be equal to the minimum cross-sectional area A_{min} of the main duct. This was calculated to be 3.333 m² to keep the air velocity no higher than 6 m/s. (In practice the main duct was made larger than this to allow access, but for this calculation A_{min} is the figure to use.)

Number of laterals = 9 per side, 18 in total.

$$
\begin{aligned}
\text{Size of outlet to lateral} &= A_{min} \text{ / number of laterals} \\
&= 3.333 \text{ / } 18 \\
&= 0.185 \, \text{m}^2.
\end{aligned}
$$

If the laterals are 0.4 m wide and 0.46 m deep at the start (i.e. 0.184 m²), the whole area is required to get sufficient outlet area.

STEP 6: CHECKING FAN BACKPRESSURE REQUIREMENT   The last requirement is to check the backpressure against the fans as in Table 6.2. The fans can then be specified for both airflow and backpressure. In this case, the fans would each have an airflow of 10 m³/s with a backpressure of 201 Pa; or to be safe and allow for additional guards or mesh to prevent bird entry, 225 Pa.

Fig. 6.19. Pedestal fans for stores where no ventilation is installed. (Courtesy of Martin Lishman, Lincolnshire, UK.)

6.4.2 Pedestal fans

Where bulk stores have no ventilation system installed, 'pedestal' ducts fitted with fans, as shown in Fig 6.19, can be installed. The system consists of a vertical galvanized steel duct with a suction fan mounted on top. The duct is placed on the floor of the store and potatoes are built up around it so that they support the duct in a vertical position. There is no need for ancillary ducting. When the fan is running it sucks air into the bottom of the duct and discharges it into the airspace above. This draws air into the pile from the surface, ventilating the potatoes as it does so. These tend to be used in small stores, where the installation of a floor ventilation system would be uneconomic or where a building is rented. Their use is mainly to dry the crop and prevent subsurface condensation. By manually opening and shutting doors, cooling can be achieved but without the benefit of automation.

The supplier of pedestal ducts with fans in the UK (Martin Lishman Ltd, Bourne, UK) recommends their use at 5 m spacing for a 2.5-m-high pile. At this spacing the 1.1-kW pedestal unit will give $0.017\,m^3/s/t$ and the 2.2-kW pedestal unit gives $0.025\,m^3/s/t$, just below and just above the standard UK ventilation rate of $0.02\,m^3/s/t$. Each unit will ventilate 42 t.

6.5 Humidification of Ventilating Air

6.5.1 Continuous humid air systems

In continuous ventilation with humid air systems, humidification of the ventilating air is essential if the crop is not to be dehydrated. The simplest method of humidification control is to use a porous evaporative humidifier, where air is forced through a water-soaked porous cellulose matrix (Fig. 6.20). The air will pick up as much water as it can absorb, but no more.

To avoid carry over of small droplets of water, the maximum approach speed of air to the humidifier is 3 m/s. At this air speed, the size of duct needed to house the humidifier is considerable. However, as the total airflow requirement to ventilate the crop is low, there is usually sufficient room in which to fit such a system.

6.5.2 Humidification of intermittent ventilation systems

The use of a porous evaporative humidifier located in the air supply duct may well be impractical for the humidification of intermittent medium- and high-rate airflow ventilation systems. To keep the approach air speed below 3 m/s requires either a very large duct or for the porous evaporative humidifier to be formed into a large box surrounding the fan. The alternative is to use a spinning disc humidifier or a pressure jet, sonic humidifier (Fig. 6.21). These need very good filtration in the water supply line to keep orifices clear of particles in the water, and pose the risk of droplet carry over, which can wet the crop. Droplet arrestors are sometimes incorporated to try to reduce this problem.

Air passing through a porous evaporative humidifier will pick up only the amount of moisture it requires to reach approximately 97% RH (Munters, 2007), so this system is self-controlling. In contrast, a spinning disc or sonic humidifier

Fig. 6.20. Water-soaked porous cellulose matrix on both sides of fan intake. (Courtesy of W. Leslie, Farm Electronics, Lincoln, UK.)

Fig. 6.21. Spinning disc humidifier. (Courtesy of Ray Andrews, Crop Systems Ltd, Norfolk, UK.)

needs a controlled amount of water to be supplied at any instant to ensure the majority of droplets become vapour and a minimum fall to the floor as water, which can flood the duct.

The evaporation of water droplets, produced by the disc or sonic humidifier, to a vapour will absorb latent heat from the air, causing its temperature to cool to near its wet-bulb temperature. The amount of water required per second will therefore be the difference between the moisture content of the air entering the humidifier and the moisture content of air at the same wet-bulb temperature but with RH of 97%. This weight of water (grams per cubic metre or grams per kilogram of air) will be multiplied by the airflow passing through the ventilating fan to arrive at a water flow in litres per second. As the temperature and RH of the ventilation air will keep changing as the weather changes, the water flow rate will have to keep changing. Therefore, to control the humidification, the temperature and RH of the air have to be monitored continuously and the water flow rate adjusted accordingly.

This assumes that the air becomes almost completely saturated, which may not happen. The blending chamber therefore has to have a drain to allow any excess water to escape. Humidification of intermittent ventilation systems using nozzles, as in the disc or sonic humidifier, is therefore more complicated than for a porous evaporative humidifier system.

Example: Humidification of 20 m³/s air produced by two fans

If the air entering the store is at 10°C and 50% RH, a psychrometric chart shows its wet-bulb temperature is 5.52°C and moisture content is 3.82 g/kg or 4.70 g/m³. Air at 97% RH, with the same wet-bulb temperature of 5.52°C, will hold 6.95 g/m³ and be at a temperature of 5.75°C. The amount of moisture that has to be added per m³ air/s is 6.95 − 4.70 = 2.25 g. For airflow of 20 m³/s, the amount of water to be added to the air will be 2.25 × 20.0 = 45 g/s or 0.045 l/s.

In intermittent ventilation systems for potatoes, weight loss is normally kept low by minimizing the duration of recirculation and by using high rates of air to cool crops rapidly when cool air is available. If weight loss in the crop is found to be excessive, it may be better to look at ways of reducing fan run time and redesigning the ventilation system so that lower backpressure fans can be used (Ch6.4).

6.6 Airspace Ventilation Systems for Boxes

The most common air distribution systems are:

- Overhead high-speed jet distribution.
- Overhead high-speed jet with 'goalpost' plenum.
- 'Goalpost' plenum under pressure or suction.
- Gable louvres with hanging recirculation fans.

6.6.1 Overhead high-speed jet distribution

In UK box stores, overhead high-speed jet distribution systems are the most common form of airspace ventilation system. This is because:

- The space between the top of the boxes and the roof is used to distribute air. No storage space is lost to provide ventilation ducts.
- The air-to-air friction between the high-speed jets and the air mass in store causes a 'roll' of air to develop (Fig. 6.22).
- The incoming air is always cooler than the crop, so as its speed reduces it falls, further encouraging the 'roll' of air necessary to induce air to flow back through the ducts formed by the pallet apertures to the blending chamber.

In principle, overhead ventilation with cold air is wrong, as the top surface potatoes are the first to be cooled by the cooling airstream. Once cooling ventilation stops, natural convection takes over and warm air from potatoes below will rise and may condense on these cooler surface potatoes. The risk of this happening

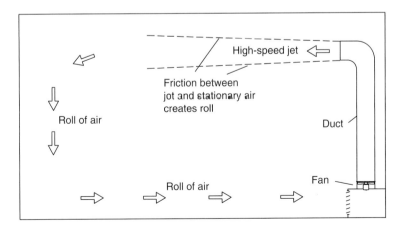

Fig. 6.22. High-speed jet causes a roll of air within the store.

is indeed ever present. However, in practice, if the cooling air temperature can be controlled so that the surface potatoes are never significantly cooler than those below, then such condensation can be prevented.

6.6.2 Air entrainment by jets

When a jet of air flows through the stationary air mass (Fig. 6.22), turbulence and friction occur at the interface. The force caused by this friction results in the stationary air mass starting to move. A jet of air will therefore induce an airflow movement many times greater than the airflow volume of the jet. Air jets therefore serve to induce a beneficial air movement within stores, which becomes a roll. The location of exit louvres has little effect on the formation of the roll, so can be in any part of the building cladding.

The units providing the high-speed jets are of three types:

- Ambient-air cooling with air blending.
- Floor-mounted refrigeration units.
- Combined ambient-air cooling with air blending and refrigeration (Fig. 6.23).

All are fitted with vertical ducts to discharge cool air at high level.

Fig. 6.23. Ambient-air/fridge cooling unit. (Welvent Unit, Warden Farming Co. Ltd, Gainsborough, Lincolnshire, UK.)

Box 6.4. Air jet hitting an obstruction

Air jets are often used in airspace-ventilated box potato stores to induce a rolling movement of air within the store. Should the jet hit an obstruction early in its trajectory, the jet can be severely deflected (Fig. B.6.2). Likely obstructions are portal frames or purlins. Such potential collisions should be avoided during the design stage. If the problem is found after the store is built, a deflector placed under the obstruction can be fitted retrospectively to minimize the effect.

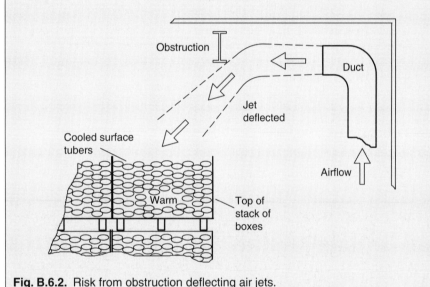

Fig. B.6.2. Risk from obstruction deflecting air jets.

Uniform air distribution throughout the stack of boxes (Box 6.4) depends on:

- Jets flowing from the overhead ducts having sufficient momentum to reach the end of the stack of boxes.
- Jets from the overhead ducts being angled so that they distribute the cool air over the whole width of the store.
- Boxes being of the same height so that their pallet apertures form a continuous duct from the front of the stack to the back. This ensures that a box of a different height does not obstruct the airflow returning from the far end of the store through the ducts formed by the box pallet apertures.

With high-speed jet air distribution systems, airflow (Fig. 6.24) can be low in the rear corners of the stack (Burfoot, 1997; Xu *et al.*, 2002). This confirms observations by store managers that these areas often experience slow drying, condensation, disease development and poor temperature control. Airflow in these sectors can be improved by jetting a small portion of the air to the corners of the store (Fig. 6.23). By suspending additional large-diameter, low-backpressure axial fans above the crop (Box 6.5), both the throw of the jets and the creation of the beneficial 'roll' of air can be increased.

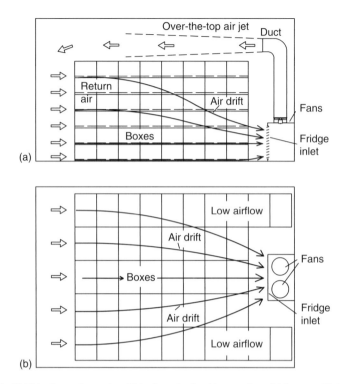

Fig. 6.24. Drift in the return air within the stack of boxes in a fridge-ventilated store: (a) side elevation of store; (b) plan of store.

Box 6.5. Use of ducts to extend the range of the jet

If a horizontal fabric or insulated steel duct is used to extend the length of the jet, the beneficial air-to-air friction is lost. While the air delivered to the far end of the store will flow back to the other end of the store, the amount of store air set in motion will be very much less. Additionally, the air within the duct will be colder than the warmer air above the potatoes, so condensation may form on the duct and drip on to the crop below. Insulation on the duct may reduce, but not completely eliminate, the potential for condensation forming on the duct. The better solution is therefore to install more axial fans to increase the 'roll'.

6.6.3 Overhead high-speed jet with 'goalpost' plenum

The best solution to the problem of low airflow in the rear corners of the stack of boxes mentioned above is to fit a 'goalpost' plenum for the boxes to be stacked against (Figs 6.25 and 6.26). It is termed a 'goalpost' plenum as it looks like a football goal post when the store is empty of boxes. This plenum reduces the tendency of the air returning through the stack of boxes to converge (Fig. 6.24) as it

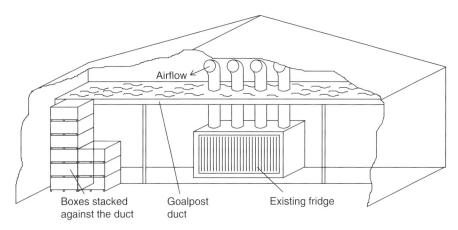

Fig. 6.25. Goalpost duct built over fridge to minimize the problem in Fig. 6.24 of air drift through stack of boxes.

Fig. 6.26. View of fridge and front face of duct from inside goalpost duct. (Wroxham Home Farm, Norfolk, UK.)

approaches the intake of the fridge or blending box, and encourages the air to flow along all the ducts formed by the box pallet apertures. Recent observations (Joe Macgrath, Norfolk, UK, 2005, personal communication) have shown more uniform temperatures and consistent fry colours with this system. The duct does, however, take up space that could have been used to store more boxes.

6.6.4 'Goalpost' plenum under pressure or suction

The use of 'goalpost' or open-fronted plenums helps fans to direct air along the ducts formed by the box pallet apertures. With typical UK 1-t boxes, 915 mm high overall, the air will be passing within 0.4 m of the potatoes at the centre of the box. The combination of the turbulence created by this airflow together with natural convection will create as good ventilation of potatoes as any airspace ventilation system can do. With the recent safety recommendations (BPC, 2005a) suggesting that permanent access systems should be provided to allow sampling of the top potatoes, such plenums can provide the dual function of providing a crop access platform combined with improved air distribution.

Plenums can be under pressure or suction. If the store already has an ambient-air cooling/refrigeration system, it can be housed within the plenum, which will be under suction due to the fans in the fridge drawing air in through its cooling coils (Fig. 6.25). Additional fans can be added to the top of the goalpost duct if the airflow of the fridge/ambient-air system is found to be insufficient for rapid drying and good temperature control. If no existing cooling system is in place, it can be pressurized, with the incoming air coming from an air blending system.

6.6.5 Gable louvres with hanging recirculation fans

A considerable number of potato stores in the UK are ventilated using extractor fans with automated louvres fitted to one gable and automated inlet louvres at the other. Small fans with socks hanging below them run continuously (Box 6.6), both to keep the air in store moving to dry any condensation on the crop and to minimize temperature differences between top and bottom of the stack of boxes (Fig. 6.27).

Air distribution is not entirely predictable as any 'roll' of air created by incoming ventilation may reverse in direction during ventilation (Box 6.7). In addition, as ambient air enters the store without first going through a blending chamber, the incoming air can be very much colder than the temperature of the potatoes, which can result in subsurface condensation when natural ventilation re-establishes air movement. If the fan blows air into store, a conventional blending box can be fitted. If the fan extracts air from the store, an inlet blending system (Fig. 6.28) can be fitted.

These systems are often complemented by the use of portable positive ventilation systems such as drying tents. These can be used to either pre-dry potatoes prior to them being loaded into store or force ventilate stacks of boxes built in a manner which allows them to be positively ventilated within the store.

6.7 Airspace Ventilation Systems for Sacks and Big Bags

The practice of storing potatoes in sacks is common in countries where the cost of labour for handling the sacks is low. Many sack stores like this are to be found in India. The stores may be up to five floors high, and have fridge cooling coils located at the top of the building (Fig. 5.11, Fig. 6.29). The floors are made from hollow section steel, to take the weight of the sacks and to be porous to the cooling air. Access

Box 6.6. Stratification in box potato stores

In a box potato store without any fans operating, natural convection from the stored potatoes causes the air in the store to stratify, with warm air in the top and cooler air below (Fig. B.6.3a). The boxes higher up the stack will therefore be warmer than those below. Where a recirculation system is not part of the ventilation system, small fans fitted with plastic hanging socks are often fitted to continually mix the air within the store and so reduce temperature differences (Fig. B.6.3b).

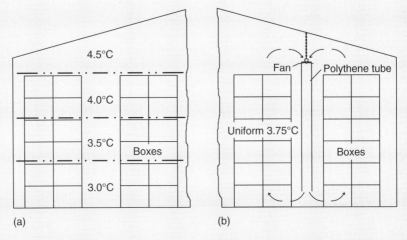

(a)   (b)

Fig. B.6.3. (a) Air stratification in store; (b) remedy, continuously running sock fan reduces stratification.

Fig. 6.27. Extractor fan-ventilated store with sock fans to mix store air. (Deveron Potato Growers Ltd, Banff, Aberdeenshire, UK.)

Box 6.7. Entry of air into a building

When air enters a building, its trajectory is determined by two factors. One is the velocity, or momentum, with which it enters the store; the other is the density of the incoming air relative to the air within the store (Fig. B.6.4). The resulting direction of airflow is the resultant of these two vectors. If the incoming air is very much colder than the air within the store, it will tend to drop vertically as it enters. If it is the same temperature as the air within the store it will travel horizontally across the store. Between these extremes it will fall at an angle.

In the example shown, when ventilation starts initially, the jet will fall vertically as the air is very much colder than the air in the store. However, as the store air is cooled to the incoming air temperature, the inlet air direction will revert to a horizontal jet. Since inlets, not outlets, determine airflow movement within stores (Randall, 1975), the induced roll in Fig. B.6.4 will start anticlockwise, but will then revert to a clockwise circulation when the air in the store has cooled to that of the incoming air temperature. It is not a good design feature for the pattern of ventilation within the store to change as the difference between inlet air temperature and store air temperature alters. It is better to ensure that air flows in the same direction at all times. This is achieved either by ensuring that the inlet air speed in the direction that the air has to go exceeds 5 m/s (Randall and Battams, 1979) or that the air is introduced at floor level so that it rises due to natural convection due to the heat from the potatoes.

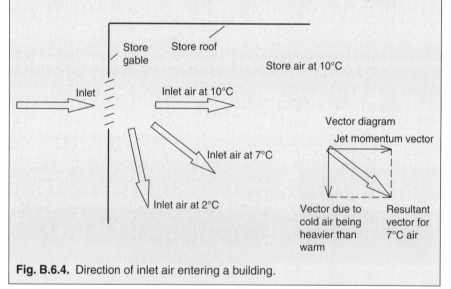

Fig. B.6.4. Direction of inlet air entering a building.

to the individual floors is by ladder. Porters carry the sacks up the ladders, or use a 'dumb waiter'-type lift and stack them in groups on the porous floors (Fig. 6.30).

While the authors have limited personal experience of these stores, warm potatoes loaded into the cold store are likely to experience subsurface condensation in the top bags. This may be minimized by rapidly batch cooling the crop to the temperature of the store using positive ventilation prior to putting them into their final storage location. Good air movement through the bags themselves, rather than just between the stacks of bags, will also be beneficial.

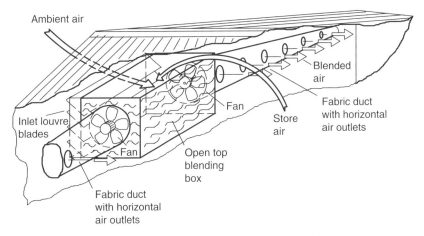

Fig. 6.28. Blending system suitable for retrofitting to existing inlets.

Fig. 6.29. Ventilation of multi-storey refrigerated sack stores. (Courtesy of W. Leslie, Farm Electronics, Lincoln, UK.)

6.8 Positively Ventilated Systems for Boxes

There are three basic methods of positively ventilating potatoes in boxes:

- Letterbox.
- Suction wall.
- Portable systems.

These can all be linked to ambient-air cooling or refrigeration systems. Portable systems are discussed in Ch2.8.

Fig. 6.30. Sacks on porous floor in multi-storey refrigerated sack store. (Courtesy of W. Leslie, Farm Electronics, Lincoln, UK.)

6.8.1 Letterbox systems

Traditional letterbox system

The conventional letterbox is the industry standard for positive ventilation (Fig. 6.31). Air is drawn via inlet louvres into an air blending chamber running the length of the letterbox duct. Recirculation louvres are fitted on the store side of the blending chamber opposite the inlet louvres. Fans mounted horizontally on the top of the letterbox duct draw air into the duct. Air is forced from the duct under pressure from the fans via rectangular slots, similar in shape to a letterbox for posting letters, hence its name, at a height to match the pallet apertures of the boxes. The air flows along alternate pallet apertures, between box layers 1 and 2, 3 and 4, and 5 and 6. The air flows downwards through layer 1 and upwards through layer 2, as foam bungs, manually put in place, prevent air flowing out the end of the pallet apertures. Pivoted flaps on the letterbox duct outlets allow air to be directed to all, or just some, of the rows and layers of boxes. A walkway is provided within the duct to allow the pivoted flaps to be reached to be shut or opened. When loading the store, the entire fan output can be directed to the first crop in, giving very rapid drying. Later in the harvest, the first rows may be dry so the airflow can be directed to the boxes of potatoes last into store. Airflows within the ducts made by the pallet apertures should not exceed 6 m/s.

Upward flow only

If boxes are to be ventilated vertically upwards only, they should be designed so that:

- The base of the pallet aperture is close-boarded to prevent air escape downwards.
- The boxes have a gap in the ends of the boxes to let air escape sideways from the rows (Fig. 6.32).
- The rows of boxes are separated by a small gap to allow air exiting from the sides of the boxes to escape.

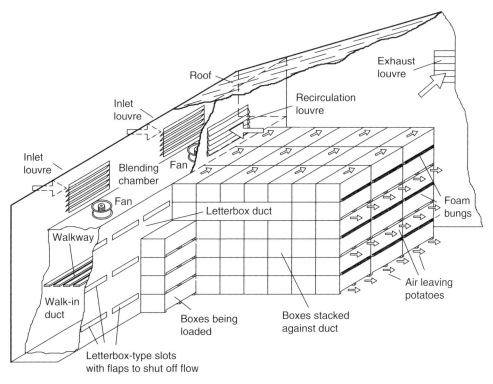

Fig. 6.31. Traditional letterbox duct with blending chamber above duct.

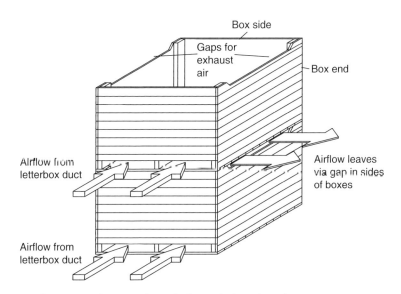

Fig. 6.32. Boxes to suit upward-ventilation-only letterbox duct.

If the duct is to ventilate a six-high stack of boxes, there will be six outlets (Fig. 6.33), one near ground level and the rest located one above another, one box height apart. Boxes for upward flow letterbox systems will cost more than those designed for upward and downward flow, due to the extra timber needed for the solid pallet base. When positive ventilation stops however, the warm zone will be above the cool zone, which avoids any risk of condensation once natural ventilation starts. This system is very suitable where high airflows are desired, as the air flowing in each duct only ventilates one layer of boxes, so the air speed along the pallet aperture ducts should not exceed 6 m/s. These systems are common for onion and bulb drying.

Upward and downward flow

In the UK, boxes are normally ventilated in pairs of layers, with the air flowing vertically upwards in the top layer and the air flowing vertically downwards in the bottom layer. Each duct outlet therefore ventilates a pair of layers of boxes in any row. Usually there are three openings per row, one above another, allowing three pairs of layers (six boxes high) to be ventilated. Only the ducts, formed by the pallet apertures carrying the ventilating air, are blocked with foam bungs. The alternate ducts have no foam bungs fitted, so allow exhaust air to escape.

The cost of boxes is lower in this dual-flow system, but ventilation control has to be more sophisticated. Since cooler potatoes will overlay warmer ones in the bottom boxes of each pair of layers, ventilation-air temperature should always be less than 4°C cooler than the warmest potatoes in the boxes.

With flow rates of 0.04 m³/s/t, the pallet aperture will have sufficient cross-sectional area to allow rows of ten boxes to be ventilated from the letterbox duct without exceeding the 6 m/s duct air speed. In practice, with the type of boxes used

Fig. 6.33. Letterbox to suit upward-ventilation boxes. (Courtesy of Flach and Le-Roy Ltd, Huntingdon, Cambridgeshire, UK.)

in the UK, leakage through gaps between the boxes is such that airflow declines with distance from the letterbox duct. A row five boxes out from the duct will dry in 4 to 7 days, while one of seven boxes may need to be turned if the crop is very wet to ensure that the boxes at the end of the rows dry rapidly. Rows of ten boxes should only be considered if the fan's airflow can be directed at individual rows by closing off other rows that are already dry, or where boxes are perfectly square and precision made, so that they fit closely together.

Sidewall letterbox
Another form of letterbox duct is the sidewall unit, which simply recirculates air within the store through the boxes (Fig. 2.17). This can be used to warm or dry potatoes as a batch process.

Open wall letterbox
The open wall letterbox, which takes three rows of 1.83-m-wide boxes, six high, sits within a 6.1-m bay of a building, with the sidewall of the building acting as the rear of the duct (Fig. 6.34). The front of the duct is open, with horizontal baffles fitted across the 'wall' to close off alternate box pallet apertures. Row length is restricted to five to seven boxes, as, unlike with traditional letterboxes, individual blowing of separate rows is rarely practised. While this could be done it would mean entering the duct to block off pallet apertures above head height. In wet harvesting years, when fan drying capacity is most needed, a 'wall' of boxes one or two boxes out from the duct can be stacked against any unused open wall letterboxes to rapidly dry them prior to moving them to their final location. This involves double handling that is unnecessary with traditional letterboxes.

Fig. 6.34. Open wall letterbox with socks on fans to aid uniformity of airflow through boxes. (Ordens Farms Ltd, Aberdeenshire, UK.)

Open wall letterboxes can have the fan either blowing or sucking; in the boxes used in the UK, the leakage of air is so great (Box 6.8) that the difference is too small to measure. In suction mode the diffuser sock shown in Fig. 6.34 is not fitted (Fig. 6.35).

Achieving uniform air distribution from letterboxes

The airflow exiting from outlets in a letterbox duct will follow the same pattern as that for underground ducts. More air will come from an outlet the further it is from the fan. Diffusers are sometimes put in to reduce this variation. If fans blow vertically downwards, canvas diffusers can be used, with the ratio of total hole outlet area to fan cross-sectional area selected from Table 6.4 (Fig. 6.36). The use of a tapered duct with greater hole area would seem a better idea, but is only partially successful due to the tendency of the air to keep flowing in a direction parallel to the duct (Fig. 6.37). This will only be an improvement if this flow of air can be deflected sideways (Carpenter, 1972).

Box 6.8. Air leakage measurement in boxes

Measurements made when boxes designed specifically for positive ventilation were carefully stacked tightly together in an experimental store had air leakage of 31 and 35% in the 2 years they were tested (Pringle *et al.*, 1997). Such care in stacking is not practical in commercial stores so the figure of 50% leakage is more likely in practice.

Fig. 6.35. Suction letterbox for open slatted boxes with plastic sheeting to seal sides. (Courtesy of D. Dickson, Doune Park, Banff, Aberdeenshire, UK.)

Fig. 6.36. Pressurized plenum with fabric socks fitted to fans to aid uniformity of air through boxes. (Wood Farm, Norwich, Norfolk, UK.)

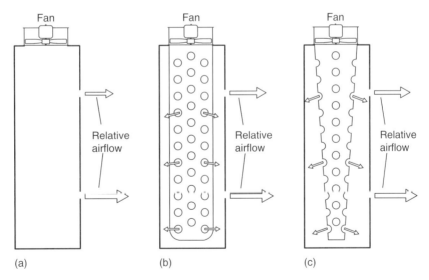

Fig. 6.37. Side section of letterbox ducts with arrow size showing effect of socks on uniformity of airflow from outlets: (a) plain letterbox; (b) letterbox with parallel-sided sock; (c) letterbox with tapered sock.

Fig. 6.38. Suction wall system where a fan sucks air from a plenum formed by the gap left between two rows of boxes and sealed with a canvas cover. (Courtesy of F. Pirie, ex FJ Pirie Pirie and Co. Ltd, Ayrshire, UK.)

6.8.2 Suction wall systems

Suction walls (Fig. 2.19) have been a traditional form of positive ventilation for cooling vegetables in porous crates or boxes for at least a century (Thompson *et al.*, 1992). Their use in the UK for ventilating boxes, under the trade name 'Aspire' or 'Boxer', is a more recent development. The basic system has two rows of boxes with a plenum between them, with a canvas sheet stiffened by battens pulled over the top and down the front face of the plenum. Either the rear of the plenum connects with an aperture in a duct, which is under suction from an extractor fan drawing air from the duct, or a portable extractor fan is placed at the front of the stack. Air movement is horizontal (Fig. 2.19, Fig. 6.38).

In selecting this system for positive ventilation a number of factors have to be considered:

- Boxes to be used in a suction wall system should be designed to suit the system, with gaps in the ends of the boxes positioned to ensure that the air travels through the potatoes in the box, not over or under them (Fig. 6.39).
- Extending the system to positively ventilate two rows of boxes each side of the plenum instead of one row each side is unlikely to be successful. Boxes are often not exactly the same height, so lining up gaps in open slatted boxes that are 25 mm wide, with those of a neighbouring box, will rarely be successful. The ducts, formed by 100-mm-high box pallet apertures, provide greater latitude to any misalignment and potential obstruction to airflow.

- Since boxes are wider than they are deep, the ventilating air path is more than twice the length with suction wall compared to letterbox ventilation.
- As with a letterbox system, a cooling front will develop, but this will be vertical rather than horizontal. As with dual-direction letterboxes care is needed to keep the temperature step in the cooling front to a minimum.
- Recent safety advice (BPC, 2005a) warns of the dangers of walking on boxes within one box distance of the edge of a stack. This system either precludes walking on the boxes or requires a harness safety system to be used when inspecting or sampling potatoes.
- The canvas spanning the plenum between the two rows of boxes can appear safe to walk on in the low light of a potato store, but hides a 5.6-m vertical drop on to the concrete below; warnings signs on the canvas are essential but may not be sufficient to prevent accidents happening. Its use should therefore be avoided unless acceptable safety provisions are installed.

Most growers using this system rely on there being gaps between boards in the end of boxes due to the timber shrinking following box construction. This is true of many but not all boxes. The largest gaps in traditional UK 'Angus' boxes (Fig. 6.40) are at the top and base of the ends, so the majority of airflow will be over the surface of potatoes, not through them.

The width of the plenum is usually sized to fit the diameter of the fan or to allow personnel access. The width can however be reduced, so long as the air speed entering the suction duct does not exceed 6 m/s. Since the air entering the rows of boxes either side of the plenum can enter via the front, top or back of the row, the gap between adjoining rows can be quite narrow, while still keeping the maximum air speed at or below 6 m/s.

Fig. 6.39. Box suitable for suction wall ventilation. (G. Mackie, Darnabo, Aberdeenshire, UK.)

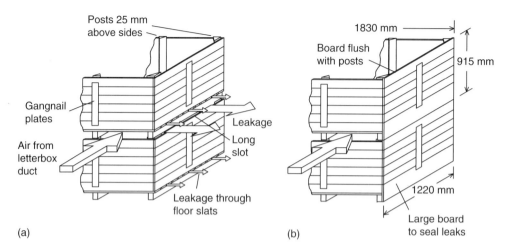

Fig. 6.40. (a) Traditional 'Angus' box; (b) closed-ended type box best for letterbox ventilation.

6.8.3 Portable systems

Portable versions of positive ventilation systems are discussed more fully in Ch2.8. They can, however, form part of a permanent ventilation system. Wedderspoon tents (Fig. 6.41) are used for initial drying of boxes in some cross-flow ventilated stores, while suction wall and Posi-vent® systems are usually linked in with the ventilation and refrigeration systems (Figs 2.18 and 2.19).

6.8.4 Single-stage positive ventilation systems

Single-stage positive ventilation is almost always used in bulk storage systems and commonly used in letterbox ventilation systems for box stores (Fig. 6.42a). One fan draws air in from outside, and forces air through ductwork and lateral ducts and subsequently through the potatoes themselves. However, a two-stage ventilation system can be used in preference in some circumstances.

6.8.5 Two-stage positive ventilation systems

A two-stage positive ventilation system uses two fans in series, one to bring the air into the store and one to blow the air through the potatoes (Fig. 6.42b). This allows two high-volume fans with a low backpressure capability to be used rather than a single fan with a higher backpressure capability. Not only can these fans be cheaper to purchase, but their running costs and the amount they heat up, and therefore dry, the air can be less. In a comparison carried out (Pringle, 1989) between a traditional letterbox system using a fan with the relatively high backpressure capability of 365 Pa compared with a two-stage system using fans each with a backpressure capability of 120 Pa, the capital cost of the two fans was 80% that of the single fan while their running costs were 70%.

Fig. 6.41. Tent drying system with operator protected by a travelling safety harness. (Deveron Potato Growers Ltd, Banff, Aberdeenshire, UK.)

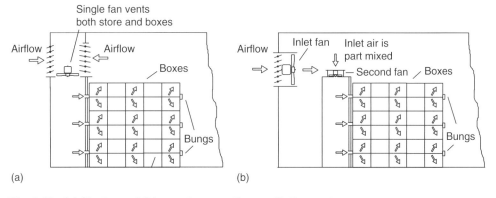

Fig. 6.42. (a) Single- and (b) two-stage positive ventilation systems.

A second benefit from a two-stage arrangement is that a certain amount of blending occurs, as ambient air is first brought into store, which causes it to mix with the store air. The second fan then takes the mixed air and forces it through the potatoes in the boxes. So long as the incoming air mixes with the store air before it comes into contact with potatoes and before the second fan draws it in, the system will achieve a degree of blending of incoming air.

For more controlled blending, a blending system on an inlet louvre can be fitted by the arrangement shown in Fig. 6.28.

6.8.6   Box designs for positive ventilation

For letterbox and suction wall systems, boxes should be designed to suit the system installed. A guide to the structural requirements of UK 1-t potato boxes is to be found in BS 7611:1992 (BSI, 1992). Boxes to suit letterbox ventilation systems should be close-boarded on their ends, from the top of the corner posts to the bottom of the bearers (Fig. 6.40).

For suction wall systems, the ends of the boxes should be open-slatted. However, the lower 200 mm of the ends should be close-boarded, to prevent air flowing sideways through the pallet apertures (Fig. 6.39).

Only with the Wedderspoon tent does box design not matter.

If the decision is to install a letterbox or suction wall system, yet only conventional 'Angus' type boxes are available, these can be modified (Fig. 6.43) or phased out over time and boxes bought thereafter can be specified to conform to the new design.

6.8.7   Airflow rates with positive ventilation systems

Airflow for positive ventilation systems should be designed at a *minimum* to give 0.02 m³/s/t through the actual potatoes. Since leakage with conventional Angus boxes has been measured to be as high as 75% (Pringle, 1989), the minimum airflow required is 0.08 m³/s/t to achieve the 0.02 m³/s/t rate through the crop. Boxes designed for use with positive ventilation systems lose approximately 50% of the air coming from the fan. The bow (outward bend) in the sides of the boxes

Fig. 6.43. Angus box modified to suit positive ventilation.

caused by the pressure of potatoes plus the difficulty of stacking boxes precisely at a height of 5.6 m and at speed means that there will always be gaps between boxes. If growers only have Angus boxes, the installed airflow should be 0.08 m³/s/t so that once the ventilation-system-specific box is phased in, a good drying airflow approaching 0.04 m³/s/t will be available.

Design of ventilation systems for flexible management

Air movement through the crop should be started immediately after the first few tonnes have entered store. At the start of loading, the airflow capacity per tonne in store will be very high, as so few tonnes are in store, and drying will be very rapid. Once the store is nearly full, the airflow rate per tonne will be much lower, approaching the design airflow rate, so drying will be considerably slower.

In wet harvesting years, the available ventilation should be capable of being spread over the entire crop already loaded into store. This will maximize the rate of drying, though it will mean that it will take longer for the individual stores to be closed up ready for cooling.

In dry harvest years, stores can be loaded in turn as the requirement for large quantities of air will not be so great. Stores can be loaded and closed up sequentially, allowing cooling to be started early.

Comparison between airspace and positive ventilation systems

There is much debate as to when positive ventilation is required and when airspace ventilation is adequate. In the view of the authors, positive ventilation is recommended for:

- Removal of respiration heat and rapid drying for early harvested seed potatoes.
- Over-wintered seed potatoes where stores are routinely opened to access seed for sale. Such stores are susceptible to condensation through air ingress and may need to be re-dried at times throughout the storage period.
- For remedial blowing of diseased batches of potatoes to mummify rots and limit spread of rotting.

If the airflow used in airspace ventilation is sufficient, its distribution uniform and the crop free from soil, the two systems become similar in performance. In a comparative study (Pringle *et al.*, 1997), once the crop had been dried after harvest, the potatoes remained dry with either airspace or positive ventilation so long as condensation on the potatoes was avoided through control of ventilation and leakage of ambient air into store was prevented.

Pre-pack and seed potatoes are normally stored at 3–4°C. If graded dry at this temperature they are likely to sustain damage (Ch10.4) and will be slow to heal any wounds inflicted. To avoid this, potatoes stored below 6°C should be routinely warmed to 8–10°C before grading. They will also heal more quickly at this temperature. If pre-pack potatoes are washed prior to grading, less fingernail damage will be done as the friction between tuber and grader is less. Pre-grading warming may therefore be omitted.

If potatoes are taken out of store and graded or dispatched at 3–4°C, and the dew-point temperature of the ambient air is warmer than the crop, the potatoes will get wet from condensation. The wet potatoes will then be vulnerable to disease. If the crop is in bags the condensation and disease may not be seen until the bags are opened. Warming prior to removal from store is good practice (Ch6.10).

6.9 Sprout Control using Chloropropham for Processing and Ware Potatoes

CIPC is the most common chemical applied to crops in store to control sprouting (Ch3.8) The European maximum residue level (MRL) for this chemical on potato tubers has recently been set at 10ppm (EC, 2004). Store managers must therefore ensure that this limit is met by strict adherence to approved application procedures.

6.9.1. When treatment is justified

Regular inspection of the crop, usually on those tubers 300mm below the crop surface, will reveal when eyes are starting to open. Treatment should then take place. Between two and five treatments may be required per season, the actual number depending on length of storage, the temperature that the potatoes are stored at and the dormancy characteristics of the variety stored.

6.9.2 Treatment formulations

In Europe the CIPC is mixed with a carrier or solvent. With some CIPC/solvent formulations, fans within the store cannot be run for fear of a spark causing an explosion, so dispersion of fog relies solely on natural convection. With other formulations, fans can assist the distribution of the fog. The formulation label will indicate what is allowed. The fog will not only be deposited on potatoes, it will also be deposited on to the fabric of the store, the boxes, fans and refrigeration equipment. Prior to fogging, measures should be taken to minimize the contamination of fans and fridge components by protecting equipment with easily removable plywood or steel panels.

6.9.3 Conditions for optimum treatment

CIPC residues on the tubers, especially in box stores, can be very uneven. Temperature gradients in a stack of boxes alter the deposition pattern (Briddon *et al.*, 1999; Xu and Burfoot, 2000) with colder potatoes on top and warmer ones at the base resulting in the highest proportion of CIPC deposited on the top box. Crop temperatures should therefore be as uniform as possible. If fans are allowed to be used, these help to distribute the chemical, but they should be run at low speed if this is possible. Best uniformity of application is achieved if a dry fog is produced, which requires the ambient RH during application to be low (SAM, 2005). Blowing high-humidity warm ambient air, along with the CIPC, into a store that is colder is likely to result in the moisture-laden air condensing on the potatoes during the treatment process. The moisture from the products of combustion from burning of petrol in the fogger makes such condensation even more likely.

6.9.4 Application procedure

Prior to applying the chemical, the air within the store should be recirculated for an hour or two to ensure that all parts of the crop are as near to the same temperature as possible. The dew-point temperature of the ambient air should be checked to ensure that it is below the temperature of the stored crop. If fans cannot be run to assist CIPC distribution, they should be enclosed with plywood or sheeting to prevent their contamination with CIPC crystals. The cooling coils of a fridge should also be blocked off. Treatment can now start.

6.9.5 Fogging process

With the doors of the store closed, the application equipment, often mounted on a vehicle for mobility, is positioned close to the store and a stainless steel pipe connected from the fogger through a port installed in the outer wall of the store (Fig. 6.44). The fan and burner unit of the fogger is started and CIPC injection

Fig. 6.44. Thermal chloropropham fogger with stainless steel flexible pipe inserted into store.

begins. With bulk stores using a no-fan formulation, the hot chemical is fed into either the main duct or the mixing chamber. As the chemical exits the generator, it cools to the temperature of the store and forms microscopic crystals that make up the fog. The fog rises through the crop by natural convection, being deposited on the crop as it goes. If fans can be run, use them to aid uniform distribution. If fans are variable speed, they should be run at low speed. If there is more than one fan, the lower air speed can be obtained by running only one fan.

With box stores, the most basic system allows the chemical to be fed into the side of the store, so that the hot vapour rises into the headspace then falls down through the stack of boxes. This tends to result in the highest levels of CIPC on the top boxes and the least on the potatoes in the boxes at the base of the stack. This effect is greatest if there is no air circulation.

Alternatively, the fog can be introduced into a stack (Fig. 6.45) so that the chemical rises up thorough the stack. Application in box stores results in 30% of the applied chemical reaching the target (Duncan, 2006) and 70% being lost to the atmosphere and deposited on boxes, store surfaces and any exposed fans and equipment.

6.9.6 Action following treatment

Most chemicals require the store to stay closed for 24 h, although some require only 12 h. Potatoes having been treated cannot be processed for between 21 and 28 days, depending on the chemical label. As the ethylene produced from burning petrol during the fogging process tends to trigger increased respiration and affect fry colour, growers would like to open their stores within 8 h of treatment.

Investigations to date to find out whether CIPC contaminates wash water suggest that the chemical is more likely to stick to the soil on the potatoes than to pass into the wash water (Duncan, 2006).

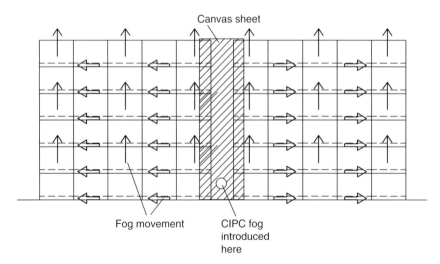

Fig. 6.45. Box stacking system using canvas sheet to minimize variation in application of chloropropham (CIPC).

Fogging is a specialized operation, with risk from particulate clouds forming an explosive mixture as well as the possibility of poor distribution of chemical. It should therefore be carried out by a competent contractor (Duncan and Kraish, 1999).

6.10 Warming in the Store

The best place to warm potatoes prior to grading is within the store itself. In store, as the RH is between 90 and 98%, the dew-point temperature will always be slightly below the temperature of the crop. Warming can therefore be carried out without the warmed air causing condensation to form on the crop (Ch3.5). If the air used to warm the potatoes has been warmed using an electric heater or an indirect gas or oil heater, the temperature of the air will increase but its dew point will not.

6.10.1 Warming of boxes of potatoes within the box store

With an airflow of $0.08\,m^3/s/t$ and a leakage of warmed air from boxes of approximately 50%, warming of potatoes from 5 to 10°C takes about 11 h (Pringle, 1993b). A proportion of the heat will enter the potatoes while some will dissipate within the store. This will have to be removed by ventilation or refrigeration, if the store is to be kept at a fixed temperature. If the batch of potatoes is small compared with the mass of potatoes stored, the increase in store temperature will be small. A fixed or portable letterbox, or a tent with recirculation, can be used to warm batches of potatoes in this fashion.

6.10.2 Warming bulk bins or a bulk store

Potatoes in small 120-t bulk bins are easily warmed using a space heater blowing warm air into the recirculating airstream. Warming the day's out-take in large bulk stores is less precise. Closing all but the last few laterals in the bulk store and recirculating warmed air through an amount of potatoes equal to a day's grading can achieve a satisfactory result.

6.10.3 Warming chamber between store and grading area

Where it is possible to have a warming chamber between a box store and the grading area, this provides the best option. One door of the chamber opens to the store while the other opens to the grading area. With the grading area access door closed, boxes are stacked in the chamber for warming. The store access door is closed, and the potatoes are warmed without introducing any outside air. The grading side door is then opened (Fig. 9.15) and the boxes removed for grading. This set-up ensures the dew-point temperature of the air for warming is below the temperature of the potatoes, while keeping the warm air in the warming chamber, so it does not leak heat into the main store.

6.10.4   Removal of cold potatoes from store for warming elsewhere

Potatoes removed without warming from cold store are very likely to get wet from condensation (Fig. B.9.3). If the potatoes are then put into a room or grading area and left to warm up naturally, the moisture on the potatoes will slow the warming process. The room should be kept at the desired final temperature of the crop during this period; usually 10°C. Warming will normally take 3–4 days.

6.11   Summary

Ventilation is the principal operation carried out on the stored crop. The subject of this chapter is how to ensure the right amount of air flows uniformly through the stored potatoes. The principal conclusions are as follows.

- Ventilation systems usually consist of an inlet, blending chamber, fan, distribution system and exit louvres.
- For buildings operating in areas where snow can be severe, inlets are often placed horizontally under the building eaves.
- Air outlets are usually placed as high as possible, to allow warm exhaust air to exit.
- In exposed sites inlets and outlets should be on the same side of the building (i.e. within the same pressure zone) to prevent uncontrolled wind-induced ventilation.
- Bulk stores may be ventilated using a main central duct discharging air into porous laterals on either side or by single tapered ducts each with their own fan.
- System-designed porous concrete or timber floors provide both uniform airflows to bulk crops and allow unimpeded store emptying.
- Potatoes, unlike grain, offer minimal resistance to airflow so the majority of the resistance the fan 'sees' arises from high-speed air passing through inlets, blending chamber, ducts and outlets.
- To allow the use of low-backpressure fans, air speeds in ducts should be restricted to 6 m/s.
- Uniformity of airflow through the crop is improved by keeping air speeds in ducts low, avoiding the Venturi effect, using tapered ducts, restricting lateral outlet area and restricting duct or floor outlet area.
- While restricting duct or floor outlets improves uniformity of air through the crop, the increased backpressure reduces the total airflow supplied by the fan.
- Fan air capacity is proportional to tonnage stored. Its backpressure capability is calculated from the sum of the resistances to airflow as the air passes through the building apertures, ductwork and crop.
- Axial flow fans whine like aircraft engines; if residences are near their speed should be kept low, intakes can be muffled and earth bunds erected.
- Media-type air humidification systems do not have to be controlled and minimize the likelihood of droplet carry over.
- Spinning disc or sonic jet humidifiers need sophisticated water filtration and control as well as droplet arresters to prevent droplet carry over.

- Overhead airspace ventilation systems in box stores require well-directed jetting of air to ensure uniform ventilation of all boxes. Poor uniformity can be overcome by fitting assistor fans to improve the roll of air.
- The use of a 'goalpost' plenum in airspace-ventilated stores can improve the uniformity of airflow, uniformity of temperature and sugars, and prevent condensation in boxes at the rear corners of the store.
- Airspace ventilation in multi-storey sack stores allows air to bypass, rather than penetrate, the product. Condensation in bags could be reduced by cooling with positive ventilation prior to storing and/or increasing the air circulation.
- Letterbox positive ventilation systems can be designed to ventilate upwards only, or to ventilate both up and down. The former allows higher airflow rates, reduces the likelihood of condensation when ventilation stops, but requires more expensive boxes.
- UK-style potato boxes are lower in cost than the more sophisticated boxes used in The Netherlands, so air leakage when used for positive ventilation is greater. Row length in the UK is normally restricted to seven, while in The Netherlands rows of ten to 12 boxes are common.
- Boxes to be used with letterbox systems should be designed to suit; standard boxes will leak 75% of the air supplied and are therefore inefficient.
- Suction wall positive ventilation systems should be restricted to one row of boxes on each side of the suction plenum and should be used for boxes designed for sideways ventilation.
- Two-stage ventilation systems, by using two low backpressure fans in series, can be cheaper to buy and run than systems with a single, higher backpressure, fan.
- Positively ventilated and airspace-ventilated box stores can perform equally well if airspace-ventilated stores have sufficient airflow, potatoes are free of soil, the store is kept sealed against air leakage and doors are kept shut. Where doors are opened regularly or air leakage occurs, positive ventilation will help to dry out any condensation that this causes.

7 Store Refrigeration

Topics discussed in this chapter:

- Refrigerated compared with ambient-air cooling.
- Principles of refrigeration.
- Direct expansion refrigeration systems.
- Coefficient of performance.
- Two-stage water/glycol systems.
- Chilled-water cooling/humidification systems.
- Sizing a new refrigeration plant.

7.1 Refrigeration Systems

7.1.1 Sealed store

In a refrigerated store, air being circulated through the potatoes is routed through a heat exchanger cooled by refrigeration. When refrigerated cooling is taking place, louvres and doors in the store are sealed. This contrasts with ambient-air cooling systems, where heat from the potatoes leaves the store with the ventilating air.

7.1.2 Principle of direct expansion refrigeration systems

The refrigeration system used for this is floor-mounted, has a vertical duct to transfer air to the top of the store and a horizontal outlet to discharge air horizontally across the boxes. A photograph of such a unit is shown in Fig. 6.23. Box 7.1 introduces some terms used in refrigeration.

A direct expansion (DX) refrigeration system consists of two heat exchangers, an evaporator (cooling coils) inside the building and a condenser on the outside (Fig. 7.1). These are connected together by pipework to make a circuit. The

---

Box 7.1. Refrigeration terminology

Refrigeration systems use a *heat exchanger* in the form of a *cooling coil* or *evaporator* to remove heat from the store. A second heat exchanger, called a *condenser*, is placed outside the store to dissipate the heat removed to atmosphere. The fluid within the pipework of the refrigerator is called a *refrigerant*, which has the property of being able to change phase easily from a liquid to a gas and back again.

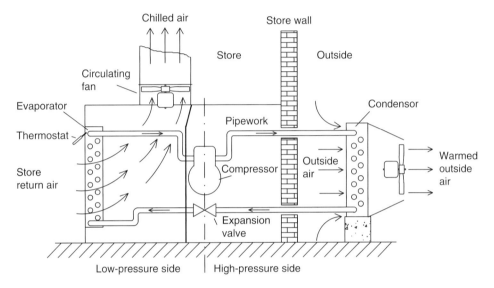

Fig. 7.1. Diagram of a floor-mounted refrigeration system.

refrigerant is forced clockwise round the circuit by a compressor. The liquid refrigerant (Box 7.2) is forced at high pressure through an orifice (expansion valve) into the evaporator, which is at low pressure. At this low pressure the liquid refrigerant starts to change phase into a gas, provided it can obtain the latent heat it needs to become a gas. This heat is obtained from the store return air passing through the evaporator. The liquid refrigerant evaporating within the evaporator cooling coils therefore cools the store return air as it passes between the coils of the evaporator.

Once the refrigerant reaches the outlet from the evaporator it has totally changed from a liquid into a gas. The compressor draws in the gas from this low-pressure side of the circuit and compresses it prior to forcing it into the condenser located outside the store. At this high pressure the gas starts to change phase into a liquid, but has to get rid of latent heat to do so. To remove this heat, fans force air from outside the store through the condenser. The refrigerant gas therefore changes phase from a gas to a liquid (i.e. condenses), within the condenser, giving out heat as it does so. This heat warms the outside air passing through the condenser. The refrigerant in liquid phase is then forced through the expansion valve once again to repeat the cycle.

> **Box 7.2.** Change of phase in refrigerants
>
> When a fluid changes phase from a liquid to a gas, heat is required. The heat absorbed in the phase change is hidden, termed latent, as it does not change the temperature of the fluid, only its state.
>
> The refrigerants used in refrigeration systems are selected for the efficiency with which they change from liquid to gas at the design temperature ($3.5°C$ for potatoes). In the heat exchanger (cooling coils) of a DX refrigeration system, the refrigerant *evaporates* from a liquid to a gas, taking in 'latent heat of vaporization' from the air passing through the heat exchanger as it does so. This is why the heat exchanger is called an evaporator. In the heat exchanger outside the building, the refrigerant *condenses* from a gas to a liquid, giving out its 'latent heat of vaporization' as it does so. This is why it is called a condenser.

This process effectively pumps heat from inside the store via the refrigerant to outside, so removing heat from the potato crop and store.

7.1.3 Control of direct expansion refrigeration systems

DX refrigeration systems are either controlled by a thermostat fitted to the evaporator coils (Fig. 7.1) or by the sensors placed in the crop, which are connected back to the control box. The sensor monitors the evaporator 'return air' temperature. When the store is set to cool using the refrigeration system, the air circulation fans are timed to start up and run for a few minutes every hour. If the return air temperature is higher than the crop set-point temperature, the thermostat or sensor will bring on the compressor. The compressor will then run until the crop set-point temperature is reached.

7.1.4 Temperature differential between return air and evaporator cooling coils

As the cooling coils of the evaporator are cooler than the store return air, heat is transferred from the warm air to the cooling coils (Fig. 7.2). The temperature differential (TD) between the return air and the evaporator coils is usually $6°C$. The return air does not cool by this amount, instead cooling by $2.5°C$ when the crop is newly harvested and warm and by $1.5°C$ once the crop is cooled to $3–4°C$. The temperature difference between the air entering and exiting the evaporator is called the air-on/air-off temperature difference. This should not be confused with the TD discussed above.

The design TD determines the RH of the air circulating through the potatoes (Table 7.1). The greater the TD, the more moisture will condense out of the air on to the cooling coils. This will reduce the amount of moisture in the air, so its RH will be lower. To minimize crop weight loss, the TD should therefore be low. This

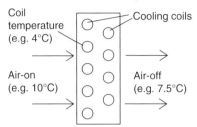

Cross-section of evaporator

Coil temperature (e.g. 4°C)

Cooling coils

Air-on (e.g. 10°C)

Air-off (e.g. 7.5°C)

Temperature differential (TD) = 10 − 4 = 6°C

Air-on/air-off temperature difference = 10 − 7.5 = 2.5°C

Fig. 7.2. Temperature differential of an evaporator.

Table 7.1. Evaporator temperature differential (TD) determines store relative humidity (RH). (From Dossat, 1981.)

Design TD (°C)	RH (%)	Size of evaporator	Cost of unit	Crop weight loss
4–6	95–91			
6–7	90–86			
7–8	85–81	Increasing	Increasing	Decreasing
8–9	80–76			
9–10	75–70			

requires the evaporator to be large, resulting in it being more expensive than an evaporator with a higher TD. Potato store evaporators are therefore large in comparison to those used to chill frozen or packaged produce.

7.1.5 Fins on the evaporator

The evaporator consists of a series of pipes, or coils, which loop back and forth across the width of the unit. As pipework on its own has a relatively small heat exchange surface area, the evaporator coils are fitted with vertical aluminium plates, or fins, 4–6 mm apart, to increase the heat exchange area. Being vertical, any moisture in the air that condenses out due to the air being cooled by the evaporator will run down the fins to the drain in the floor. If however the fins are too close together, surface tension can slow the rate that water runs off and the water can turn to ice. This can block up the fins and prevent the fridge from working (Fig. 3.23). For high-humidity storage atmospheres, fin spacing at minimum should be 4 mm with 6 mm being preferable. These fins are delicate and can be damaged by impact or overzealous pressure washing. A uniform airflow across the evaporator is also essential if icing is to be avoided.

7.1.6   Defrost systems

The closer the store set-point temperature is to freezing, the more likely the evaporator will be to ice up. For conventional DX systems, a minimum set-point temperature should be 3.5°C. Most refrigeration systems use off-cycle defrost to melt any ice that has formed. This simply shuts off the compressor for a quarter of its running time, to let the evaporator warm up to allow any ice to melt. In sizing a refrigeration unit, this requirement should be taken into account.

Alternatives to off-cycle defrost include electric heating elements beside the evaporator or valves in the refrigeration system to allow the compressor to be operated in reverse, as a heat pump. This latter approach is called hot gas defrost. Both electric heaters and hot gas defrost systems melt the ice more rapidly than off-cycle, but cost more and add heat to the store. They should be installed if stores are to be operated below 2.5–3.0°C.

7.1.7   Refrigerants

Hydrochlorofluorocarbon refrigerants such as R22 were the industry standard until they were required to be phased out in the year 2000 under the Montreal Protocol of 1987 to protect the Earth's ozone layer. At the time of writing the new industry standard is hydrofluorocarbon R134A. This too is ozone-depleting but to a lesser degree than chlorofluorocarbons (CFCs). The recently introduced R410A has zero ozone-depleting potential so may take over from R134A. Other refrigerants less damaging to the ozone layer include ammonia and hydrocarbons like propane and iso-butane. Other non ozone-depleting refrigerants may be developed over time.

To ensure ozone-depleting chemicals do not escape to atmosphere, refrigeration equipment should be leak-tested twice per annum by an approved contractor (EC, 2000).

7.1.8   Coefficient of performance

The cooling capacity of a refrigeration system is quoted in kW. The electrical power required to power the refrigeration system compressor, air circulation fans and any secondary chilled water distribution systems is also measured in kW. The coefficient of performance (COP) of the refrigeration system is obtained as:

$$\text{Coefficient of performance} = \frac{\text{Cooling capacity of refrigeration system (kW)}}{\text{Electrical consumption of compressor, fans and pumps (kW)}}$$

For overhead cool-air distribution type refrigeration systems, a COP of 2.0–2.5 is common.

The refrigeration system is therefore a heat pump working in reverse, using a small amount of electrical energy to pump a larger amount of heat energy from the crop inside the store to the air outside.

7.2   Two-stage Water/Glycol Systems

While single-stage DX refrigeration systems are most commonly used in potato stores, two-stage water/glycol systems are also used, the glycol being added to the water as antifreeze. The water/glycol system uses DX refrigeration, but the evaporator is located in a tank of water and cools the water/glycol mixture in the tank. The chilled water/glycol mixture is then circulated through one or more heat exchangers located within the potato store. The chilled water/glycol mix is in a closed circuit, with the store air flowing over the heat exchanger.

7.3   Chilled-water Cooling Systems

Chilled-water refrigeration systems also use a DX refrigeration system to cool a tank of water. However, in this case the chilled water is trickled over a porous medium made from moisture-resistant cardboard or plastic chips (Fig. B.3.7). The store air flows through the medium, and is both cooled and humidified by the chilled water. The air and chilled water are in intimate contact.

This system is used where weight loss is to be minimized by ensuring RH in store is maximized. Its use is mainly with systems which continuously ventilate potatoes with humidified air (Ch.3.7) or where vegetables such as winter cabbage or turnips are to be cooled for long-term storage.

7.4   Combined Refrigeration/Ambient-air Cooling Systems

Combined refrigeration/ambient-air cooling systems are commonly used in the UK. They require well-sealed louvres to prevent air leakage during the cool storage period, but they allow ambient-air drying and ventilation of the crop during wound healing and for cooling when cool ambient air is available. This requires less electricity than refrigeration. When cool outside air is not available, refrigeration can be used.

Such a unit is shown in Fig. 7.3a and b. The unit is similar to that in Fig. 7.1 but has motorized air recirculation louvres fitted in front of the evaporator and motorized intake louvres in the rear to the right of the diagram. In refrigeration mode (Fig. 7.3a), the louvres in front of the evaporator are open, allowing the return air from the store to be cooled as it passes through the evaporator prior to being drawn vertically up the duct by the main cooling fans. All louvres to the outside are closed.

In ambient-air cooling mode, the motorized air recirculation louvres in front of the evaporator are closed and the motorized intake louvres to the right of the diagram are open. This allows air to be brought into store and exhausted via louvres in the building gable.

The system also allows air blending (Fig. 7.3b) by partially opening both air recirculation and air intake louvres. In the figure they are open 80% and 20% respectively, which would occur if the ambient air were considerably colder than

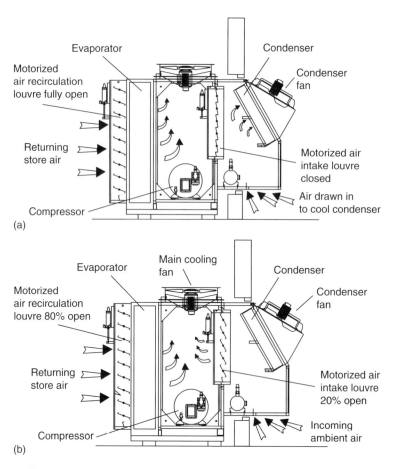

Fig. 7.3. Floor-mounted fridge: (a) with louvres set for refrigeration mode; (b) with louvres set for ambient-air cooling. (Courtesy of W. Leslie, Farm Electronics, Lincoln, UK.)

the desired duct air temperature. Figure 7.4 is a photograph showing this system, mounted on a skid unit, being brought into the store during installation. The fans in view are those used to cool the condenser.

7.5 Comparison between Direct Expansion, Two-stage Water/ Glycol and Chilled-water Systems

Table 7.2 compares the operating characteristics of DX, two-stage water/glycol and chilled-water systems.

7.5.1 Direct expansion units

DX refrigeration units are often selected for being relatively low in cost and modular in construction. Floor-mounted units mounted on a skid unit are favoured in preference

Fig. 7.4. Floor-mounted fridge unit being installed. (Courtesy of W. Leslie, Farm Electronics, Lincoln, UK.)

Table 7.2. Comparison of direct expansion (DX), two-stage water/glycol and chilled-water systems.

System	Temperature differential (°C)	Air-on/air-off temperature difference (°C)	Store relative humidity (%)	Suits
DX	6	2.5	92–95	Single store
Water/glycol	6–0	2.5–0	92–95	Multi-store units
Chilled water	6–0	2.5–0	95–98	Both

to roof-mounted units in box stores as they can be placed in position by forklift and draw air through the boxes of potatoes. Being at ground level they can be serviced even when the crop is in store. They can incorporate ambient-air cooling as described above. To install, they need only a hole left in the store wall for the condenser to project outside. Once the fridge is in place the hole is sealed. A removable wall fills the gap. Alternatively, the unit comes in two sections, with the evaporator and refrigeration unit inside the building connected by two pipes to the condenser mounted outside. DX units used to store potatoes over a period of 6–8 months can produce a potato firm enough to satisfy the demanding requirements of supermarkets (Table 7.3).

7.5.2 Two-stage water/glycol systems

Two-stage water/glycol systems have the benefit that the TD across the cooling coils can be reduced proportionally from 6°C to 0°C. The lower the TD the higher the store RH will be, since less moisture in the air will condense on the cooling coils. The rate of cooling will however reduce as TD reduces, so in intermittently ventilated systems the cooling will be stopped at a preset value. In continuously ventilated systems, there will almost always be some cooling taking place to remove respiration heat.

The system suits multi-store units as the one refrigeration system can provide cool water for a series of stores. This can provide a lower capital cost system than having DX systems in each store, if all the units are working simultaneously. In stores that are emptied sequentially, individual DX units may provide the lower operating costs. In the UK most growers have a separate DX refrigeration system per store rather than a water/glycol multi-heat-exchanger system.

7.5.3 Chilled-water coolers

Chilled-water coolers are the preferred system for the long-term storage of vegetables but can also be used for potato storage. In one unit installed at the Potato Marketing Board's packhouse at Sutton Bridge, Lincolnshire, UK, a chilled-water unit using a Baudelot evaporator was installed in a 2700-t box store and compared with a conventional DX unit located in an identical, adjoining store. The chilled-water cooler had an airflow of $0.01\,m^3/s/t$ and the fans ran continuously, the air being distributed through the stack of boxes via their pallet apertures from a pressurized 'goalpost' style duct, the height of the boxes. The DX unit was intermittent in operation, with an airflow of $0.025\,m^3/s/t$, with air from the distribution fans being discharged over the top of the stack of boxes.

Over 2 years of monitoring (Mawson *et al.*, 1992, 1993; Statham and Cunnington, 1993), where the store was operated for 31 weeks and 35 weeks respectively, the results were as shown in Table 7.3.

The wet cooled store did reduce weight loss in both years, but the financial value of the saving in product for sale and the marginally improved quality of the wet cooled stock of potatoes did not compensate for the 15–20% additional capital cost of the wet cooling system compared with the DX cooler or the potential risk in the slower drying of the newly harvested crop.

Table 7.3. Comparative weight loss in a direct expansion (DX)-cooled, compared with a wet-cooled, 2700-t refrigerated store over 31 weeks and 35 weeks, respectively.

	Weight loss (%)		
Store	1991–1992	1992–1993	Power consumption (kWh)
DX-cooled	4.1	5.3	145,000
Wet-cooled	3.4	4.8	149,000

7.6 Common Faults and Maintenance

Refrigeration systems tend to run with few problems. The most likely is for one of the switches that monitor refrigerant pressure, either the high-level or the low-level switch, to cut out and stop the compressor working. If this occurs, the control system should also shut off the recirculation fans. In some systems this is not the case and the fridge appears to be working because the circulation fans are operating, but the compressor is stopped. Leakage of refrigerant is one cause of these switches tripping. There is usually a light on the control box to indicate that the switch has tripped.

Icing up of the evaporator with well-designed units is rare. It is best avoided by keeping store temperatures above 3.5°C, and preventing store air RH rising to near 100% due to the ingress of warm humid ambient air through leaks or open doors. Another cause of high store RH, sprout growth, will increase the risk of icing if it is not halted.

7.7 Sizing a Refrigeration Plant

Refrigeration systems have to be able to remove all the heat that enters from outside or is generated within the building, as well as being able to cool the crop after harvest or cool the stored crop back down to the set-point temperature should the crop temperature rise for some reason. These heat loads include (Fig. 7.5):

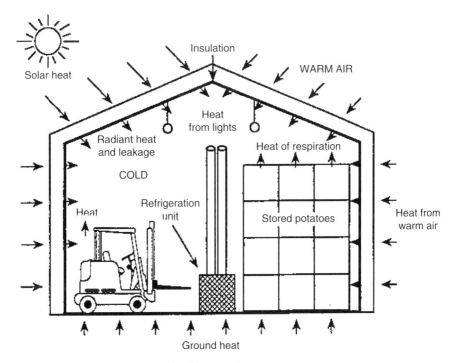

Fig. 7.5. Heat gains in a sealed refrigerated store.

- Solar heat gain.
- Heat transfer through walls, roof and floor from the warm air and ground outside.
- Warm air leaking into the building through gaps in the store fabric.
- Heat of respiration from the stored crop.
- Heat from forklifts, lights or other equipment operating in the store.
- Heat from circulation fans.
- Heat removed from the crop during cooling or after a breakdown.

7.7.1 Example calculation

The best method of illustrating how a plant is sized is to do an example. The one chosen is to size the refrigeration plant for a 930-t pre-pack store similar in design to, but smaller than, that used as the basis for the costing exercise in Ch13.

The 24-m-long by 18-m-wide building, with 6.5-m eaves height and 15° pitch roofs, holds 930 t of potatoes in boxes. The building is medium in colour, and has roof, wall and floor U-values of 0.25, 0.32 and 0.36 W/m²°C, respectively. Air leakage is assumed to be 0.5 air changes per hour. Two air distribution fans of 2.0 kW each run continuously to circulate the store air. Potato respiration heat output is 10 W/t at the storage temperature of 3.5°C (Ch1.5).

The maximum heat load is probably in early summer for a fully loaded store, when the average external temperature over the 24 h is taken as 18°C, while the store set-point temperature is 3.5°C.

7.7.2 Heat gain by conduction through the building fabric

A high level of insulation and good sealing of the store will help keep the plant size to a minimum. The higher the outside temperature, the larger the plant cooling capacity will need to be. A plant to keep potatoes well into the late spring/early summer will therefore be larger than one to store crop until early spring.

The heat transfers through walls, roof and floor are as in Table 7.4, using the equation:

$$H_F = U \times A \times \Delta T,$$

where

H_F = heat flow (W)
U = U-value (W/m²°C)
A = area of building fabric (m²)
ΔT = temperature difference between outside and inside of the building (°C).

A supplementary temperature increment is added for solar gain (Ch5.4, Table 5.1).

The maximum total heat transfer through the building fabric and floor is therefore 7.97 kW.

Table 7.4. Heat transfers through store fabric.

Item	U-value (W/m² °C)	Area (m²)	Solar gain (°C)	ΔT (°C)	ΔT + solar gain (°C)	Heat flow, H_F (W)
North gable	0.32	139	0	14.5	14.5	645
West wall	0.32	156	3.3	14.5	17.8	889
South gable	0.32	139	2.2	14.5	16.7	743
East wall	0.32	156	3.3	14.5	17.8	889
Roof	0.25	447	8.3	14.5	22.8	2548
Floor	0.36	432	0	14.5	14.5	2255
Total						7969

7.7.3 Heat gain due to air leakage

The heat gain due to leaking air is found using the equation:

$$H_L = \rho \times C_p \times q \times \Delta T,$$

where

H_L = heat from air leakage (W)
ρ = density of air = $1.23 \, \text{kg/m}^3$
C_p = specific heat of air = $1.005 \, \text{kJ/kg}$
q = airflow = 0.5 air changes/h
ΔT = temperature difference outside to inside = 14.5°C.

If the volume of the building is $3336 \, \text{m}^3$, then

$$\text{Heat from air leakage} = 1.23 \times 1.005 \times (3336 \times 0.5) \, / \, 3600 \times 14.5$$
$$= 8.30 \, \text{kW}.$$

7.7.4 Heat gain due to electrical equipment operating in store

$$\text{Approximate heat from two 2-kW circulating fans} = 2 \times 2.0$$
$$= 4.0 \, \text{kW}.$$

7.7.5 Heat from potato respiration

The metabolic heat from the stored crop is constant and is determined by the store's set-point temperature (Fig. 1.18):

$$\text{Metabolic heat from potatoes (kW)} = \frac{\text{Mass of potatoes} \times \text{Respiration rate (W/t)}}{1000}$$

$$= \frac{930 \times 10}{1000}$$

$$= 9.30 \, \text{kW}$$

7.7.6 Heat to be removed to cool the crop after harvest or after a breakdown

In addition to the above heat loads, the plant must have capacity to cool the crop after harvest and have a reserve capacity at the warmest time of the year that will enable it to cool the crop should it warm up for some reason. This could be because the plant stopped working for a day or the power supply failed.

A reasonable estimate of crop cooling capacity for box potato stores is to size the plant to cool the crop by 0.5°C per day to avoid any risk of subsurface condensation. If the crop enters at 18.5°C and is cooled to 3.5°C, cooling will take 30 days. Only at the later stages of cooling, when the store and crop temperature is significantly cooler than outside, will the plant have to cope with both cooling of the crop and removal of heat entering the store through the fabric and through gaps.

In early summer, the plant can be sized to cool the crop by 0.5°C over a period of 24 h, while still coping with the heat load from warm outside temperatures and solar heat gain.

$$\text{Rate of heat removal to cool the crop} = \frac{m \times 1000 \times c \times \delta T}{t},$$

where

m = mass of potatoes in store (t)
c = specific heat of potatoes (3.80 kJ/kg)
δT = temperature reduction over cooling period (°C)
t = time of cooling period (s).

As the store holds 930 t of potatoes,

$$\text{Rate of heat removal} = \frac{930 \times 1000 \times 3.80 \times 0.5}{24 \times 3600}$$

$$= \quad 20.45 \text{ kW}$$

What has not been included in this calculation is the cooling required to cool the potato boxes which, at 100 kg per box, is a significant load.

7.7.7 Total heat to be removed by the fridge plant

The total heat gain that the plant will have to deal with is the sum of the above heat loads.

Heat gain through fabric	7.97 kW
Heat from warm air leakage	8.30 kW
Respiration heat of potatoes	9.30 kW
Heat from the equipment	4.00 kW
Heat removed from crop	20.45 kW
Total cooling load	50.02 kW

If the refrigeration plant can only work for 83% of the time, to allow off-cycle defrosting of six periods of 20 min each, the plant will have to be increased in size appropriately:

$$\text{Plan size to allow for off-cycle defrost} = \frac{\text{Cooling load}}{0.83}$$

$$= \frac{50.02}{0.83}$$

$$= 60.27 \text{ kW}$$

Therefore

Final cooling capacity $= 60$ kW or 65 W/t.

The specification can be looked at for different times in the year, at cooling, during winter storage and in early summer, and the plant size adjusted to the highest demand. What refrigeration engineers fear most is thermal runaway: where the crop starts to heat up for some reason and its respiration rises to an extent that the refrigeration system cannot regain control of the store temperature. The only solution then is to remove the crop from the store.

7.8 Summary

In the UK refrigeration is increasingly being installed as an addition to ambient-air cooling, to satisfy supermarket demand for blemish-free washed potatoes and to slow dormancy break caused by a decade and a half of warm autumns. If fossil fuel-generated electricity is taxed or rationed to limit global warming, greater use is likely to be made of alternative disease and sprouting reduction measures such as better drying, condensation control, more sophisticated control of cooling and use of chemical sprout suppressants. This chapter has introduced the following concepts.

- Refrigeration systems are only operated when store louvres and doors are shut completely.
- While ambient-air cooling systems remove warm air from the building with the ventilation air, refrigeration systems cool the store air with a heat exchanger (cooling coils) and dump the heat outside the building using a second heat exchanger.
- Refrigeration systems consist of a circuit of pipework, with the refrigerant fluid expanding to a gas within the internal heat exchanger (evaporator) and condensing in the heat exchanger (condenser) outside the building. The evaporation of the refrigerant fluid absorbs heat while condensation emits heat.
- Direct expansion (DX) refrigeration systems use a single circuit of pipework to cool the store air and discharge heat to outside.
- Two-stage water/glycol systems use a DX system to cool a tank of water and then pass the cooled water through a series of heat exchangers to cool the air within the stores.

- Chilled-water coolers also chill a tank of water, but then produce a spray or trickle of chilled water which comes in direct contact with the store air. The air is therefore both cooled and humidified by the water.
- The coefficient of performance of a refrigeration system is the cooling capacity in kW of the system divided by the energy consumption of the fridge compressor, water pumps if present and air circulation fans.
- DX cooling systems are favoured in the UK over water/glycol and chilled-water systems due to their low cost and ability to produce a firm potato.
- The required cooling capacity of a refrigeration plant is calculated by adding the heat loads on the building to the cooling required to reduce the crop temperature.
- An approximate cooling capacity for box potato stores in the UK is $65\,W/t$.

8 Store Environmental Monitoring and Control

Topics discussed in this chapter:

- Commonly used sensors.
- Information provided by sensors.
- Control of ambient-air cooling.
- Control of continuous ventilation with humidified air systems.
- Control of combined ambient air/refrigeration systems.
- Controls to prevent condensation during drying and wound healing.
- Controls to minimize energy use.
- Advanced sensors.
- Store logging systems.
- Modem connections for comparing store performance.

8.1 Internal Store and External Air Monitoring

Sensors most commonly used in potato stores are:

- Temperature sensors.
- Air wet- and dry-bulb temperature sensors.
- Air RH capacitance film sensors.

8.1.1 Temperature sensors

Temperature sensors can be solid-state thermistors (i.e. *thermal resistors*), resistance thermometers, or thermocouples. In the first two, their resistance when connected into an electric circuit changes with temperature. In the third, a voltage is produced proportional to the temperature difference between the ends of a pair of wires, joined at each end, and made from different metals. UK storage manufacturers prefer thermistors, as they are low in cost, have a high resistance so that connecting leads can be different lengths yet have minimal effect on sensor accuracy; they are also reliable. Thermocouples are favoured by those carrying out experimental work as they can be cut to the length required for the experiment, provide good accuracy and are low in cost so can be discarded if necessary after use. Some European ventilation suppliers use thermocouples as standard. Resistance thermometers are only used where high accuracies are required.

Temperature sensors accurate to ±0.2°C are normally specified for potato storage applications. Thermistors are inserted into 4-mm stainless steel tubes and sealed to exclude moisture. Thermocouples can be similarly 'potted' or can be left as an exposed, welded tip (Fig. 8.1).

Connecting wires have to be long enough to reach from the point being monitored back to the control box/data logger. Cable lengths of 60 m are not uncommon. Cables must be kept clear of elevators and forklifts, so are commonly hung from the store roof, becoming accessible to staff for insertion into the potatoes only when the crop or boxes are loaded into store.

8.1.2   Capacitance film air relative humidity sensors

Air RH is increasingly being measured using capacitance film sensors (Fig. 8.2). They give a read-out directly in percentage RH. However, when used outside they can fail completely, possibly due to condensation forming on the sensor or, where the store is near the coast, due to salt spray from the sea. When used inside the store, they are particularly prone to dust, so should be protected with a fine filter. Ventilation manufacturers using these sensors should ensure self-checking routines are installed in the monitoring software to inform the store manager when a fault has occurred.

The recent availability of plug-in replacement heads allows routine replacement of the vulnerable sensing portion. Non-plug-in units need to be disconnected each year and sent to the supplier for recalibration.

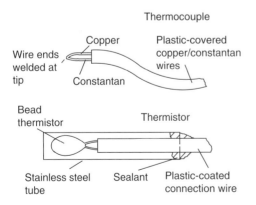

Fig. 8.1. Thermocouple and thermistor temperature sensors.

Fig. 8.2. Capacitance film sensor to measure air relative humidity.

8.1.3 Wet- and dry-bulb temperature sensors

Wet- and dry-bulb sensors, fitted with a fan to draw air over the sensors (Fig. 8.3), can be used to determine RH from a table or formula (CIBSE, 2006). Like the capacitance RH sensors, these too have limitations. The reservoirs of sensors located outside will freeze in subzero conditions, sometimes causing the capillary action of the wick to fail once the ice melts. Busy store managers can forget to top up the reservoir. When used within the store, dust in the air can collect on the wet wick, causing it to become coated in a wet paste. Filtration of the aspirated air will reduce but not totally prevent this problem. The monitoring software should again be designed to identify such failures.

8.1.4 Purpose of store monitoring

Store monitoring must serve a purpose. What monitoring tells the store manager is described in Table 8.1.

Variations in temperature of the crop in different parts of the store, near cold walls or by leaky doors can cause poor fry colours or localized condensation. Temperatures across the store should optimally be within 1°C of each other, less if possible. A large temperature difference between the top and base

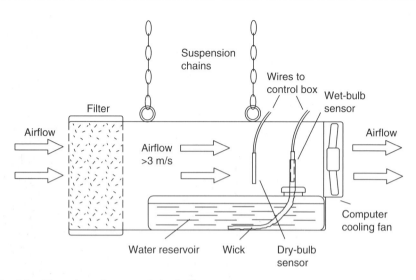

Fig. 8.3. Air aspirated wet- and dry-bulb temperature sensor.

Table 8.1. Information obtained from store monitoring.

Measurement	Aspect measured	Indicates	Indicates in addition
Temperature	Crop/external temperature difference	When cooling fans should be operating	
	Differences throughout the pile or the stack of boxes	When air should be recirculated to even out variations	Poor uniformity of air distribution
	Excessive difference up a pile	Need to recirculate air to reduce difference	That the ventilation inlet low-temperature setting may be too low
Relative humidity (RH) or wet-bulb and dry-bulb temperature	RH in store	Low store RH may indicate a leaky store	
	RH outside	Potential for outside air to condense on the crop in store (see Ch8.6)	

of a cooling front is likely to reduce with further cooling, but is undesirable over long periods when no ventilation is occurring. Recirculation will reduce this differential. Since there is little a store manager can do about store RH other than to keep doors shut, the value of internal store RH measurement is primarily to indicate store leakage.

8.2 Control of Crop Cooling and Cool Storage in Bulk Stores

8.2.1 Intermittently ventilated bulk storage systems

Bulk potato storage ambient-air cooling systems are almost always provided with air blending. Blending allows the temperature of the cooling air to be controlled so that it is never more than 4°C (Box 3.7) cooler than the crop temperature. It also allows ventilation when the ambient air is below freezing. Blending control is illustrated by the plot in Fig. 8.4. As the ambient air temperature (T_a) cools, ventilation starts at point A when T_a crosses the 'ventilation initiation line', 1.0°C below the crop temperature line. Ventilation continues as T_a cools, until it meets the low temperature limit 4°C below the crop temperature line. At this point (B) the intake louvres will start to close and the recirculation louvres open to blend the incoming cold air with the warmer store air (Fig. 8.5). The duct temperature sensor then controls the position of the two louvres to ensure that the duct temperature T_d is always 4°C below crop temperature T_c. As the ambient air temperature rises, blending will stop at C and the inlet louvres will return to being full open. As the ambient air continues to warm, ventilation will stop at D when T_a meets the ventilation initiation line.

If no low limit is put on the temperature of the incoming air, the base of the pile can become very much cooler than the top. This presents a condensation risk. As the crop sensors are in the top of the pile or boxes, air cooler than the top of the pile but warmer than the base could be blown into store, resulting in condensation forming on the potatoes near the floor.

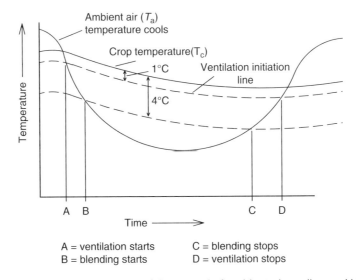

A = ventilation starts C = blending stops
B = blending starts D = ventilation stops

Fig. 8.4. Graphical representation of the control of ambient-air cooling and blending.

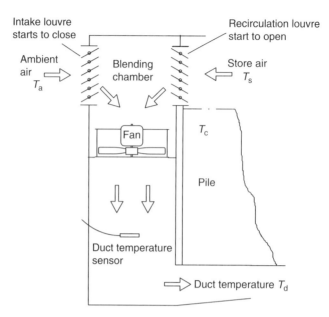

Fig. 8.5. Control of intake and recirculation louvres to achieve blending of inlet air.

8.2.2 Continuous, low-rate, humidified-air ventilation system for bulk storage

Continuous, low-rate, humidified-air ventilation systems use a wet-bulb sensor fitted after the intake louvres (Fig. 8.6) for controlling whether the air is suitable for cooling. This works on the premise that ventilation air will approach its wet-bulb temperature once it passes through a humidifier (Box 3.11), so even if air is actually warmer than the crop, it can still effect cooling. We can illustrate this by an example.

Example
Assume the ambient air is at 10°C, 75% RH, while the crop temperature is at 9°C. The wet-bulb temperature of this air is 7.8°C. If the 10°C, 75% RH air is used to ventilate the crop, as it leaves the humidifier it will be near 7.8°C and so will effect cooling of the crop.

As with intermittent ventilation systems described above, the system is fitted with blending, to ensure that the ventilation air is never too much cooler than the crop. To avoid the risk of blowing air warmer than the potatoes into the bottom layer of the pile, a temperature sensor is put into a litre bottle of sand located in the main duct. This varies in temperature in a similar manner to the potatoes at the base of the pile. Ventilation is stopped if the wet-bulb temperature of the air entering the main duct is warmer than the sensor in the bottle.
    An alternative to using the bottle filled with sand is to have a sensor located in the base of the pile, but this has to be put into position as the pile is being filled. It cannot easily be put in once the store has been filled.

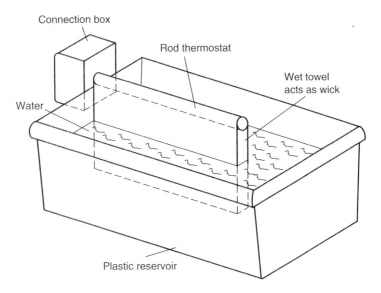

Fig. 8.6. Inlet air wet-bulb sensor for continuously ventilated, humidified-air ventilation systems.

8.3 Control of Crop Cooling and Cool Storage in Box Stores

The control system installed is dependent on the ventilation configuration (Fig. 8.7) used in any particular store. These include:

- Direct entry of ambient air into store with air either blown into store (Fig. 8.7a) or sucked out of store (Fig. 8.7b).
- Entry of ambient air via a blending system, with air either blown into store (Fig. 8.7c) or sucked out of store (Fig. 8.7d).

The air can be brought into store and forced through the crop by means of:

- An airspace ventilation system where air circulates around the boxes.
- A single-stage positive ventilation system.
- A two-stage positive ventilation system.

8.3.1 Direct air entry, airspace ventilation systems

In the most basic of ventilated box stores, ambient air enters the store directly from outside (Fig. 8.7a and b). The air is either blown into or sucked out from the store. In the latter, fan heat is discharged to outside. In the former it blows into store, raising the inlet air temperature by about 0.1–0.2°C, slightly reducing its cooling potential.

Direct air entry systems have a number of limitations:

- They may result in air very much colder than the crop flowing over warm potatoes, which may result in subsurface condensation in the crop, particularly in the top boxes.

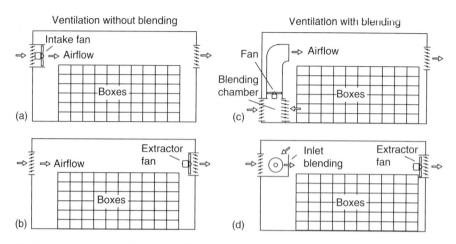

Fig. 8.7. Box store ventilation and blending configurations: (a) ventilation without blending, air blown in; (b) ventilation without blending, air sucked out; (c) ventilation with blending, air blown in; (d) ventilation with blending, air sucked out.

- If controls are added to prevent ventilation with very cold air, valuable air available for cooling may be lost.
- The incoming air path is unpredictable as its direction is dependent on its speed and its temperature compared with that of the store (Box 6.7).

To modify existing direct entry ventilation systems, the frost thermostat (Fig. 8.8) fitted to the control box can be altered on a weekly basis to prevent the incoming air from being too much colder than the crop.

If fans blow into the building, it may be possible to add a blending system to the existing ventilation system. If fans draw air out of the building, an inlet blending system can be specially constructed and fitted to match the size of the inlet louvres (Fig. 6.28).

The availability of blending prevents the problem that can occur where ventilation is prevented initially due to cool air being unavailable, only to be prevented subsequently due to ambient air being too cold to introduce into store.

8.3.2   Blended air entry, airspace ventilation systems

As with the blending system for a bulk store, the system (Fig. 8.7c and d) allows a maximum crop/duct differential to be set so that the cooling air is never cooler than the crop by more than a fixed value, usually 4.0°C. This will minimize the chance of subsurface condensation forming on the potatoes, especially those in the top boxes.

8.3.3   Single-stage positive ventilation systems for boxes

Single-stage positive ventilation (Fig. 6.42a) systems for boxes fitted with blending are controlled in the same way as for bulk storage ventilation systems.

Fig. 8.8. Frost thermostat. (Courtesy of Danfoss Ltd, Denham, Middlesex, UK.)

8.3.4 Two-stage positive ventilation systems for boxes

In two-stage positive ventilation systems (Fig. 6.42b), two sets of fans in series are used, one set to bring the air into the store and one to blow the air through the potatoes. Where ambient air enters the store directly, not via a blending chamber, the incoming air is blended to some extent with the store air. The second fan then takes the partially blended air and forces it through the crop. So long as the incoming air is forced to mix with the store air before it enters the second fan, the system will achieve a degree of blending of incoming air.

8.4 Controlling Combined Ambient-air/Refrigeration Cooling

Where refrigeration is fitted, most control boxes allow the store manager to select from:

- Ambient-air cooling only.
- Refrigeration only.
- Ambient or refrigeration cooling.

The third setting instructs the controller to try to cool the store using ambient air if cool air is available and using refrigeration if it is not. Since this system is 'blind' to time of day or to a cool spell of weather coming soon, in the UK it tends to use refrigeration more often than is needed.

A restriction may be put on the refrigeration to allow it to operate only during the times when low-tariff electricity is available.

Where control boxes are linked by modems to the World Wide Web, long-range weather forecasts can be utilized to decide when refrigeration can be delayed when a cool spell of weather is approaching. Such weather information could include ambient dew-point temperature information to allow growers to plan when to empty their stores without the risk of them becoming wet with condensation.

8.5 Ensuring Uniformity of Temperatures in Store

8.5.1 Control of recirculation in intermittently ventilated bulk stores

Recirculation of air in box stores is often controlled by a time clock so is routinely carried out. Recirculation reduces air temperature stratification in store and temperature differences in the potatoes. It does however add to cumulative weight loss. A better approach is only to recirculate when stratification has developed, subsurface crop temperatures differ by more than 0.5°C, or temperatures across the stack of boxes vary by more than 1°C. Recirculation is best restricted to night time, as solar heat gain tends to warm the headspace air during the day. Recirculation with warm headspace air needlessly warms the crop, so should be avoided.

To monitor the small temperature differences involved to initiate or monitor recirculation in the top surface of the pile or box, pairs of sensors should be specially selected that give the same temperature reading. These should be strapped together so they can be used to give a reliable subsurface temperature difference (Fig. 8.9). To initiate recirculation, a maximum acceptable differential should be set, for

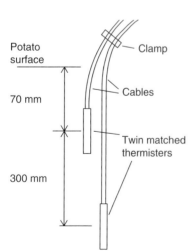

Fig. 8.9. Paired sensors to measure subsurface temperature differentials.

example 0.5°C. A table of suggested values is given elsewhere (Table B.3.1). These are arbitrary figures, which may need adjusting with experience.

If warm ambient air enters the headspace of stores through leaks or open doors, recirculation ventilation of the pile can lead to warmer headspace air condensing on the cool potatoes at the base of the stack (Ch3.5). Prevention of this problem is simply achieved by monitoring the dew-point temperature of the headspace air, and comparing this with the temperature of the potatoes at the base of the pile. So long as the dew-point temperature of the headspace air is, at minimum, 1°C below the lowest temperature of the base of the pile, the 1°C providing a margin of safety, recirculation can be carried out without the risk of condensation. The same control can be used to warm potatoes without causing condensation.

An alternative control is only to allow recirculation to be carried out if the headspace air is at, or below, the temperature of the crop at the base of the bulk store. Air that is cooler than the crop will not condense on warmer potatoes. This requirement will restrict the times when recirculation is possible more than when the dew-point method is used.

8.5.2 Roof-space heating for condensation control

Condensation can form on the roof or on the coldest potatoes due to sudden falls in ambient air temperature (Ch3.5). Figure 8.10 shows the output from a sample tuber fitted with a skin resistance sensor (Fig. 8.11) located in the top boxes of a well-sealed, continuous, low-rate, humidified-air ventilated refrigerated store in December 1991. The high resistance at the start shows that the tuber is dry. The reduction in resistance shows that condensation is almost certainly occurring. A white mould was subsequently seen to develop on the surface tubers, which suggests that condensation did occur.

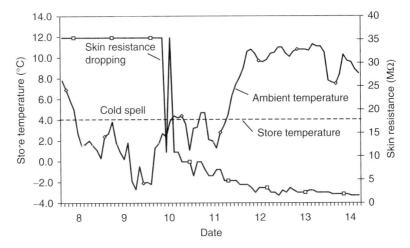

Fig. 8.10. Condensation on potatoes in a well-sealed store. (Ambient temperature courtesy of UK Met Office.)

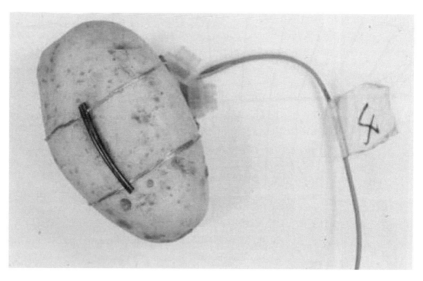

Fig. 8.11. Skin resistance sensor fitted to a tuber.

The ambient air temperature* had cooled to 6°C below the crop storage temperature, causing heat to be lost from the headspace air and condensation to form on the coolest potatoes. The problem is easily overcome by replacing the lost heat using roof-space heaters.

While many stores have roof-space heating fitted, store managers rarely know if they ever come on. Control is often by thermostats set on the heaters themselves and these are often accessible for checking only once the crop is in and the thermostats come within reach of staff walking on the crop.

A better arrangement is for the heaters to be controlled by the main control box. Heating should be triggered when the headspace air becomes colder than the top of the crop by about 2.0°C. This differential should be fine-tuned from experience. An interlock is needed in overhead cooling air distribution systems to ensure the heaters do not come on when cooling is in progress.

There should be indication as to when the roof-space heaters come on, as their operation is likely to be during the middle of the night when most store managers are in their beds.

8.5.3 Store heating to maintain crop temperature during cold weather

When ambient temperatures fall with the onset of winter, the store insulation and crop respiration heat may be insufficient to maintain the required set-point

---

*The ambient air temperature plotted in Fig. 8.10 came from a meteorological station 50 miles (80 km) away, which may account for the coldest weather not coinciding with the condensation starting. A second explanation is that the door was opened, which allowed warm air into store, but this was ruled out by the store manager.

temperature, particularly in processing stores where temperatures of 8–12°C are required to maintain fry colour.

Roof-space heaters can be used not only to prevent condensation forming on the crop but also to maintain store temperature. For crop heating, they should be switched on in conjunction with the recirculation fans in the fridge or ambient-air recirculation system.

Controllers for processing crops, which need to be kept at a required crop set point, should be able to vary between a lower and upper limit. Above the upper limit cooling should be started; below the lower limit, warming should be started. The band should be sufficiently wide to ensure that heaters and cooling systems do not start competing with each other, resulting in hunting.

8.6 Control of Ventilation During Drying and Wound Healing

When crops are ventilated with ambient air immediately after harvest, the aims are to:

- Prevent respiration heat from causing subsurface condensation.
- Dry surface moisture on potatoes and dry adhering soil.
- Stop ventilation if ambient air is likely to rewet the crop (i.e. crop temperature below dew-point temperature of the ventilating air).
- Maintain a warm crop temperature to speed wound healing and to prevent subsequent condensation during the next warm weather front (Ch3.5).

Ventilation is normally continuous immediately after harvest, reducing to periodic recirculation once the crop is dry and as crop respiration rate reduces. The ventilating air will vary in temperature, sometimes being warmer and sometimes being cooler than the crop. Blowing warm air on to the crop can result in moisture from the air condensing on the potatoes, undoing the benefit from drying. Limits should therefore be incorporated into control boxes during this period to prevent condensation occurring.

8.6.1 Control using relative humidity monitoring

Using dew-point temperature
Dew-point control of the ventilating air consists of monitoring the ambient air's RH, calculating its dew-point temperature and ensuring that the potatoes being ventilated are never below this temperature. A safety margin of 1–2°C is usually included to allow for inaccurate RH measurement. If this condition is met, the control box can be set to ventilation without fear of moisture in the ventilating air condensing directly on to the crop.

Using air moisture content
An alternative to dew-point temperature control is to use air moisture content as the factor being compared. In this case the controller calculates the saturation moisture content of air at the temperature of the potatoes. If the moisture content of the ambient air, determined using the ambient RH sensor, is below this value,

the air will not become saturated when it meets the potatoes, so condensation on the crop will not occur.

Example
If the potatoes are at 10°C, the saturated air moisture content at their temperature will be 0.0076 kg/kg dry air. Ambient air at 20°C, 45% RH has a moisture content of 0.0065 kg/kg, so moisture in the air will not condense on the crop directly; while air at 15°C, 75% RH with a moisture content of 0.0079 kg/kg will.

Both these control strategies act in a similar manner to prevent ventilation, intended to dry the crop and remove respiration, from causing condensation on the crop.

To avoid the complexity of dew point or moisture content control, some growers prefer the simpler system of avoiding the use of ambient air for drying, relying on a refrigeration system to do both the drying and the cooling. This scenario is discussed in Box 8.1.

8.6.2   Maintenance of crop temperature

If a period of cold weather occurs during the drying and wound healing period, the crop will be cooled to below the average temperature for the time of year. The dew-point temperature, or air moisture content, control will increasingly restrict ventilation when warmer temperatures re-establish. To prevent this happening, a low temperature set point, 4°C below the temperature of the crop, should be incorporated into the control system to switch louvres from ventilation to recirculation. This will maintain the crop temperature at near the average ambient temperature for the time of year, while allowing ventilation with ambient air to be maximized. In addition it will speed the rate of wound healing.

8.6.3   Control without the use of relative humidity measurement

A more basic but inferior method of control to the one above has the fans and the louvres switched to manual and controlled by two thermostats. The frost override thermostat (Fig. 8.8) switches ventilation to recirculation when the temperature of ambient air falls too low and a second, specially fitted high temperature thermostat switches ventilation to recirculation when the ambient air temperature rises too high. In between, the fans ventilate the crop continuously. A time clock can be fitted to operate the fans in recirculation mode on a periodic basis if ventilation is not taking place.

To operate such a system with existing control boxes, the frost thermostat is adjusted to 4°C below the temperature of the crop entering the store, while the high temperature thermostat is set at 4°C above the crop temperature.

Compared with 'dew-point temperature' or 'air moisture content' control, this approach is not guaranteed to keep the crop free from condensation. The store manager can contribute to improving its reliability by monitoring weather forecasts which include dew-point temperature information.

Box 8.1. Relative cost of drying potatoes using ambient air and refrigerated air

---

Some growers would rather dry their potatoes using their refrigeration system than use ambient air. This saves having to fit an ambient-air cooling system.

Ambient-air drying system

If ambient air is assumed to be at 15°C, 70% relative humidity (RH) over 10 h of the day and the air leaving the store is assumed to be 15°C, 95% RH, then the moisture-carrying capacity of air will be 0.0101 − 0.0074 = 0.0027 kg/kg (CIBSE, 2006) or 0.0033 kg/m³. If each tonne of potatoes has 20 kg* of moisture on the surface of the potatoes and in the attached soil, and the airflow supplied is 0.02 m³/s/t stored, it will take 6060 m³ air/t to remove this moisture, which at 0.02 m³/s/t will take 84 h, or 8.4 days with 10 h of drying per day, to dry the crop.

Fridge system

With a fridge system, again having an airflow rate of 0.02 m³/s/t, if the air leaves the wet crop at 17.5°C and RH of 95%, and the fridge cools this air to 15.0°C, 0.00153 kg of moisture will be removed per cubic metre of air passing over the fridge cooling coils. To remove the 20 kg of moisture, this will require 13,072 m³ of air or 182 h, 7.6 days, of continuous refrigeration and air recirculation to dry the crop.

Comparative energy used

A 1000-mm, six-bladed fan producing 9.4 m³ air/s at a backpressure of 120 Pa requires 2.4 kW of power or 255 W/m³ air per second. If the airflow per tonne is 0.02 m³/s, the fan electrical energy use per tonne per second is 5.1 W.

If instead of a fan blowing in ambient air, the refrigeration system is used to dry the crop, the difference in enthalpy or heat content of air at 17.5°C, 95% RH and 15.0°C, 100% is 6390 J/m³. If the airflow is 0.02 m³/s/t, 0.02 × 6390 = 128 J of heat will have to be removed per second, i.e. 128 W. If the coefficient of performance (COP) of the refrigeration plant is 2.5, then the fridge will use 128 / 2.5 = 51 W of electrical power per tonne of potatoes per second.

From the above:

- In the ambient-air drying case, if the fans blow ambient air for 10 h/day and then recirculate store air for the remaining 14 h, the fan will consume 5.1 × 8.4 × 24 × 3600 / 1000 = 3700 kJ/t of potatoes.
- In contrast, the fridge will consume 51 × 182 × 3600 / 1000 = 33,415 kJ/t of potatoes.

The fridge system therefore uses nine times the energy used by the fan to dry the crop. If heat needs to be added to prevent the crop being cooled down, this will increase energy costs further. If crop cooling along with drying is wanted, this will reduce the benefit ratio from ambient-air drying. So long as the louvres are very well sealed, so that they do not contribute to air leakage over the winter, the availability of ambient air for drying will reduce overall energy consumption.

---

*This assumption is based on there being 9860 potatoes, average diameter 57 mm, in a tonne, with a 0.2-mm-thick layer of water on their skins and in the attached soil.

8.6.4   Control of warming of the crop for reconditioning or prior to grading

Dew-point control or air moisture control can also be used when warming potatoes. This is more fully discussed in Ch6.9.

8.7   Control Measures to Minimize Energy Use

8.7.1   Minimizing fan operation with airspace-ventilated systems for box stores

In airspace-ventilated stores, where incoming air flows over the top boxes, the fans will run for very different times depending on where the controlling crop sensors are placed. Sensors may be:

- In the roof headspace.
- Approximately 70 mm below the top surface of the potatoes in the top boxes.
- Deep down, near the centre of the top boxes.
- In the top and base layers of the stack of boxes.

When cooling starts, the roof headspace sensors will cool rapidly. Second to cool will be the sensors 70 mm below the surface of the potatoes in the top boxes. Last to cool will be the sensors deep down in the top boxes and in the lower layers of the stack.

With UK rates of ventilation (0.02 m³/s/t), it will take 3–4 min of fan running to flush the whole store. Ambient-air fan operation, and therefore energy use, will be minimized if the air in the store is flushed with cool ambient air, then the fans stopped. Heat exchange between crop and air will take place by natural convection. Since there is no forced air movement, the rate of cooling is likely to be slower than when fans are operating continuously. If fans are run continuously when cool ambient air is available, the cooling rate will be a maximum, but a proportion of the fan energy used will be wasted moving air from one part of the store to the other, without necessarily passing through the potatoes. A compromise has to be found.

8.7.2   Cooling efficiency of the box storage ventilation system

The aim must be to ensure that when the fans are running, they are removing heat from the crop. In a bulk store, all the ventilating or recirculating airflow passes through the crop, so that heat transfer is high. In airspace-ventilated stores a proportion of the airflow passes through potatoes in the boxes, but unlike bulk stores, a proportion will circulate over the potatoes and through gaps between boxes. If heat transfer to the ventilating air is low, the ventilating air will exhaust from the building little warmer than the temperature at which it enters. The optimum is for the exhaust air to leave at the same temperature as the crop. This brings in the concept of cooling efficiency.

The efficiency of the cooling process can be assessed, in part, by monitoring the temperature of the air exiting through the exhaust louvres and the temperature of the air entering the store from the blending chamber or air inlet. The cooling efficiency can be obtained (Box 8.2) by dividing the temperature difference between

---

Box 8.2. Cooling efficiency of ventilation

$$\text{Cooling efficiency} = \frac{\text{Exit air temperature} - \text{Inlet or duct air temperature}}{\text{Crop temperature} - \text{Inlet or duct air temperature}} \times 100.$$

For example, if the crop is at 10°C, the duct inlet air is 6°C and exit air is 7.5°C, then

$$\text{Cooling efficiency} = \frac{7.5 - 6.0}{10.0 - 6.0} \times 100 = 38\%.$$

---

the air leaving and entering the store by the temperature difference between the crop and the ventilating air at entry. If all the air flows through the potatoes, cooling efficiency will approach 100%. If half the air bypasses the potatoes, the cooling efficiency will be 50%. What the efficiency measure does not take into account is the cooling by evaporation that takes place during ventilation. It will however give some indication as to how much air is flowing through potatoes and whether fans should be run continuously or in short bursts.

8.7.3 Energy use in cooling using ambient air compared with refrigeration

If cool ambient air is available, ambient-air cooling uses considerably less energy than that needed to run a refrigeration system (Box 8.3). In continental climates ambient air for cooling is usually more regularly available than in maritime climates like the UK (Ch3.6), where there may be weeks of mild weather when air suitable for cooling is unavailable. The availability of cool air below 4°C, particularly in the autumn, appears to be reducing in recent years.

In making an assessment as to whether to wait for ambient air or to resort to refrigeration, it is important to take the following aspects into account:

- Cooling costs in the UK using ambient-air and refrigerated cooling are approximately £1.50/t and £6.00/t, respectively, for 7 months' storage; 1.2% and 4.6%, respectively, of the value of a crop selling at £130/t.
- If refrigeration is delayed to the period of the night-rate (Economy 7) tariff, usually 00.30 to 07.30 hours, the cost of electricity drops to approximately 67% of the standard day rate.
- If the crop could sprout and lose value for want of temperature control, it is better to use refrigeration if it is available rather than try to save money by relying on ambient-air cooling instead.
- Energy costs are likely to rise in future.

8.8 Sensors to Provide Fuller Information of Store Climate

Temperature, RH and wet- and dry-bulb sensors provide only partial information of the store microclimate. A number of other sensors (Table 8.2), when fitted to a

Box 8.3. Cost of ambient-air cooling compared with fridge cooling

Ambient-air cooling uses fans to draw cool ambient air, when it is available, through the crop to cool it. The cool air is free, only the electricity required to run the fan has to be paid for. In contrast, a refrigerated store is sealed, and the refrigeration system has to remove the heat from the recirculating air using a compressor, which is the main power user in a refrigeration system. This uses significantly more energy than just running a fan.

With any refrigeration system, the temperature difference between the air entering the evaporator (air-on temperature) and the air leaving the fridge (air-off temperature) is constant, usually about 2.5°C. If the crop is at 10°C, the air entering the fridge cooling coils (air-on) will be about 10°C and the air-off temperature will be 2.5°C cooler, i.e. 7.5°C. (This difference will reduce by about a degree as the crop nears 3.5°C, but it will not vary in the same way ambient-air ventilation does.)

Ambient-air cooling usually starts automatically when ambient air temperature drops to 1–1.5°C cooler than the crop. Cooling will normally continue until the ambient air falls to 4°C below the crop. After this point, blending should start to ensure that the differential between the cooling air temperature and crop is not excessive as this could result in condensation on the crop.

The rate of cooling with ambient air is therefore likely to be similar to using the fridge. The problem with ambient-air cooling is that there may well be no cool air available when it is required.

Presuming that there is air available, the power rating of the fans and fridge will be as follows.

Ambient-air cooling
As in Box 8.2, the power used to supply a ventilating fan operating against a back-pressure of 120 Pa is about 5 W/t of crop stored, regardless of the ambient air temperature.

Refrigeration
Using a refrigeration system to cool a crop from an initial temperature and relative humidity of 10°C, 95% to air at 7.5°C, 95% requires 6.43 kJ of enthalpy to be removed per cubic metre of air. At an airflow of 0.02 m^3/s/t, this comes to 6430 × 0.02 = 129 J/t. As in Box 8.2, if the coefficient of performance (COP) is 2.5, the electrical energy needed to power the fridge cooling is 129 / 2.5 = 51.6 W/t.

The fridge cooling costs are therefore 51.6 / 5 = 10 times that using ambient-air cooling.

data logger, can provide a complete picture of what is occurring in store. These additional sensors include:

- Carbon dioxide sensors.
- Tuber skin resistance sensors.
- Simulated potato to monitor tuber condensation.
- Simple solar heat gain indicator.
- Anemometer.
- Wind vane.

Table 8.2. Sensors to provide fuller information of store microclimate.

Measurement	Aspect measured	Indicates	Indicates in addition
Carbon dioxide in store air	Percentage CO_2 level in store	Risk of increasing sugar levels in crops and danger to staff	When or if store air flushing is required
Tuber skin resistance	Electrical resistance of skin of a tuber	Condensation forming on the tuber or a wet tuber drying	Tuber skin being desiccated by ventilation and recovering when ventilation stops
Simulated tuber condensation sensor	Condensation forming on a simulated potato	Condensation forming on tubers	
Solar sensor	Temperature increase in glass container due to solar radiation	If roof-space warming could be due to solar radiation	If no solar heat gain, roof-space warming due to warm air entry
Anemometer	Wind speed	When increased leakage of air into store is likely	Cause of crop condensation or store warming/cooling
Wind vane	Wind direction	Some wind directions will result in more air leakage than others	

8.8.1 Carbon dioxide sensors

A carbon dioxide monitor should be installed in all processing stores and in any store that is very well sealed. The information that it provides will indicate when and if flushing of the store is necessary. This avoids routine flushing, which may often be unnecessary, and the associated risk of the ventilating air causing condensation to form on the crop.

8.8.2 Tuber skin resistance sensors

The tuber skin resistance sensors shown in Fig. 8.11 have been used extensively in experimental work (Pringle and Robinson, 1996) to determine when condensation is occurring in a store and why it is occurring, and to help determine whether condensation has contributed to the subsequent development of disease in the stored crop.

The sensor used (Fig. 8.12), measures the electrical resistance of the tuber skin between the two wires 30 mm apart girdling a sample tuber, placed within a mass of potatoes. It was developed from the idea of a large grid, invented by Rasmussen (1989), to measure the drying of crop in store, and was miniaturized to allow detection of condensation in different layers of a box or pile. Details of its electrical circuitry are given in Box 8.4.

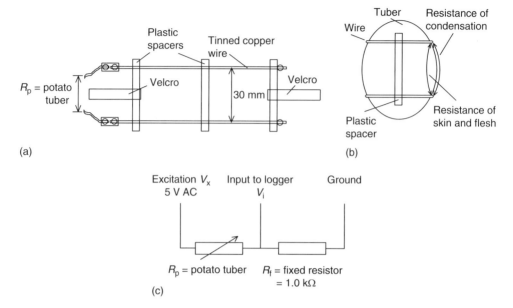

Fig. 8.12. Skin resistance sensor – technical details: (a) sensor laid flat to receive tuber; (b) sensor wrapped round tuber; (c) electrical circuit for skin sensor.

Box 8.4. Skin resistance sensors

The measurement of skin resistance is best achieved by applying an AC voltage across a bridge (see Fig. 8.12). The AC voltage is required to avoid the problem of polarization when a DC voltage is used.

The equation for resistance of the potato skin, R_p, is:

$$R_p \text{ (k}\Omega\text{)} = X \times R_f / (1 - X),$$

where

R_f = resistance of fixed resistor (1.0 kΩ)
$X = V_i / V_x$
V_x = excitation voltage
V_i = input voltage into logger.

Example

If $V_x = 5$V, $R_f = 1.0$kΩ and a voltage $V_i = 2.1$V is monitored at the data logger, then

$$X = 2.1 / 5 = 0.42$$

and

$$R_p = 0.42 \times 1.0 / (1 - 0.42) = 0.72 \text{ k}\Omega.$$

Some data loggers can provide the required AC excitation voltage. If it is not available, a separate AC voltage source should be used. A fixed resistor can be clipped across the sensor wires to check the sensors are working before they are wrapped round the tubers.

The results need careful interpretation. While the sensor does show drying or wetting of the tuber surface, it also shows skin cells losing moisture during ventilation and recovering once ventilation stops. This complication has restricted its use to those carrying out experimental work and has stimulated the concept of a simulation potato, which has no skin to confuse the output.

8.8.3   Simulated tuber condensation sensor

A sensor that monitors surface moisture and condensation, but not skin resistance, would solve the problems of using real tubers discussed above. It would not indicate how rapidly the skin of potatoes dries after harvest, but it would monitor condensation on the crop thereafter.

For such an instrument to be successful, it must have the following properties:

- Its thermal mass should be similar to that of the potatoes it is monitoring, as its thermal lag, related to its mass, plays a significant part in why condensation forms on stored tubers.
- The sensing grid should not come into contact with surrounding tubers.
- Air within the crop must be free to circulate across the sensing element to evaporate moisture when microclimate conditions change.

A sensor to this specification was developed (Martino and Gow, 1994). It used wet flour dough packed tightly within a plastic cylinder to mimic the thermal characteristics of a real potato and used an etched copper grid condensation sensor to monitor condensation. In limited trials its performance was comparable with the skin resistance sensor but without the confusion of variable skin resistance. A number of companies have developed their own versions. A modified version of the condensation sensor concept, utilizing newly available film sensors used for measuring condensation on building surfaces, is shown in Fig. 8.13.

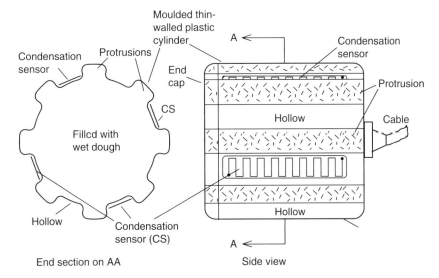

Fig. 8.13. Simulated tuber condensation sensor.

8.8.4 Solar radiation sensor

In monitoring the performance of the whole store, it is important to know why the headspace temperature of a store is rising. This could occur when the store door is left open on a warm day, allowing warm air to drift into store. Alternatively, it could occur on a sunny day with the door tight closed. A simple sensor indicating the effects of solar radiation is therefore required. A temperature sensor mounted within an inverted jar, or similar translucent container, with no base fitted, is all that would be required. If fitted on the external south-facing side of the store, any reading of the sensor in the jar which exceeds an external, shaded, store sensor by a margin of 3–4°C could be assumed to indicate that solar heating was taking place. This sensor would then contribute to the store performance analysis.

8.9 Sensors to Aid Store Management

8.9.1 Condensation risk warning

The placement above the store door of a large digital display repeater showing external dew-point temperature compared with internal crop temperature or a traffic-light-style warning light (Fig. 8.14) will warn forklift drivers and store staff that condensation on the crop may occur if store doors are opened. The forklift

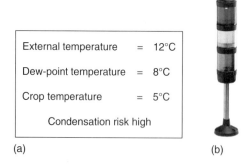

(a)

(b)

(c)

Fig. 8.14. (a) Display screen, (b) stack light and (c) LED above door to warn when an open door could allow ambient air to condense on the stored crop.

driver will soon learn at what time of day he can open the door without risk of condensation forming on the crop inside. When circumstances like a waiting lorry require the door to be opened regardless of the warming, it will serve to ensure the door is left open for the shortest possible period.

8.9.2 Plant operation sensors

In any store data-logging system, indication of fan, fridge or louvre operation is usually taken from the controller. If a maintenance engineer or store manager has switched the isolator of any piece of equipment to off, then the log of data will suggest that the equipment is running when in practice it is not. The same problem arises if fans burn out, louvres jam or refrigeration high- or low-pressure sensors trip.

One approach to this problem is to fit temperature sensors or motion sensors to each piece of rotating equipment. This can lead to considerable quantities of additional wiring and more items to be monitored. The other approach is to monitor all aspects of the store climate and identify any failures in equipment through the failure to achieve the desired result. Should a crop fail to cool when the fridge is running, for example, this may indicate that the fridge compressor motor has tripped. Whole-store monitoring is simpler than fitting multiple lengths of wire, but it does need good analysis to ensure faults are quickly rectified.

8.9.3 Carbon dioxide monitors

In many stores, particularly processing stores, fans are regularly run to flush the store to prevent carbon dioxide levels rising to levels that could spoil the fry colour of the potatoes. This may well be unnecessary as there may be sufficient air leakage to prevent this happening. The flushing may also risk causing condensation on the crop, and may be better avoided if unnecessary. By installing a carbon dioxide sensor in store, flushing can be carried out only when actually required. If the level is logged, this can indicate for future use when high levels of carbon dioxide are likely.

8.10 Logging Systems

The more basic data-logging systems, where internal and ambient temperature, crop temperature and plant operation are monitored but ambient RH, skin moisture, crop condensation or solar heating are not, allow recording of:

- Crop temperatures and crop temperature differences.
- Ambient-air cooling or fridge operation.
- Louvre position.

These are usually recorded at hourly intervals. By the end of a week, these data accumulate to a mass of figures. Experience suggests that most raw logged data of this nature are rarely looked at.

Logged data may be required as part of the traceability requirements of the company purchasing the crop and can be extremely useful if a problem arises. The data can be examined to find out:

- When equipment stopped working.
- Where excessive crop temperature differences may have caused condensation to form on the crop.
- Whether the cooling equipment is working as it should.

Information displayed on a daily basis therefore should not attempt to show all the data collected, but only those data that can give warning of problems. These could include:

- Average crop temperature for the last week.
- Subsurface temperature differences across the potato pile or stack of boxes.
- Temperature differences between top and base of pile or stack.
- Duration that ambient air was below crop temperature over the last 24 h.
- Ventilation run-hours over the last 24 h.
- Refrigeration cooling hours over the last 24 h.

If an external RH sensor is fitted, the display can also indicate:

- The number of hours of ambient air suitable for use for ventilating the crop during drying and wound healing over the last 24 h.
- When ambient air could have caused condensation on the crop if the door had been left open.

If an internal RH sensor is fitted, the display can also indicate:

- Whether the crop temperature is lower than the store air's dew-point temperature so that it could have caused condensation on the crop.

If skin resistance sensors are fitted, these can also indicate:

- The rate of drying of the crop.
- If increase in resistance, either ventilation is taking place or crop is drying or both.
- If decrease in resistance, either ventilation has stopped or condensation is taking place.

If simulated tuber condensation sensors are fitted, these display:

- When and where any condensation on the crop occurred.

If a solar radiation sensor is fitted, this can help determine whether roof-space warming is due to:

- Solar heat gain.
- Warm ambient air leaking into store.
- Roof-space heaters being on.

An example of store data analysis is shown in Fig. 8.15 (Pringle and Robinson, 1996). Figure 8.15a shows ambient air temperature, ambient dew-point temperature and crop temperature. The crop was cooler than the ambient air dew-point temperature on 16 and 17 October but not cooler than the internal store dew-point temperature,

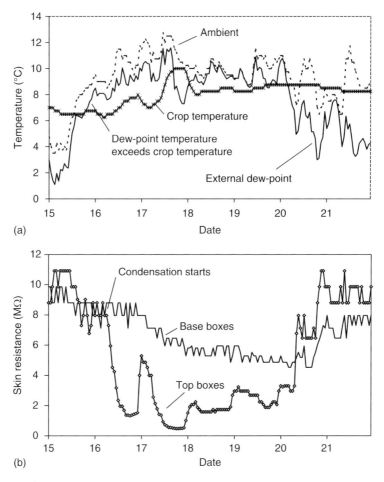

Fig. 8.15. Condensation on the stored crop due to air leaking into store: (a) crop temperature versus ambient air dry-bulb and dew-point temperature; (b) drop in skin resistance indicates condensation occurring on stored crop.

the latter being omitted from the graph for clarity. Figure 8.15b* shows that condensation occurred on the potatoes in the top boxes first, then in the base boxes later during this period. Ambient air must therefore have drifted into store during this period, causing the condensation to occur. Subsequent examination of the store found a louvre blade jammed open when the louvres were closed.

8.11 Data Storage, Use and Access

By monitoring a store, the store manager has access to all the data collected, but not how it compares with other stores. His store may be using more energy than

---

*Skin resistance measurement was carried out using a short pulse of DC voltage so resistances monitored were higher than true resistances measured when using AC voltages.

other stores without his knowledge. By fitting a modem and telephone line to his data logger, or the personal computer used as the controller/logger, he can make the logged data available to a grower group manager. Comparative energy usage or ambient-air cooling/refrigeration mix can be compared between the stores within the group. Reasons for excess energy usage can then be found.

One UK company, Proctors Ltd, encourages growers to fit modems as standard and to contract into a scheme whereby store performance is checked at 08.00 hours each morning. Any fault or problem is usually found before any damage can be done. Often the problem can be sorted remotely. If a service engineer is required to visit the store, he goes there knowing what the problem is, so he can take the tools or equipment he needs. This can save long drives to remote locations only to find that a special part is required.

An additional benefit from such a system is that the store manager can access store records through his palmtop computer or mobile phone when he is away from home. The increasing power of these devices means that the usefulness of this information can be improved over time.

The information collected can be integrated into the traceability and quality assurance systems discussed in Ch12, providing information that is often the missing part of the jigsaw when seeking explanations for problems. The knowledge gained should result in better store management in the future.

8.12 Summary

Electronic environmental control and monitoring systems are an essential part of potato storage and have the potential to improve store management, identify potential savings in running costs and reduce the cost of repairs. The main aspects to consider are the following.

- Sensors used routinely in potato storage include temperature, RH and wet- and dry-bulb temperature.
- Sensors can be used for control only, for store monitoring, or for both.
- There are two main modes of control: (i) condensation control during drying and wound healing; and (ii) cooling using ambient air or refrigeration. A supplementary control is the 'recirculation of air through a pile' and 'warming of crops' while avoiding condensation.
- Temperature sensors indicate when cooling should be occurring, when excessive crop temperature differentials have occurred and when recirculation should take place.
- Ambient-air cooling of intermittently ventilated bulk stores is controlled by monitoring crop and external temperature and controlling blending to keep the cooling airstream within 4°C of the crop temperature.
- Continuous, low-rate, humidified-air ventilation systems are controlled using wet- bulb sensing located just inside the intake louvres.
- Control of ventilation during crop drying and wound healing using ambient RH sensing will stop ventilation when the moisture in the ventilating air would cause condensation on the stored crop.

- A more basic condensation control system during crop drying and wound healing restricts incoming air temperature to ±4°C of crop temperature.
- The precise location of sensors in box stores that are used for control can have a significant impact on ambient-air cooling fan running times and energy costs.
- Box stores differ from bulk stores in that a proportion of the ventilating or recirculating air goes over or between boxes, so does not pass through the potatoes themselves.
- The concept of cooling efficiency for box stores is useful for determining how much air is going through potatoes and whether intermittent ventilation or continuous ventilation during cool weather is the more efficient in energy use and effectiveness.
- The use of additional sensors such as skin resistance sensors, simulated tuber condensation sensors, carbon dioxide monitoring, solar radiation indicators, anemometers and wind vanes can greatly assist troubleshooting in stores.
- Fitting of modems to store data loggers or personal computer-based controllers allows ventilation and refrigeration equipment suppliers to remotely monitor store equipment and save on any repair costs.
- The value of data logging of store climate is greatly enhanced if data can be partially analysed automatically and then compared, using modems and the World Wide Web, with data from other stores. It can also be used for traceability.

9 Store Management

Topics discussed in this chapter:

- Store hygiene.
- Store preparation for loading.
- In-field sampling.
- Loading into store.
- Ventilation for drying and wound healing.
- Cooling.
- Inspection of crops in store.
- Energy management.
- Application of sprout suppressants.
- Reconditioning processing crops.
- Warming crops prior to grading or dispatch.
- Procedures for dispatch.

9.1 Introduction

This chapter synthesizes the information contained in the previous chapters to provide a strategy as to how managers should fill and run their stores. To avoid excessive repetition, links to other chapters are included to explain why particular recommendations are made. The emphasis is on what actions store managers can carry out with the equipment they have. Discussion about what alternative equipment they could use is to be found in the appropriate other chapters.

9.2 Store Hygiene and Equipment Inspection

At the end of each winter storage period, significant quantities of soil and dust carried in with the crop during the previous harvest remain within the store, in lateral ducts, on floors, on ledges and walls, on boxes and machinery. As this dust is likely to contain pathogens that can contaminate the crop about to be harvested, the store should be thoroughly cleaned.

Removal of soil and dust from store will minimize the risk of disease being carried over from one year to the next and is particularly important if parts of the crop were diseased in the previous season. Cleaning of the store must be completed in time for the new harvest to begin.

9.2.1 Vacuuming the store

Vacuuming (Fig. 9.1) is much preferred to brushing as it removes both heavy and light dust particles. Brushing removes the heavy large particles while dispersing the remaining lighter fraction containing spores and the resting bodies of bacteria, only for them to settle back on to the floor and sheeting rails (BPC, 2001b).

While store preparation is mentioned separately from store cleaning, it is often best to do the cleaning and inspection at the same time. The process of vacuuming dust from louvres, fans and equipment will reveal bent blades, frayed sensor wires, damaged fan impellers, loose bolts, rat-chewed cables and gaps in the store fabric. The action of cleaning forces staff to inspect the equipment at the same time.

9.2.2 Washing the store

While vacuuming the store is normally sufficient to remove the risk of infection from the previous year's crop, washing may be justified for high-health pre-basic

Fig. 9.1. Vacuuming the store is preferable to brushing. (Slackadale, Aberdeenshire, UK.)

seed storage (BPC, 2001b). While composite panel stores and Styrofoam sheet-insulated stores can be washed with a pressure jet washer, spray foam insulation can be damaged if jet pressure is too high. Sprayer pressures should therefore not exceed a pressure of 3.4 MPa (34.5 bar) if spray foam-insulated stores are to be cleaned (Ch5.7).

9.2.3 Fogging the store to disinfect it

Fogging equipment like that used for CIPC treatment can also be used to disperse a disinfectant in the store after it has been vacuumed and washed. This is usually only carried out for pre-basic seed storage.

9.2.4 Cleaning the boxes

If box tippers do not rotate the boxes through a great enough angle, some tubers can be left in each box. Internal box bracing can make this problem worse by trapping tubers under the bracing. If some tubers remain, they may contaminate the new crop with last year's decayed tubers. Increasing the rotation of the tipper is a simpler solution than removing a handful of tubers from hundreds of boxes by hand. Unless very high-health seed is being stored, removal of soil and crop residues is usually sufficient to minimize any disease carry over from one year to the next.

Cleaning boxes with a water spray hand lance is a major undertaking. If it is done properly it can take a day to wash 30 boxes. While it is difficult to justify financially washing all boxes every season, it is prudent to mark with chalk any particular boxes that have stored rotting stocks, which may contain an abrasive film of sharp soil particles intimately mixed with, for example, dry rot spores or other pathogens. These should be cleaned thoroughly before re-use. Where cleaning is to be carried out every year, a conveyor-based automatic washing system can be installed.

Boxes can be left outside to allow ultraviolet light to kill pathogens living on the timber. However, in the UK climate, box life was reduced by 40% when stored outside rather than inside (Pringle, 1993a).

9.3   Store Preparation

9.3.1   Store cleaning and inspection

Some weeks before the crop is loaded into store, the store fabric and doors should be examined to find any damage, holes or problems. Ventilation equipment and the refrigeration system should be inspected and operated, sensors calibrated, and control and logging systems checked. This can be combined with the cleaning of the store as both cleaning and inspection can be done simultaneously.

To check for gaps in the store fabric, the store doors should be shut during daytime and any daylight showing through doors, cracks or louvres identified. Large holes can be sealed using sheet insulation, while smaller gaps can be filled with polyurethane foam, squirted into the gaps using spray-cans, or an opaque mastic or silicon sealant sold in tubes, injected using a sealant gun.

9.3.2 Calibration of internal store sensors

Crop sensors should be checked against a calibrated thermometer to ensure that they have not been affected by moisture over the past year and are still reading accurately (Fig. 9.2). Where possible, all the sensors should be bundled together with a probe of an accurate thermometer or handheld electronic thermometer (Fig. 9.3) which reads to 0.1°C and is accurate to ±0.1°C. If the sensors are in steel tubes, they can be partially dipped in a container of water to just below the top of

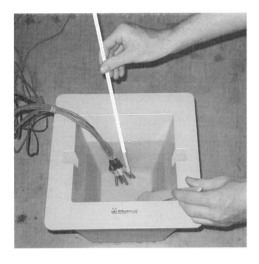

Fig. 9.2. Calibrating store sensors against an accurate thermometer. (Courtesy of Potato Council, Oxford, UK.)

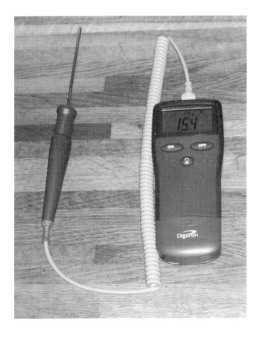

Fig. 9.3. A portable electronic temperature probe.

the tubes to ensure that they are all at a uniform temperature. Liquid paraffin can be used instead of water if the water could damage the sensors.

The temperature reading on the store control box or monitor is then read off and compared with the reading on the reference thermometer. A correction table should be made up to show the offset of each sensor. Any offset is usually constant over the range of monitoring. However, it is best to calibrate sensors using liquid at a temperature within the normal store operating range (e.g. 5–14°C in the UK). If the temperature of the water or liquid paraffin is being increased gradually by the addition of warmer liquid, the liquid should be stirred well to ensure that the probes are all at the same temperature.

If it is impossible to bring all the sensors together, the next best alternative is to take the handheld electronic thermometer round each sensor, measure the temperature and compare it with the reading on the control box (Fig. 9.4). Two people will be required to do this, as the sensor may be a considerable distance from the control box.

9.3.3 Location of crop sensors

In both box and bulk stores, sensors should be located in the top and bottom of the stack, with more sensors located in the top of the crop than at the base (Fig. 9.5). With the overhead throw systems commonly used in box stores, excessive temperature differentials between cool air jet and crop surface can result in

Fig. 9.4. Switchgear boxes for two separate stores with control boxes to left and right.

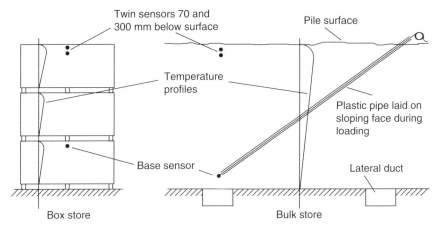

Fig. 9.5. Suggested location for temperature sensors in box and bulk stores.

condensation in this top layer. If this differential exceeds 0.5°C in the top 300 mm of the crop, this indicates that the incoming cool air is likely to cause condensation in this top layer. The use of two sensors (Fig. 8.9), selected as being within ±0.1°C of each other, always used as a pair, will give valuable feedback about the likelihood of condensation if located under the path of the cooling jets, particularly where the jet starts to lose momentum and flow over the top surface of the potatoes.

In bulk stores, sometimes the sensor to measure the base of the pile is not put in the pile itself but in a 1-l plastic bottle filled with sand, placed in the ventilation duct. This saves having to install the sensor in the crop as it is being loaded. The thermal mass of the sand means that its temperature will approximately follow the temperature of the potatoes in the base of the pile. The alternative is to lay a tube down the face of the pile during loading (Fig. 9.5) so that the base sensor can be inserted once the store is full.

The base temperature of the pile is monitored by the controller to ensure that the ventilating air dew-point temperature never exceeds this value.

9.3.4 Calibration of store's external sensors

Checking the external and blending duct temperature sensors and thermostats is usually complicated by their out-of-reach location. Calibration is carried out on one sensor at a time and usually needs two people, plus possibly two mobile phones if distances between monitor and sensors are large. The external temperature sensor and the duct temperature sensor are checked by comparing them with the handheld electronic thermometer in the same way as for checking internal sensors individually. However, as they are usually fixed to the gable of the building and in a blending chamber, respectively, they may require a ladder or forklift cage to enable them to be reached. A risk assessment should be carried out prior to undertaking this task. While this calibration is time-consuming and difficult, it is vital as a 1.0°C error in these sensors can result in considerable loss of potential cooling time.

When carrying out these checks, ensure that the sun or any exterior floodlight cannot affect the temperature of the external sensor.

To check the frost override thermostat (Fig. 8.8), rotate the switch-on/off temperature setting knob to increase the set temperature until you hear a click and then turn it down until you hear another click. Turn the knob back and forth to find the setting between clicks. Check the temperature on the thermostat dial and compare it with the handheld electronic temperature sensor. These should be the same if the thermostat scale is accurate.

If the frost thermostat is being set at 0°C and it has an exposed sensing element (Fig. 8.8) which is waterproof, then the frost thermostat can be set by immersing its sensing element in an ice/water mix. As mains voltage is fed to the thermostat, the mains power should be switched off during this operation if there is any chance of moisture coming into contact with a live terminal. The knob is rotated until it is in a position between the two clicks, and this position marked for future reference.

External RH is measured by the control box using either a capacitance film sensor (Fig. 8.2) or wet- and dry-bulb sensors (Fig. 8.3). Capacitance sensors should either be fitted with new sensing heads annually or the whole device sent to the supplier for calibration, although kits of salt solutions can be supplied to users to carry out the checks themselves.

The thermistors of a wet-bulb sensor are checked with the wick removed using the electronic thermometer. The wick is then washed in warm soapy water and rinsed in clean distilled water to ensure it takes up water properly. It is then fitted over the wet-bulb thermistor and suspended in the reservoir of distilled water.

RH sensors can be checked for approximate accuracy using a whirling hygrometer (Fig. 9.6) and compared with the reading on the control box display. As whirling hygrometers are at present available only with glass thermometers, they should not be used inside the store in case they break and the glass or mercury contaminates the crop.

Capacitance film RH sensors used inside the store should be protected from dust using a filter. Wet- and dry-bulb RH sensors similarly need good filtering and regular cleaning to prevent the wick clogging up with dust.

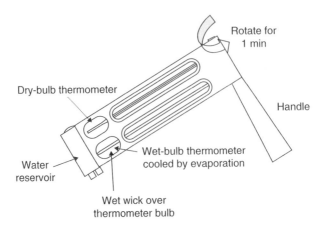

Fig. 9.6. Whirling hygrometer with glass thermometers.

9.3.5 Checking ventilation and refrigeration plant operation

Checking that the frost thermostat stops fans and shuts louvres
when fans and louvres are switched to manual

To check that the store is adequately protected against being ventilated with freezing air, the fans are started and the louvres opened using the manual switches on the switch panel beside the control box. The adjustment knob of the frost thermostat is rotated until the arrow of its pointer is greater than the ambient air temperature; a click should be heard. The fans should stop and the louvres shut. If the frost thermostat works only when the switches are set to auto, there is no frost protection when fans are set to work on manual. This is a serious flaw in the control system and means that if the fans are switched to manual and unintentionally left on over a frosty night, the potatoes could suffer frost damage.

Checking automatic ambient-air cooling

Automatic control equipment usually consists of one electronic control box per store. This box is usually dwarfed by a much larger steel box containing relays, timers and motor switchgear, which have manually operated switches on the front of the box to turn electrical ventilation or refrigeration equipment on and off and to switch control from manual to automatic. Fan and fridge hours-run meters and indicator lights are also fitted to the front of the box to indicate plant operation. In Fig. 9.4, two stores are being controlled separately. One is controlled by the small control box on the left and the large switchgear box to its right. The second store is controlled by the control box on the far right, with its switchgear box to its left.

To continue checking the ventilation system, the switches on the switchgear box and the automatic control box should be set to automatic cooling. Disconnect all the store sensors but one, and warm this remaining sensor with the heat of your hand. If this can be done beside the control box, you can watch the temperature of the sensor rise. When the store sensor temperature rises to 1–2°C higher than ambient, the fans should come on and the louvres open. If the sensor is allowed to cool back down, the fans will run on for 10–15 min or whatever period the control box suppliers have set. This delay is designed to stop fans hunting, or switching on and off rapidly. A second delay, also to prevent hunting, may prevent the fans starting again if the sensor is warmed immediately after it has cooled.

Checking blending

The ventilation supplier will probably check operation of the blending system using fixed resistors that plug into the ambient, duct and crop temperature sensing ports. Unless these are left for you to use, it may not be easy to check whether the blending system is working properly. Blending should start when the crop/duct differential exceeds the set value, usually 4°C. Experience has shown that in many cases blending will not operate when the ambient air is below 0°C, as the frost override thermostat located in the duct is often cooled to below 0°C by unmixed freezing air and stops ventilation altogether (Pringle, 1989). This can only be found by monitoring the blending process when ambient air is below freezing. It can be avoided by placing the frost thermostat after the fan, so that the air is well mixed before it reaches the sensor.

Checking dew-point control

Where dew-point temperature control on the ventilating air during crop drying and wound healing is provided, the system can be checked by immersing the crop temperature sensor in water below the dew-point temperature of the ambient air. The controller should switch the louvres to recirculation ventilation for as long as the sensor is in the cold water. If the louvres do not close, the most likely problem is an inaccurate low reading by the RH sensor.

Checking roof-space heater operation

The operation of the roof-space heaters can be checked when no cooling is taking place, but the controller is in cooling mode. The headspace temperature sensor should be placed in water a few degrees below the temperature of the single crop sensor used to check that cooling was working properly. When the headspace sensor exceeds the set differential, usually 2°C below the crop sensor, the roof-space heaters should come on.

9.4 In-field Sampling

Potatoes can be harvested and loaded into store at rates of 10–25 t/h. Long transport distances may mean that crops may not start coming into store until a couple of hours after lifting starts, whereupon 50 t may have already been harvested. Should a problem be found in the crop as it enters the store, it will be too late to divert it elsewhere, especially as another lorry with 24 t of potatoes may soon be due.

It is vital therefore to sample crops in the field a day or two ahead of harvest. These need to be washed in a bucket of water if the evaluation is done in the field. A plastic-coated egg basket is useful for this purpose (Fig. 9.7). Markets can then be found for crops showing rots or blight, so that they can be sold straight off the field.

The agronomist or fields person works ahead of the harvester, assessing skin set and disease. Skin set indicates whether the crop is ready to harvest. The disease assessment will help to assess the best market for the crop. If there is more than 1–2% rots or physical disorders, such as skin surface cracking, or disease is so bad that the crop is not worth storing, the crop can be bypassed or harvested for immediate sale or stock feed.

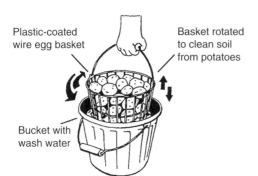

Plastic-coated
wire egg basket

Basket rotated
to clean soil
from potatoes

Bucket with
wash water

Fig. 9.7. Wire basket for washing samples in the field.

When sampling to a standard, the number of randomly selected tubers required to find defective tubers in a load is shown in Table 12.2. A typical specification for a pre-pack crop is given in Table 12.3.

9.5 Loading into Store

Potatoes being loaded into store:

- Will have damaged skins and cuts.
- May be wet.
- Are likely to include soil, weeds, clods, stones and haulm.
- Will include undersize and oversize tubers, i.e. out-grades.

The objective is to load into store a crop free from trash, with soil removed so that subsequent ventilation is unimpeded, and this achieved with minimum abrasion of tuber skins and impact damage. The options on how best to achieve this are described in Ch2. By the time store loading starts, the grower and store manager will have committed themselves to either separating and cleaning the crop going into store at the steading or putting boxes into store 'as dug'.

9.5.1 Loading into a bulk store

The equipment used for loading bulk stores is described in Ch2. The key management aims are as follows.

1. If blight or rots are likely to cause progressive crop breakdown in store, sell crop off the field or keep separate from the main store.
2. Remove any rots, mother tubers or haulm that are present.
3. Adjust cleaning equipment to remove as much soil as possible while minimizing damage to the crop.
4. Stop harvesting if crop starts to smear with soil or crop temperature falls below 9°C.
5. Move loading head constantly to avoid the formation of soil cones.
6. Load pile in layers (Fig. 9.8) to minimize amount of roll-back of tubers.
7. Have fans running immediately when the crop starts to enter store (Fig. 9.9).
8. Open up lateral ducts as they are covered with crop (Fig. 9.10).
9. Keep surface of pile as level as possible to ensure uniform ventilation.
10. Soil cones reduce ventilation to tubers within the cone, which slows drying and cooling of tubers in this area. This can result in localized rotting. Pockets of potatoes with blight or rots can start to rot down (Box 9.1), causing hot spots which can lead to large-scale rotting.

9.5.2 Loading into a box store

The same principles (1) to (4) above should be applied for storage in boxes. However with boxes, instead of harvesting having to be stopped if smearing is occurring,

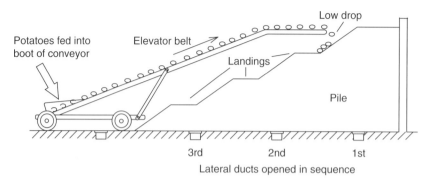

Fig. 9.8. Loading bulk pile in layers to minimize drop damage and tuber roll-back.

Fig. 9.9. Axial flow fans for a bulk store. (Courtesy of Farm Electronics, Lincoln, UK.)

empty boxes can be put on trailers for loading direct from the harvester. Smearing will be less and the boxes can be blown with air to dry the soil so that when the potatoes are put over the cleaner later the soil falls off. It may take a week of drying to get the soil to this stage.

Fig. 9.10. Slide located at entry to a lateral of bulk store photographed from inside of main duct (courtesy of A. Johannson, Norrköping, Sweden).

Box 9.1. Formation of hot spots in bulk stores

Hot spots in potatoes can occur both in boxes and in bulk storage, but their consequences are particularly serious in bulk storage. The problem starts with a tuber starting to rot due to a blight or soft rot infection that invaded the growing tuber (Fig. B.9.1a). Rotting may occur whatever the store management or may be exacerbated by tubers remaining warm and wet after harvest. The rotting tuber collapses under the weight of potatoes above (Fig. B.9.1b), resulting in an anaerobic mass of exudates though which ventilating air cannot penetrate. The metabolic heat from rotting exudates results in warm moist air rising from the hot spot and condensing on the cooler sound potatoes above (Fig. B.9.1c). They become warm and wet, causing further rotting to occur and further breakdown of tubers. The hot spot increases in size with more and more tubers collapsing, resulting in the surface of the pile starting to slump (Fig. B.9.1d). This, and the exudates flowing into the laterals and main duct, may be the first indication of rotting that the store manager sees.

Prevention, by hot box testing at-risk crops and diverting them elsewhere and rapidly drying newly loaded potatoes, is the only cure for hot spots. Temperature monitoring of the crop should give early warning of hot spots occurring, so that the store can be emptied before rotting has progressed too far. Processing stores, kept warm to keep sugar levels low, are the most prone to hot spots developing.

Continued

Box 9.1. Continued

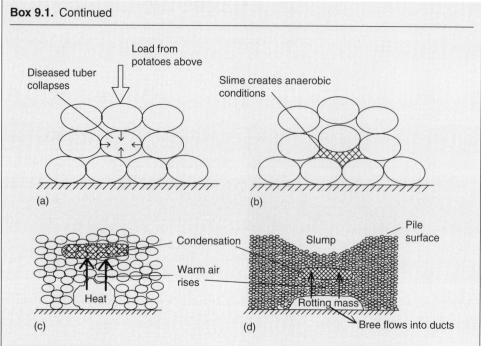

Fig. B.9.1. Sequential development of a hot spot in a bulk pile: (a) tuber rots to a bacterial slime; (b) slime contaminates other tubers; (c) heat from rots causes warm air to rise and condense on tubers above; (d) collapsing tubers cause slump.

Box 9.2. Dealing with problem batches of crop

Store managers are often advised by agronomists to treat batches of crop in store in different ways. If the store has a single airspace, it is not possible to treat any part differently from the rest, except to provide more air to some potatoes than to others.

   Were it possible to cool one part of the crop more quickly than the rest to slow the multiplication of a certain disease, then convection currents from the warmer potatoes in store can result in the cooler part of the crop being soaked with condensation (Ch3.5). This unintended result can make disease more, rather than less, likely.

If any rots are present, it may take up to 6 weeks to dry the rotting tubers to prevent them from causing sequential rotting. It is important to evaluate whether it is worth saving such a crop or whether it is better to sell or dispose of the crop straight from the field (Box 9.2).

   Boxes loaded into store should be stacked in such a way that ensures ambient air or recirculated store air can be forced through the newly loaded boxes within hours of being harvested. While this may be easily achieved with positive ventilation systems, the airflow in airspace-ventilated systems will bypass the first boxes loaded unless steps are taken to prevent this.

In stores holding a single variety, boxes should be stacked in a wall across the direction of airflow (Fig. 9.11a). The recirculating air in this arrangement has to flow through the boxes to return to the intake of the fan. In seed stores, where boxes are stacked in rows to ease access to specific varieties, the only approach is to try to prevent air bypassing the rows. This can be done by building temporary stacks of boxes across the store to prevent air bypassing the rows (Fig. 9.11b). Any boxes containing particularly wet potatoes can be used for this purpose, as they will get more ventilation than those in the longer rows.

As the store becomes full, it is common, though not particularly good practice, to fill the end of the store with boxes placed at right angles to the main rows of boxes. If this is done, spaces should be left between every second or third box to allow air to pass between these boxes prior to entering the pallet apertures of the boxes in the rows (Fig. 9.12).

9.5.3 Application of chemicals

As stated in Ch2.4 and Ch2.5, chemical treatments must be based on a fully justified crop risk assessment, with the chemical used targeted at the disease of concern. It is preferable to concentrate on non-chemical disease control measures, such as rapid drying of crops, gentle handling, removal of respiration heat and minimization of temperature differentials of crops in store.

If liquid chemicals are being applied, every effort should be made to dry off as much liquid as possible before the potatoes go into store. Ventilation immediately

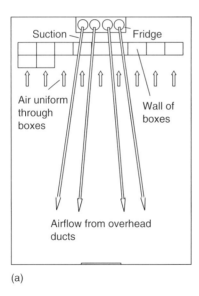
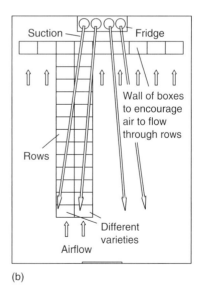

Fig. 9.11. Recommended box stacking arrangement for airspace ventilation systems to encourage air to flow through boxes: (a) single variety store (e.g. ware store); (b) multi-variety seed store.

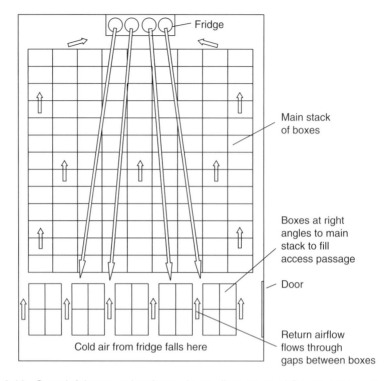

Fridge

Main stack
of boxes

Boxes at right
angles to main
stack to fill
access passage

Door

Return airflow
flows through
gaps between boxes

Cold air from fridge falls here

Fig. 9.12. Gaps left between last boxes in, to allow return airflow to enter main stack pallet apertures.

after loading will help to remove the remaining moisture. While the chemical is likely to prevent development of the target disease, any remaining liquid may well make the development of a non-target disease more likely.

9.5.4 Duration of loading period

Stores should be sized so that harvesting equipment can load them within 5–8 days. This minimizes the risk period when doors are open and ambient air or ventilation can cause condensation to form on the crop when it is still warm and disease multiplication can be rapid (Box 9.3).

9.5.5 Loading bags

In hot countries, storage in open-weave hessian bags placed in general-purpose cold stores is popular (Fig. 6.30). To avoid subsurface condensation within the bags, it is best to pre-cool the bags at intake, using a forced airflow to reduce their temperature to within a few degrees of the cold store temperature, before the bags are stacked on the ventilated floors. Prior to being loaded into store, the potatoes should be given time

---

Box 9.3. Condensation during store loading

Harvesting and store loading in a Scottish seed store was interrupted for 2 weeks due to rain. The grower decided to start cooling the half-filled store during this period as it was dry and the wounds had cured. Soon after harvest resumed, the farmer discovered that the first half of the crop into store was now wet. A few weeks later it had developed severe silver scurf. The later harvested crop was dried using ambient air and was unaffected by disease.

    The warm humid air from the newly lifted potatoes had condensed on the cool, dry crop, thoroughly rewetting the latter. This provided perfect conditions for silver scurf spores present to infect the crop and any infection already in the crop to multiply. The result was a highly contaminated, low-value stock of seed.

---

to heal their skins and should be stacked in such a manner that plenty of air passes through the sacks to remove respiration heat and any associated condensation.

9.6 Drying and Wound-healing Period

From the start of store loading, ventilation should be switched on and the store temperature maintained so that it tracks the temperature of the crop in the ground (Fig. 9.13). The reasons for this are:

- Ventilation removes respiration heat and so prevents temperature differences developing which can cause subsurface condensation.
- Keeping potatoes in store at the same temperature as newly harvested crop minimizes the possibility of convection currents forming, which can result in condensation forming on the cooler potatoes.

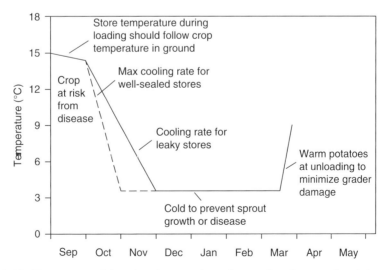

Fig. 9.13. Target over-winter store temperatures for seed and pre-pack potatoes in the UK.

- Ventilation dries wet crops and removes moisture from adhering soil.
- Keeping the potatoes at harvest temperature, normally 10°C and above in the UK, ensures wound healing is rapid and allows ambient-air ventilation with little risk of crop condensation.
- Rapid healing of wounds helps minimize weight loss in the crop by preventing evaporation of moisture from wounds.

At loading, differences between sections of a pile or potatoes in stacks of boxes should not exceed 3–4°C, while differences in temperature of surface and subsurface potatoes should be restricted to 0.5°C (Box 3.7).

Drying is fastest at warm temperatures as warm air has greater water-carrying capacity than cool air. Positive ventilation dries crops more quickly than airspace ventilation.

Once the skins of tubers are dry, continuous ventilation can be replaced by periodic recycling of air through the crop to remove respiration heat and ensure that the temperature of the crop is uniform throughout. The tubers should be dry throughout the pile or box, rather than just the surface layer. In positively ventilated systems, tubers nearest the incoming drying airstream will dry first, while those next to where the air leaves the box may take up to 1–2 weeks to dry. In airspace-ventilated stores the top layer of the top boxes will dry first but potatoes lower down in the boxes and the stack of boxes may still be wet. Extensive recirculation may be necessary to dry these deeper potatoes. Unnecessary ventilation however increases evaporative weight loss and loss of skin bloom, so should be undertaken only when necessary.

Wound healing is fastest at 20°C, and reduces to almost zero at 7°C (Ch1.5). To evaluate the speed of wound healing, the manager can cut a few tubers in half, place them on the corner post of boxes or on a piece of board and visually monitor the progress of wound healing of the exposed flesh.

If crop temperatures at harvest are 18°C or above, cooling can be started before wound healing has finished. Particular care has to be taken to prevent temperature differentials forming in boxes if the crop is still respiring at a high rate (Box 9.4).

9.6.1 Drying crops using fridge rather than ambient air

Drying of potatoes is almost always carried out using the ambient-air ventilation system rather than the refrigeration system. Ambient-air drying uses considerably less energy. However, some stores do not have ambient-air ventilation so have to make do with the refrigeration system. On occasions where the air outside is foggy, the refrigeration system can be used to continue drying when drying using ambient air would have to stop.

Drying crops with doors open and compressor off using fridge fans for air circulation

There are two ways of drying crops using a refrigeration system. The first is to open the door of the store and run the air circulation system with the compressor switched off. So long as the store is not too sheltered, the wind will change the air

Box 9.4. Condensation during ventilation drying

The four plots in Fig. B.9.2 show what can happen when a store is ventilated soon after harvest with fans set to manual. There was therefore no control on inlet air temperature. Ambient air temperature varied considerably over the period (Fig. B.9.2a), with unblended, ventilating air being blown over the top of the boxes. As the weather was very wet, ventilation was often stopped. This caused the relative humidity (RH) in store to rise into the nineties (Fig. B.9.2b). The air was blown over the top of the stack of boxes, causing large temperature differences in the top layer of boxes. On one occasion, on the night of 12 September, the temperature

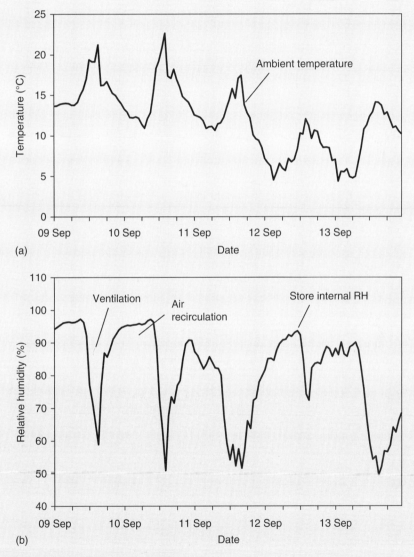

Fig. B.9.2. Ventilation-induced condensation in an airspace-ventilated box store:
(a) ambient temperature; (b) store relative humidity (RH); (c) box temperature;
(d) tuber skin resistance.

Continued

Box 9.4. Continued

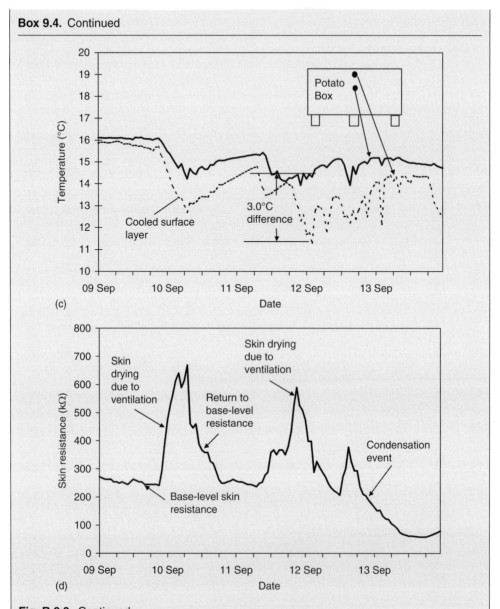

Fig. B.9.2. Continued

difference between the top tubers and those 350 mm below was 3°C (Fig. B.9.2c). At this point the top boxes became wet due to condensation (Fig. B.9.2d). This was evident when the skin resistance of a sample tuber fell below the base level of 270 kΩ. The previous peaks in skin resistance were due to air drying the top potatoes during ventilation.

The crop, which was harvested with no sign of disease or rots, eventually started to rot after a series of 'ventilation-induced' condensation events.

While this was an extreme case, the store manager was completely unaware that there was a problem and did not know how it was caused.

within the store while the air circulation system distributes the air through the crop. This is a manually controlled system, with the store manager opening and closing doors as weather conditions dictate. There are no automatic controls to prevent ventilating the crop with freezing air or air with high dew-point temperature, which may condense on the stored crop. There is therefore considerable risk in this approach.

Drying crops with doors shut and compressor operating

Keeping the store doors closed and running the fridge in cooling mode can dry crops. The moisture evaporated from the wet crop by the circulating air will subsequently condense as the air passes over the cooling coils. The fridge therefore acts as a dehumidifier. This is a high-energy, expensive way to dry potatoes (Ch8.6), but will dry the crop even if the outside air is foggy. However, the evaporation of moisture from the potatoes will also cool the crop, which may conflict with keeping the crop at the same temperature as the potatoes being lifted. Adding heat using gas or electric heating can prevent such cooling, but this adds further to fuel use and cost.

9.7   Cooling

Once the crop is dry and wound healing complete, or at least in progress, cooling can begin. As mentioned previously, cooling should only start once the doors have been shut for the winter. The rate of cooling is determined in part by the store equipment and external climate and in part by the store manager. Two alternative rates of cooling are recommended, one for well-sealed stores and one for stores that are less well-sealed or which are regularly opened to source potatoes, usually seed (Fig. 9.13). The fast cool helps dormancy to be maintained and slows the multiplication of any established disease. The slow cool aims to prevent any ambient air that does leak into store from condensing on the crop and causing disease to develop owing to the presence of moisture.

9.7.1   Rate of cooling dependent on store equipment design

The rate of cooling is dependent on the:

- Availability of cool air where cooling uses ambient air.
- Rate of airflow per tonne of crop stored ($m^3/s/t$).
- Uniformity of airflow passing through the crop.
- Cooling capacity of refrigeration system per tonne of crop (W/t).
- Defrost period per day to prevent ice building up on cooling coils, usually six 20-min periods with off-cycle defrost.

Store managers have to work with the system they have, not what they would like. The aspects listed above are largely outside store managers' control. The only one that the manager can influence is the uniformity of airflow passing through the crop, and this is achieved by stacking the boxes to optimize airflow though each and every box.

9.7.2   Rate of cooling determined by store manager

Store managers can influence the rate of cooling to an extent. They cannot speed cooling for individual, at-risk batches of crop (Box 9.2), nor can they cool crops at the same rate as vegetables are cooled (Box 9.5).

They can restrict cooling:

• To a maximum of 0.5°C/day to minimize the possibility of temperature differences occurring in the crop, particularly in box stores.
• For stores regularly opened to ambient air, where cooling should not overtake the natural cooling of the atmosphere as winter approaches.

Too rapid cooling can cause temperature differences within boxes to occur in box stores. The store manager should monitor crop temperature differences during cooling and slow the rate of cooling should temperature differences within boxes exceed 0.5°C. By setting the crop/duct differential to a maximum of 4°C (Box 3.7), the likelihood of temperature differences within boxes exceeding 0.5°C is minimized. To slow cooling when using refrigeration systems, the simplest way is to switch the unit off for a period each day.

9.7.3   Minimizing weight loss during cool storage period

There will always be some crop weight loss as there is continuous breakdown of carbohydrate through respiration to provide cell nutrients and energy to maintain life. The minimum theoretical weight loss of potatoes at 3.33°C has been calculated as 0.072% per week (Hunter, 1985). These figures assume saturated air. At this rate, weight loss for a 30-week period will be 2.16%, or 21.6 kg/t of crop stored.

Even with cooling air humidified to 95–98%, cooling will always result in weight loss in the crop. The amount of weight loss can be minimized by:

• Ensuring leakage of warm air into store is minimized (Box 9.6).
• Opening doors only when absolutely necessary.
• Using personal doors for entry rather than the main store doors.

Box 9.5. Cooling potatoes compared with cooling vegetables

Store managers or agronomists sometimes try to treat potatoes like they would treat newly harvested raspberries or broccoli. They want to cool them over a few hours to 3–4°C, both to reduce their respiration rate and to protect them from developing disease. However, raspberries may have a value of £2500/t and a respiration rate of 580W/t at 20°C, while potatoes have a value of £130/t and a respiration rate of 70W/t. Potatoes may be harvested at 200t/day, while a typical harvesting rate for raspberries is 1.5t/day. It is uneconomical, impractical and unnecessary to cool at such speed. Potatoes have relatively low respiration rates, and their thick skins are their primary barriers to disease. The store manager should concentrate on keeping potato skins free from damage, aiding the healing of any wounds that do occur, and ensuring the skins stay dry during the 30 days that it takes to cool the crop down to the long-term storage temperature.

Box 9.6. Study of efficiency of energy use

A validated computer-based study on the efficiency of operation of a 1500-t refrigerated pre-pack store (Devres and Bishop, 1995) produced the following results:

- Energy use *increased* by 23, 46 and 70%, respectively, if the store temperature was held at 3, 2 and 1°C compared with 4°C.
- Energy use *decreased* by 7 and 13%, respectively, if the store temperature was held at 5 and 6°C, respectively.
- Energy use *increased* by 20% if the leakage rate increased from 0.24 to 0.6 air changes per hour.
- Reducing the speed of cooling saved little energy.

For optimum energy efficiency, stores should be held at 4°C and be sealed well.

Recirculating air through the crop in stores without humidifiers also leads to weight loss. The need to recirculate air through the crop can be minimized by:

- Setting crop/duct differential to a maximum of 4°C so that differentials >0.5°C in the crop do not occur.
- Using sensors measuring temperature differences in the crop, rather than time clocks to control when recirculation occurs.

9.7.4 Efficiency of cool storage

Comparing energy consumption with other stores or for the same store in previous years will give an indication as to whether the store is working at maximum efficiency. This has been tried on an experimental basis (Bishop, 1992). A validated computer study (Box 9.6) suggests the optimum settings for optimum energy efficiency.

With stores cooled using refrigeration only, the quantity of condensate produced from the drain under the cooling coils should match to the amount of moisture lost from the crop. Condensate in excess of the actual moisture loss from the crop may be caused by ambient air leaking into store. Excessive moisture loss in DX refrigeration systems may be caused by the TD between the cooling coils and the store air exceeding the recommended 6.0°C (Ch7.1).

9.8 Store Monitoring

9.8.1 Inspection of crops and checking correct store operation

Store management should include:

- Checking tubers in the tops of boxes or pile for condensation and disease, taking samples from specific boxes or areas of the store to assess any disease development in store.
- Daily recording of temperatures on the store diary sheet located beside the control box (Box 9.7).

Box 9.7. Store diary sheet

Even with the advent of computer-based data logging it is useful to keep a manual store record sheet beside the store controller. This ensures that:

- The store is visited daily.
- The requirement to write down temperatures, fan and fridge times forces the store manager to check that the plant is working correctly.
- Problems such as sticking louvres or burnt-out fans are quickly identified.
- On the store person's day off, others can see how the plant has been performing.
- Indications of rotting, such as the presence of insects, can be noted on the sheet.

Table B.9.1. Example store record sheet.

Marybank Farm Ltd – Potato store record sheet

Store: Pre-pack store 2   Variety: M Piper   Year: 2008/9

Date	Time (hours)	Crop temperature (°C)				Air temperature (°C)		Fans		Fridge		Comments
		Set point	Ave.	Min.[a]	Max.[a]	External	Store ave.	Cum. hours	Hours/day	Cum. hours	Hours/day	
2/10	08.30	10	12.5	11.7 s3	13.2 s9	9.2	12.8	221	–	23	–	Amb. air cooling
3/10	08.15	10	12.1	11.2 s2	12.7 s9	8.3	12.4	234	13	23	–	As above
4/10	09.30	10	11.6	10.8 s2	12.3 s7	13.2	11.8	241	7	31	8	Fridge cooling
7/10	08.10	9	10.3	9.5 s2	11.2 s7	12.2	10.5	241	0	85	54	Rots in box Row1Col3

[a]s3 denotes that sensor 3 is the coldest; and s9 that sensor 9 is the warmest.

To aid inspection of the top of boxes or pile, a fixed safe access should be provided and safe areas marked by hazard tape (Ch5.10).

Regular sampling of tubers for disease development over the storage period will not only help to determine suitable markets for the crop but may also help to identify problem areas within the store that need to be modified for the following season.

The use of a modem (Ch8.11) fitted to the control box and data logger allows cooling performance and energy use to be compared with other stores and will give an early indication of poor performance and equipment failure.

9.9 Managing the Use of Low-tariff Electricity

9.9.1 Control settings

If the store is on low-tariff electricity, the control box will usually be configured to run refrigeration only during the night and in periods during the day when cheap electricity is available. It is normal to set limits which will override these restrictions should the crop warm over a set limit, by 0.5–1.0°C for example.

9.9.2 Using crop thermal mass

Potatoes in store represent a large thermal mass. This can be used to delay cooling of the crop until ambient air cooler than the crop becomes available. The use of refrigeration can then be minimized. A judgement has to be made as to how much the crop temperature can be allowed to rise between periods of cooling, with the amount being less with short dormancy varieties than long. Any diseases present should also to be taken into account, as the warmer the crop, the faster these will multiply. As the end of the storage period approaches and ambient air temperatures start to rise, store refrigeration equipment may not have sufficient cooling capacity to enable it to bring the crop back down to temperature if the crop temperature rises too much. Energy saving through the use of the crop's thermal mass has therefore to be done with care.

9.10 Sprout Suppressants

CIPC is the most common chemical applied to crops in store to control sprouting (Ch3.8). It and other sprout suppressants must never be used on seed potatoes, as otherwise the seed will not grow. Even storing seed in boxes or stores that have been contaminated with CIPC in the past may prevent uniform sprouting after planting. It is strongly recommended that growers subcontract the application of CIPC to experienced contractors.

9.11 Reconditioning of Crops for Processing

Where fry colours are found to be poor, as indicated by a low Agtron or Hunter L value, crops can be reconditioned by warming them for a period of 2 to 6 weeks in store to improve their fry colour. In a series of experiments (PMB, 1991–1995) over

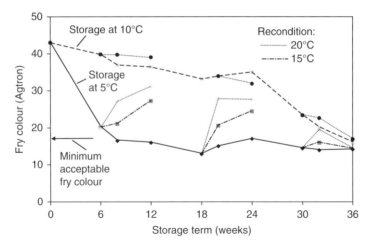

Fig. 9.14. Plots showing reconditioning of processing potatoes. (Courtesy of Potato Council, Oxford, UK.)

4 years on cv. Pentland Dell potatoes, the Agtron value of crops stored at 5°C was raised by up to 16 points when the store temperature was increased to 20°C (Fig. 9.14). Reconditioning will affect the entire crop stored in the one airspace. The safest way to warm the crop is to use roof-space heaters with recirculation ventilation switched on. If the controller has a control mode for safely warming the crop with ambient air without causing condensation on the crop, this can be used instead (Ch8.6).

Reconditioning will not work on longer stored crops, where senescent sweetening has set in.

9.12 Warming Crops for Grading and Dispatch

Pre-pack and seed potatoes are normally stored at 3–4°C. If graded at this temperature they are likely to sustain damage (Ch10.4) and will be slow to heal any wounds inflicted. To avoid this, potatoes stored below 6°C should be routinely warmed to 8–10°C. They will also heal more quickly at this temperature.

9.12.1 Warming in the store

Warming potatoes in the store allows the crop to be warmed without risk of wetting the crop. In box stores this is best done if the store has a separate warming chamber (Ch6.10), accessible via canvas doors (Fig. 9.15) from both the store and the grading area. The store-side canvas door is opened to allow boxes to be stacked prior to warming. The door is then closed and warming started. Once warming is complete, the boxes are removed using the second door opening into the grading area.

Potatoes in bulk bins can be easily warmed using a space heater. Warming the day's out-take in large bulk stores is less easy. The only way this can be done

Fig. 9.15. Canvas door of chamber for batch warming of boxes.

is by opening up the last few laterals in the bulk store, and ventilating with warmed air. This, however, prevents ventilation of the rest of the pile while warming is taking place.

9.12.2   Removal of cold potatoes from store for warming elsewhere

If potatoes at 3–4°C are taken out of store and the dew-point temperature of the ambient air is above that of the temperature of the crop, the potatoes will get wet from condensation. The warming airflow will first have to evaporate the water that has condensed on the crop (Box 9.8) and, once it has dried, only then will the crop start to warm. The warming process will be considerably longer than if the crop was warmed within the environment of the cold store and the possibility of disease development will have been increased.

9.12.3   Warming on the grading line

Radiant heat applied to tubers on a roller table for 1 min immediately after store has been shown to reduce damage (Bishop *et al.*, 2000) and has the benefit of heating only the very outside of the tuber, so being economical on heat energy. The temperature profile under radiant heat was also measured, which showed an increase in the outside flesh temperature of 8–12°C depending on tuber colour.

Box 9.8. Warming boxes of potatoes coming from cold store

A letterbox warming system fitted with an electric heater and an airflow of 0.08 m³/s/t was installed in the grading area of a multi-store complex. The heater raised the temperature of the recirculated grading-area air by about 2.5°C. The batch of potatoes became wet when removed from a cold store held at a temperature below the dew-point temperature of the ambient air. The warming system had then to dry the crop before warming started. Figure B.9.3 shows how the potatoes nearest the incoming warm air dried almost immediately and started to warm thereafter. In contrast, the potatoes in the layer where the air exited the potatoes were first evaporatively cooled by the air. Only when they were dry did they start to warm. The wetting of potatoes extended the warming process by about 9 h.

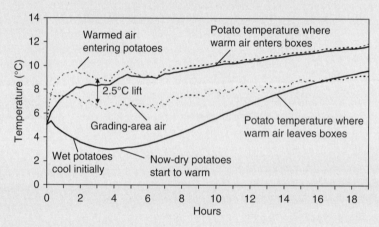

Fig. B.9.3. Warming of potatoes that had previously become wet from condensation on removal from cold store.

Similar radiant heat treatments have been used to reduce black spot (Trenckmann, 1988) but this has not been adopted commercially. Since wound healing is necessary following grading, tuber temperature should be held above 7°C optimally for 1 week, but for 3 days at a minimum.

9.13 Summary

This chapter has concentrated on what actions the store manager can take to ensure that the crop is stored in as good condition as possible. The key actions are as follows.

- A month before harvest, the store is cleaned and inspected for damage, leaks or faulty equipment.
- If boxes are used for storage, these are emptied of last year's debris.
- Temperature and RH sensors are cleaned and calibrated ready for inserting in the crop.

- Ambient-air cooling control and fridge operation are tested prior to the store being filled.
- Crops are sampled before harvest to help decide whether to store or sell straight from the field, with sales staff alerted as to what action is intended.
- Crops identified by hot boxing samples as being likely to rot in store should be sold immediately or dried separately from the main crop.
- Stores are loaded within a period of 5–8 days to minimize disease risk to crop in store at this particularly vulnerable stage.
- Minimization of damage has priority over speed of harvest.
- If liquid chemical is being applied to the crop entering store, skins should be dried immediately after application using ventilation.
- The temperature maintained in the store during loading should follow that of the temperature of potatoes being lifted from the ground.
- Harvesting should stop if soil on tubers is wet and starts to smear, or crop temperature falls below 9°C.
- Elevator heads in bulk stores should be kept moving to prevent soil cones forming within the pile.
- Ventilation should be started immediately when potatoes or boxes enter store.
- In bulk stores, crops that could rot should be kept out of the main pile; in box stores vulnerable loads can be kept separate from the main storage area and blown with high volumes of air to prevent hot spots occurring.
- Potatoes grown in hot countries which are to be stored in bags in refrigerated stores should be cooled down to store temperature in batches prior to stacking within the main store.
- During store loading and wound healing, the control box should be switched to wound healing mode to maximize ventilation while preventing condensation.
- Stores should be closed and sealed as soon as possible after loading.
- Once wounds are healed, the control box can be set to cooling. The rate of cooling of 0.5°C/day should allow cooling to take place without the risk of condensation forming on the crop.
- A record sheet should be kept of internal and external temperatures, crop temperature differences and ventilation and refrigeration run times to ensure that the store is operating optimally.
- A modem fitted to the control box and data logger allows remote monitoring by ventilation equipment suppliers and energy use to be compared with other stores.
- CIPC should be applied to crops in processing stores when eyes show signs of opening. Stores should be opened for ventilation, preferably within 12 or even 8 h of treatment.
- Crops with sugar levels too high for processing can be reconditioned by maintaining the stored crop at 15°C for a period of 15 days.
- Potatoes to be graded dry should be warmed, preferably within the storage area, to 8–10°C prior to handling to minimize damage.

10 Seed Grading and Preparation for Planting

Topics discussed in this chapter:

- Storage systems for multiple varieties and generations of seed.
- Particular requirements for seed production facilities.
- Manipulation of seed size by chronological ageing.
- Seed grading and dispatch procedures.
- Warming seed prior to handling.
- Chitting systems to maximize yield of ware.
- A chitting system for planting 160 ha.
- Pre-grading seed reports.
- Communication between seed suppliers and customers.

10.1 Introduction

Seed has special requirements for storage, grading, dispatch and pre-planting preparation that differ from the production of processing or pre-pack potatoes:

- Seed needs to be put into store so that individual boxes and varieties can be removed throughout the storage period.
- In the UK seed is usually grown in the cooler areas of the country where aphids, the transmitters of viruses, are few. In wet, cool autumns, potatoes may be harvested with considerable amounts of soil, which can impede subsequent ventilation.
- Due to the small tuber size, small voidage and presence of soil, positive ventilation to force air between tubers is often advantageous.
- Seed may be required soon after harvesting, either for export or for chitting on the customer's farm, or may have to be stored over winter for springtime delivery.
- Seed may require up to five grader screens to split it into the size ranges required to achieve uniform planting.

- By selling split graded seed, the remaining fractions may have to be returned to store until requested by another customer.
- Following grading, seed may spend weeks or even months either in transit or on the recipient's farm waiting for soils to dry or warm sufficiently for planting.
- Most seed supplied locally is required over a short period of approximately 4–6 weeks, putting pressure on the grading process.

10.2 Specific Requirements for Seed Storage

While ware growers tend to grow one or two varieties of potatoes, seed growers may grow six to 20 varieties. They often grow different generations of varieties, such as Super Elite (SE) 1 and SE2, requiring further separation. Keeping the seed lots separate is easy if the crop is stored in boxes.

Another approach is for cooperatives or companies to organize seed growing on a large scale, so that one or two farms produce all the seed of one variety and generation for the group. This allows the seed to be stored in bulk stores in 500- to 1000-t lots. This approach is used in The Netherlands, while the UK has traditionally stored seed in 0.5- to 1.5-t boxes.

Boxes are commonly stacked in store primarily to suit accessibility, rather than to ensure that the ventilating air flows uniformly through the newly harvested crop (Jackson, 2005). As access to different varieties and generations of seed is usually required throughout the storage period, boxes cannot be put in a tight stack, as is done in pre-pack stores, where only one or two varieties are grown. A gap of 100–150 mm on either side of the boxes (Fig. 10.1a) is required if a forklift is to remove a row of boxes from between two neighbouring rows. Unless the air distribution system is designed to block off these gaps, much of the airflow that would otherwise go through the pallet apertures, or the box slatted floors, will go along the gaps instead, reducing the airflow through the potatoes.

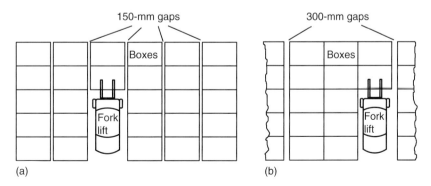

(a) (b)

Fig. 10.1. Stacking arrangements to allow removal of individual rows: (a) spaces between each row allow individual rows to be removed; (b) two to three rows tightly stacked, with boxes removed by first using a forklift side-shift, if fitted.

As ventilation of air though boxes should start immediately the crop comes into store, to prevent subsurface condensation, the ventilation system should be designed to allow this. Positive ventilation systems (Ch6.8) should therefore either be designed to allow rows of boxes to have 150-mm gaps between them, or to be only two or three boxes wide, so that the box can be moved sideways prior to the forklift reversing (Fig. 10.1b).

If airspace ventilation systems (Ch6.6) are used for storage of seed, the first rows in should be in the centre of the store where the airflow from the ventilation system is greatest (Fig. 10.2). Air through the first boxes in can be further improved by blocking off the air return path with either a wall of empty boxes or a temporary wall of boxes containing seed that requires to be dried. Poor air distribution can be compensated for to some extent by increasing the volume of recirculation airflow to above the UK figure of $0.02\,\mathrm{m}^3/\mathrm{s}/\mathrm{t}$.

In other ways the storage of potatoes for seed is similar to that for the pre-pack or processing market but with some distinct differences. No sprout suppressant will be used, as the ability of the seed to grow subsequently would be affected by its application. There are fewer restrictions on pesticide use in seed as compared with ware, so fungicide treatments prior to storage, or after grading, are common.

The crop may be pre-graded into different size grades during the storage period to remove the grading bottleneck in the spring. Minimal re-grading and inspection is then required prior to putting the seed into bags or boxes for dispatch.

In a recent survey in the UK on seed storage (BPC, 2005b), the most common size of seed store was 1000–1250 t (~25% of those sampled) with a typical loading capacity of 100–200 t/day (~60%), giving a time to fill the store of 8 days or less

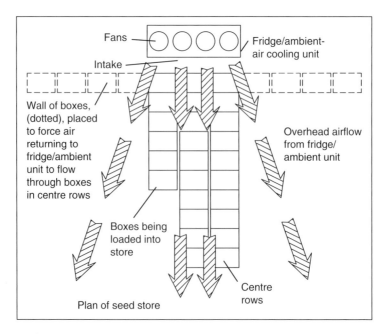

Fig. 10.2. Stacking arrangement where seed is airspace-ventilated.

(~60%). In the UK, box storage for seed is universal, with traceability, in addition to separation, being a major issue. After storage nearly 80% is warmed before grading. A failing in many large stores is that they take more than 8 days to fill, and so cannot be rapidly closed, wound healed and then cooled.

Seed should always be dispatched at a temperature above average ambient (usually 8–10°C in the UK) to prevent condensation during transit.

10.3 Altering Chronological Age of Seed to Produce Bold or Prolific Crops

Increasingly potatoes are being grown for a specific market, be it bakers, processing, pre-pack, salad potatoes or seed. Each has an optimum tuber size. The greater the proportion of the crop in the optimum size range, the greater the potential profitability of the crop. Extending or reducing the period between tuber initiation of the seed crop and the planting date of the resulting seed can increase the proportion of tubers within the optimum size range (Ch1.3).

10.4 Grading of Seed

10.4.1 Seed graders

Most of the equipment for grading seed is the same as that used for grading ware. The primary difference is the grader itself. In ware production, the grader simply removes oversize and undersize material (e.g. <45 mm and >85 mm), requiring two screens. In seed the size grades may be:

• <25 mm.
• 25–35 mm.
• 35–45 mm.
• 45–55 mm.
• >55 mm.

To achieve this number of grades, using a grader with one chain mesh screen after the other on a horizontal plane (Fig. 10.3a) would lead to a very long machine. To conserve space, therefore, graders are used that have mesh riddles (screens) stacked vertically (Figs 10.3b and 10.4) with potatoes cascading downwards through the screens, with the largest tubers being removed first and the smallest being removed last. A compromise between the chain mesh screen grader and the vertically stacked riddle grader is to use a stepped riddle grader (Figs 10.3c and 10.5), which, although longer than the vertically stacked riddle grader, allows the falling tubers to land on a rubber belt rather than cascading through the series of oscillating riddles below. While manufacturers of the step grader claim it does less damage than the vertically stacked screen grader, due to the rubber belt being a softer landing for tubers, both types of grader are popular with seed producers. As seed tubers tend to be considerably smaller than ware, the potential for damage is less than for ware.

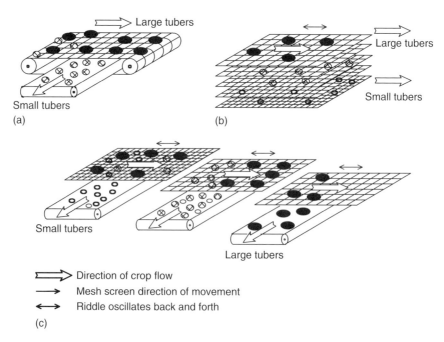

Fig. 10.3. Types of grader: (a) chain mesh screen; (b) vertically stacked riddle; (c) stepped riddle grader.

Fig. 10.4. Vertically stacked riddle grader. (Courtesy of D.T. Dijkstra, B.V., The Netherlands.)

Fig. 10.5. Step grader with riddles removed. (Slackadale, Aberdeenshire, UK.)

10.4.2 Seed grading systems

A seed grading system has to:

- Remove at the start any rots that could contaminate other tubers.
- Remove any adhering soil, stones, clods or haulm.
- Separate tubers into a range of seed sizes.
- Allow visual inspection of every tuber and removal of those which exceed the tolerance levels.
- Allow possible application of a pesticide.
- Enable tubers of different size bands to be put into boxes or bags.
- Ensure boxes or bags are sealed after inspection so that the customer gets only material that has been passed by an inspector.

After grading the seed should be ventilated to dry any moisture on the skins caused by grader damage and to remove the respiration heat associated with grading, which often results in subsurface condensation.

A typical seed grading system (Fig. 10.6) will start with a box- or trailer-fed bulk hopper feeding potatoes over a soil extractor or cleaner. Alternatively, the gentler system of a pair of box tippers discharging potatoes on to a flat belt (Fig. 10.7) can be used instead of the hopper. The cleaner will be a spiral coil (Fig. 10.8), a star cleaner or a continental web (Fig. 2.9), and will remove any soil, small stones or clods. The tubers are then fed into the boot of a fluted belt elevator, which discharges the potatoes at high level on to the grader screens. The tubers cascade down through the square mesh screens, with the largest being removed by the top screen and successive fractions being separated by the screens below. The separated fractions are fed on to the dividing flat belt conveyor. Steel dividers remove

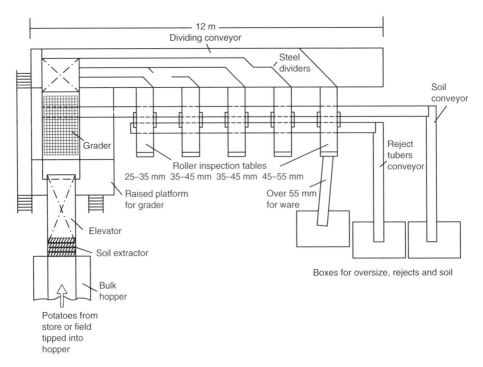

Fig. 10.6. Layout of a seed grader system. (Redrawn from illustration by D.T. Dijkstra, B.V., The Netherlands.)

Fig. 10.7. Twin box tippers feeding a flat belt. (Courtesy of Haith Tickhill Group, Doncaster, South Yorkshire, UK.)

Fig. 10.8. Cleaning coils at discharge point of bulk hopper.

the separate fractions from the belt and channel the tubers on to a series of five roller tables, which rotate the tubers so that inspection staff can see the whole surface of the tubers and remove any that are blemished or damaged. The roller tables feed the seed into boxes, big bags or sacks (not shown), each with a close graded range of seed size. As the 35–55 mm size range contains the majority of the seed, two roller tables are allocated to either 35–45 mm or 45–55 mm, depending on which size predominates.

The inspection tables are usually housed in a well-lit cabin, illuminated to 500–750 lux at the inspection table, so that inspection staff can be kept warm and free from the dust that is generated by forklift movement, tipping of boxes, filling of sacks or bags, or from any point where potatoes drop.

A soil and clod conveyor takes soil from under the grader and roller tables, while the reject tubers conveyor removes damaged or diseased tubers picked off the tables by the inspection staff.

10.4.3 Warming of seed prior to grading

Seed crops in storage for longer than their dormancy period are usually kept at 3–4°C to prevent sprouting. There are, however, risks of inducing damage to the crop if it is handled or graded at this temperature. When potatoes were subject to a standard bruising test, those at 3°C had a bruise incidence of 83%, while those at 21°C suffered only 8% bruising (Wiant *et al.*, 1951). Potatoes dropped 1.1 m at 5°C had 77% splitting, while at 8°C splitting was reduced to 38% (McRae *et al.*, 1975). Warming of potatoes to 8–10°C prior to handling or grading is therefore commonly recommended for potatoes that are below this temperature.

If the whole store is being emptied, the whole crop can be allowed to rise in temperature prior to grading. However, since seed may be graded mid-storage to reduce the bottleneck in the spring or to satisfy orders throughout the storage period, the store is normally kept at 3–4°C and batches of crop warmed prior to grading. If the warming is done rapidly overnight, this tends to dry out the skin of tubers, making fingernail-type scuff damage more likely. Giving the skins time to regain their original moisture and elasticity can reduce this problem.

The crop not only needs to be warm to prevent damage, it also needs to be warm so that any wounds inflicted during grading are healed rapidly. Periderm formation in tubers develops more rapidly at warmer temperature (Ch1.5).

10.4.4 Where damage can occur

Seed, due to its small tuber size and therefore small mass, is less liable to suffer damage during grading than ware. This allows the use of vertically stacked screen graders or step graders, which require potatoes to be elevated to high level only for the tubers to cascade downwards through riddles to near ground level. Ware graders are designed to be flat to reduce such impact damage, which is possible due to the limited number of ware size grades.

Seed is graded dry, so there is considerable friction between tuber skins and surfaces where soil has built up and dried. In contrast, ware is often graded after it has been washed, so that tubers are slippery and any soil on the grader is soft and non-abrasive.

Damage is likely to occur in:

- Bulk hoppers if the exit is constricted.
- Cleaning coils (Fig. 10.8) if the steel is grooved by hard stones or small stones jam between the coils.
- Drops into the boot of an inclined elevator.
- Drops from the inclined elevator on to a belt or grader riddle (Fig. 10.9a).
- Any change of direction from one belt to another, particularly if belt guides are located beneath where tubers land on the belt (Fig. 10.9b).
- Increased drops on to a grader riddle when the top riddle has been removed (Fig. 10.9c).
- The tail of a grader screen due to soil build-up (Figs 10.9d and 10.10).
- Drops into the boot of a bag filler (Fig. 10.9e).
- Drops into the sack, box or big bag (Fig. 10.9e).

In the experience of the authors, where grading lines have been designed, supplied and installed by the one company, damage points are few. Problems arise when growers install unmatched pieces of equipment in the line or put in pieces of board to bridge gaps or divert tubers (Fig. 10.11). Even the softest piece of rubber foam padding will build up with soil, which may then dry to the consistency of sandpaper. The skin of tubers landing on such surfaces will then tear, giving the familiar 'thumbnail' marks or scuffing that become obvious 2 or 3 days later (Fig. 10.12). This not only spoils the appearance of the seed but provides access points for disease to invade.

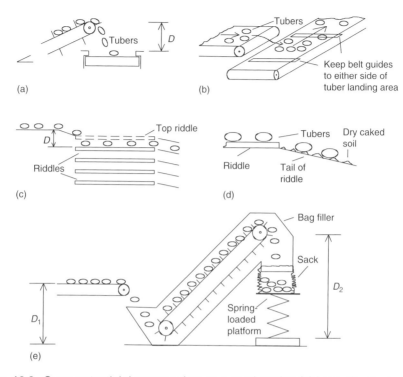

Fig. 10.9. Some potential damage points on a seed grader: (a) keep elevator incline <25° to keep drop height (*D*) low; (b) try to avoid 90° belt-to-belt transfers; (c) if remove top riddle to reduce number of tuber sizes, this increases drop *D*; (d) soil build-up on tail of riddles causes scuffing; (e) minimize drop to intake of bagger (*D*₁) and place empty sack on spring-loaded platform to minimize drop (*D*₂).

Fig. 10.10. Soil build-up on the tail of a mesh screen.

Fig. 10.11. Misuse of a board to try to reduce fall height.

Fig. 10.12. Tubers with 'fingernail' or scuff damage. (Courtesy of Potato Council, Oxford, UK.)

10.4.5 Split graded seed

Where uniformity in size of tubers in the daughter crop is critical, cup-type potato planters (Fig. 10.13) are commonly used to ensure seed is planted at regular spacings. If the cups of the planter are much larger than the smallest tubers in the batch

Fig. 10.13. Planter cups which deliver seed tubers to drill.

of seed, two tubers can be carried by the one cup and planted together in the drill. Alternatively, if the cup is smaller than the largest seed in the batch, the tuber can fall off the cup giving a miss. The results are doubles and misses, leading to a variable-sized daughter crop.

Split grading of seed, where seed is subdivided, for example into two fractions, one 35–45 mm the other 45 55 mm, allows planter cups to be sized closer to the size of the batch of seed used. Doubles and misses are minimized and a more uniform crop is obtained. For the seed supplier, it may mean having to grade the seed, take out the fraction the customer wants and put back into store the remaining fraction. This means warming the crop to be graded, putting the fraction to be sold into bags or boxes, allowing them and the remaining fraction to wound heal, then returning the unsold fraction of warm tubers back to store where they will be cooled back down to the temperature of the store.

10.4.6 Seed potato classification for specific diseases

In the UK, under the Seed Potatoes Regulations, the government sets the percentage tolerances for disease, skin blemishes and dirt (DEFRA, 2006). These vary depending on whether the seed is Pre-basic, Basic, Super Elite, Elite, A or A/S. These are shown in Table 10.1. This provides a quality assurance standard for customers, so that they know what quality of seed to expect. The tolerance for Group I diseases or pests is nil.

The individual, group and collective group tolerances work as follows. In Group III for Pre-basic stocks, for example, individual tolerances of up to 1.0% of

Table 10.1. Individual tolerances (%) and group tolerances (%) for Pre-basic, Basic, Super Elite (SE), Elite (E), A and A/S stocks of potatoes. (DEFRA, 2006.)

Specified diseases or pests; damage or defects	Pre-basic 1–4 stocks			Basic seed potatoes and equivalent SE, E, A and A/S		
	Individual tolerances	Group tolerances	Allowable % surface cover	Individual tolerances	Group tolerances	Allowable % surface cover
Group II						
Blight (*Phytophthora infestans*)	0.2	0.2		0.5	0.5	
Blackleg (*Erwinia*, Ecc and/or Eca)	nil			0.5		
Gangrene (*Phoma* spp.), dry rot (*Fusarium* spp.), wet rots	0.2			0.5		
Frost-damaged tubers	0.2			0.5		
Group III						
Skin spot (*Polyscytalum pustulans*)	0.2	5.0		0.5 (2.0*)	4.0 (5.0*)	12.5
Black scurf (*Rhizoctonia solani*)	1.0		12.5	3.0		12.5 (25.0*)
Common scab (*Streptomyces* spp.)	5.0		33.0	4.0 (5.0*)		25.0 (33.0*)
Powdery scab (*Spongospora subterranea*)	1.0		12.5	3.0		12.5
Group IV						
External blemishes or tubers other than diseased tubers whose shape is atypical for the variety	1.0			2.0	2.0	
Superficial necrosis caused by strains of potato virus Y	nil			0.1		
Group V						
Dirt or other extraneous matter	1.0			1.0		
Group II, III, IV collective tolerances		5.0			4.0 (5.0*)	

*Tolerances to be applied in the case of class A seed potatoes.

Ecc, *Erwinia carotovora* subsp. *carotovora*; Eca, *Erwinia carotovora* subsp. *atroseptica*.

The table excludes the Group I diseases of wart disease (*Synchytrium endobioticum*), eelworm (*Ditylenchus destructor*), potato cyst nematode (*Globodera* spp. infesting potatoes), and diseases and pests not established in the UK.

black scurf alone or 5.0% of common scab alone can be tolerated, but they cannot jointly exceed a maximum of 5.0%. In the same way, a tolerance in Basic stocks in Group III of 4.0% is acceptable as is a tolerance of 2.0% in Group IV, but the collective tolerance of Groups II, III and IV must not exceed 4.0%.

10.4.7 Seed chemical treatments

In a survey carried out in the UK (Heywood *et al.*, 2006), 39% of seed was treated with chemicals into store or prior to dispatch. Methods of chemical application into store are dealt with in Ch2.5. Systems of treatment of seed out of store are similar in design, but it is even more important that treated tubers should be thoroughly dried before dispatch. For seed, unlike ware, chemicals in the form of dusts can be applied.

10.4.8 Packaging of seed

Packaging for seed has a number of essential requirements. It should:

- Allow air exchange between seed and its surroundings to prevent condensation forming on the potatoes and an accumulation of carbon dioxide.
- Protect the seed from damage during transport.
- Be free from any disease spores or dirt that could contaminate the new seed.
- Be capable of being sealed to ensure the contents are not tampered with following inspection.

Within these constraints, many forms of packaging are available. The one selected will take into account whether the:

- Distance of travel is such that the cost of returning the packaging exceeds its value or is excessive.
- Packaging packs down, or nests, to reduce the cost of its return.
- Seed quality regulations allow re-use of packaging.
- Packaging can be re-used by the client to save it being returned.

Large 1-t (900 mm × 900 mm × 1800 mm) and 1.3-t (970 mm × 970 mm × 1950 mm) polypropylene bags with ventilated sides are now probably the most popular packaging used for seed transport. These collapse down into a small space when empty. They are usually only used once for seed to prevent contamination of new seed by old.

If boxes are used for deliveries, clients commonly require the boxes to be new, so that they can use them to replenish their own stock of storage boxes. Bulk transporters, commonly used for grain transport, are sometimes used for delivering seed direct to the field.

10.4.9 Dust control

As seed is almost always graded dry, dust in the grading area and on the floor of the store passageways is a major problem. Dust is categorized by particle size as inspirable and respirable. The inspirable dust is filtered out in the nose and throat, while the respirable dust can enter the lungs. The latter is therefore the more likely to lead to long-term respiratory illness. The Occupational Exposure Standard (OES) for organic dust (HSE, 2005c) is $10\,mg/m^3$ time-weighted average for an 8-h shift. A study of personal exposure to total inspirable dust in six on-farm grading areas recorded dust levels averaging $28\,mg/m^3$ (range 3 to $148\,mg/m^3$), well above the OES limit (Robertson, 1993). The OES for the respirable fraction of the dust is $5\,mg/m^3$. Only 9% of readings recorded exceeded this limit. Highest personal exposure to dust occurred at inspection roller tables and at bagging-off points. Dust from the potato grading process and dispersed into the air by forklift movement produces a most unpleasant environment in which to work, and one which commonly exceeds acceptable OES levels.

Dust control equipment, in the form of air extraction hoods, can be fitted wherever tubers are subject to drops during their passage along the grading line. Locations include the box tipper, transfer points from one conveyor to the next and where bags are filled. The air extracted is fed to a dust filter to remove the dust and allow the air to be recycled. Dust extraction systems help to reduce dust levels but rarely are fully effective. A cabin fitted with windows installed over the roller tables, supplied with a downward flow of filtered warm air, can improve the atmosphere for inspection staff. For full dust control, a steam or mist generator to moisten tubers at the start of the grading process should be fitted. This approach may increase the likelihood of disease development on the tubers, so the graded seed should be thoroughly dried using positive ventilation to prevent any disease developing.

For short periods of working, face respirators designed for dust can be used. These must fit the face well, be kite-marked to a recognized standard and have an exhaust valve to ease breathing out and to prevent spectacles steaming up.

10.4.10 Access to stores for crop to grade

Whenever seed is graded, whether prior to dispatch or to prevent spring bottlenecks, access to store is required both to remove seed for grading and to return seed that has been graded. If store temperatures are 3–4°C, the sequence in operation is optimally as follows:

- Warm seed to ≥10°C prior to grading, preferably within the store airspace.
- Grade seed and put into porous bags or boxes.
- Ventilate seed at ≥10°C for 3 days prior to dispatch.
- Dispatch seed at ≥10°C unless it is to be transported in refrigerated containers.
- Ventilate seed at ≥10°C for 3 days prior to return to store.
- Either pre-cool prior to returning to store or force-ventilate once in store to prevent condensation forming on neighbouring cold potatoes.

Every time a store is opened to access seed there is the possibility of allowing warm humid air into store, resulting in condensation forming on the crop inside (Box 10.1).

Box 10.1. Strip curtains only partly reduce air exchange between store and outside

Plastic strip curtains are often used to keep warm ambient air out of refrigerated stores while allowing forklift access to fetch potatoes for grading. This is unlikely to be very effective. With the store door closed (Fig. B.10.1a), the store is full of cold air that is denser than the warm ambient air outside. If the door is opened (Fig. B.10.1b), even with the strip curtains in place, the dense cold air will flow out of store and the less dense, warm ambient air will enter. If the crop temperature is less than the dew-point temperature of the ambient air now within the building, the potatoes in store, particularly in the top boxes, will become wet with condensation.

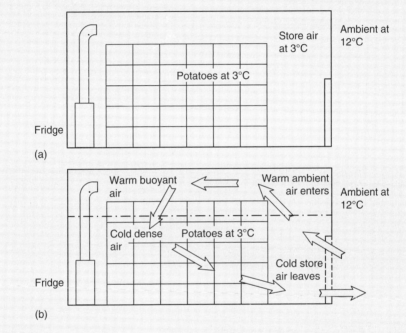

Fig. B.10.1. Cold store with: (a) main door tight closed; (b) main door open but plastic strip door present.

This also puts an additional burden on the cooling system, as the air within the store has to be cooled back down to the set-point temperature. The best routine is to open the store when ambient temperatures are as near to the store temperature as possible and to remove the day's grading over as short a period as possible. In the UK, this is early in the morning or late in the day.

This problem of doors being open is worst with large seed suppliers, where high-volume grading is taking place and a number of stores are being accessed simultaneously. There is a tendency for doors to be kept open all day so as not to slow forklift operation. If such a system is practised, it may be better to keep seed at a warmer temperature so that it remains dry rather than trying to keep it cool, where repeated wetting through condensation is likely. However, if the weather is warm, this can lead to early dormancy break and sprout development.

10.4.11 Dispatch of seed

Unless seed is being dispatched in refrigerated transport, it should be dispatched at a temperature that is near to or above the average ambient temperatures for the part of the country through which it will pass. If the seed is dispatched at 3–4°C and travels through humid warm conditions, the seed will arrive at its destination covered in condensation. If the bags or boxes are not examined on arrival, the seed can be stored wet for weeks prior to planting. This problem is particularly serious in transport by ship, as seed is usually travelling from cold to warmer climes, in conditions of very high RH (Fig. 10.14). Pre-warming the seed reduces this risk.

Seed also has to be protected from frost. The warmer seed is when it is dispatched, the longer will it take to cool to temperatures likely to cause frost damage. Where freezing conditions are likely, seed should be transported in sealed, insulated transport or wrapped in airtight insulation material to ensure ingress of ambient air is minimized.

A popular form of road transport is the Tautliner, or curtain-sided lorry. These have a flat bed and hard roof, but have canvas sides hanging from rails which open the whole length of the lorry. While very convenient for loading and

Fig. 10.14. Seed transport by ship. (Courtesy of S.G. Baker Ltd, Angus, UK.)

unloading, they are not airtight. When travelling, air will leak through the edges of the curtain and ventilate the load.

In frosty conditions, therefore, the pallets or boxes should be encased in insulation material, such as corrugated cardboard or fleece, and then sheeted with canvas or plastic tarpaulin to cocoon the load to prevent air ingress. As it is difficult to seal the pallet apertures of boxes, or 1-t or 20–25-kg bags on pallets, the canvas should be put over the floor of the flatbed prior to loading so that it can be pulled over the load to meet at the top with the canvas from the other side (Fig. 10.15). Some specialist transport companies have an insulated cocoon, which both seals and insulates the loads when conditions are frosty.

The other common form of transport is the flatbed lorry. When double sheeted, they provide better sealing than the Tautliner (Pringle and Thompson, 1987). An insulating layer such as corrugated cardboard, plastic fleece, bubble wrap or straw can be used as the insulating medium between the two sheets of canvas for the top and sides of the load. In severe weather conditions, of temperatures below −5°C, seed should not be transported in uninsulated lorries. Above this temperature, the chance of frosting potatoes on seed transported on journeys of less than 36 h is low.

Seed can also be transported loose in bulker lorries for just-in-time direct filling of planters in the field. Since planting is unlikely to be carried out in periods of frost, insulation is not required.

Containers are sometimes used for the transport of seed. The cheapest containers are simply steel boxes holding around 20 t (Fig. 10.16). How airtight

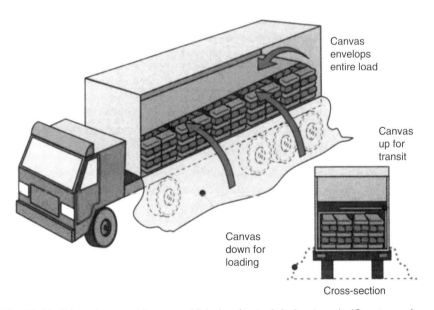

Canvas envelops entire load

Canvas up for transit

Canvas down for loading

Cross-section

Fig. 10.15. Wrapping seed to prevent it being frosted during transit. (Courtesy of Potato Council, Oxford, UK.)

Fig. 10.16. Loading big bags into a steel container. (Courtesy of S.G. Baker Ltd, Angus, UK.)

they are depends on how much damage they have sustained in use. As steel is a good conductor of heat, seed tubers are at risk from frost damage. So long as the journey is short, the seed has been well ventilated and is warm and dry, and the box is lined with corrugated cardboard, they are suitable for short journeys. Physical damage can be incurred on the tubers if drop heights during loading are excessive. For longer journeys refrigerated containers will provide a more reliable form of transport.

10.4.12 Refrigerated seed transport

If seed is shipped from seed producer to customer in refrigerated transport, problems of the crop getting wet through condensation are minimized, as the container or lorry will be sealed. Problems of the seed becoming wet only occur once the doors are opened at the destination. If the crop is below the dew-point temperature of the ambient air, condensation will occur. To avoid this, seed should be allowed to warm up during the transport period, so that the seed approaches the temperature of the ambient air at the seed's destination. This is obviously more of a problem in humid countries than in ones that have low RH. In cold weather such as in the winter in the northern USA it may be necessary to warm up the lorry prior to loading it with seed.

10.5   Chitting, or Pre-sprouting, of Seed

While manipulation of chronological age is concerned mainly with increasing the proportion of small or large tubers in a sample (Ch10.3), chitting (i.e. pre-sprouting) of seed is designed to produce a crop which bulks early and can be harvested before blemish diseases become established. While the size of investment required in buildings and equipment, and the high cost of labour in the UK, have reduced the number of growers chitting seed, a number still do so using trays, chitting sheds illuminated with artificial lights or naturally illuminated greenhouses.

Tubers should have a short (<5 mm), strongly attached green sprout (Fig. 10.17). The sprouts should not be white and long, as these are easily knocked off in the planter. To achieve such sprouts, every tuber has to be exposed to either natural or artificial light and kept at temperatures which initiate sprouting but do not result in excessive growth. The light, combined with low chitting-house temperature, suppresses growth (Krijthe, 1948).

Pre-sprouting seed does not always have a consistent outcome. For instance, the dry matter of a crop grown from pre-sprouted seed increased in one season out of four for cv. Maris Piper and in two seasons out of four for cv. Rooster, yet in the other two seasons there was no difference (Burke *et al.*, 2005).

10.5.1   Chitting in greenhouses

Traditionally, the chitting of seed in trays was carried out in greenhouses (Fig. 10.18), with high light levels partially compensating for the lack of temperature control. A 3-kW blower heating system for every 10 t of seed provides sufficient frost protection for the south-east of England. In trials carried out by one of the authors using

Fig. 10.17. Sprouts should be green and short, not white and long.

Fig. 10.18. Glasshouse chitting shed.

such a system, the temperature of the seed was maintained between 2.5 and 7.5°C for approximately 50% of the time with the remaining 50% being above 7.5°C (Bishop and Maunder, 1980). Since there is little control of temperature, sprouting can become excessive if planting is delayed by wet weather. Greenhouses are now rarely used.

10.5.2   Insulated chitting sheds with artificial light

An insulated store with lights can provide better temperature control than greenhouses. The artificial fluorescent strip lights (65 W/t) are hung on the side of, or between, stacks of potato trays and moved from one stack of trays to the next. As the capital cost of the lighting is so high, it is sensible to keep the lights on 24 h/day once sprouts start to appear and to move them twice a day to ensure that all the trays get their share of light. If lights are moved twice per day, one hanging florescent tube is needed per eight stacks of trays (FEC, 1985). To simplify moving the lights, they are either hooked on to the top tray or suspended on wires or tracking. The store is fitted with a recirculation ventilation system rated at 0.04–0.05 m³/s/t of seed, to ensure that the temperature within the chitting shed is uniform. In the trials mentioned above, the temperature in the insulated shed was between 2.5 and 7.5°C for 84% of the time, with the remaining 16% being above 7.5°C (Bishop and Maunder, 1980). The heat produced by the lights provides frost protection, but also makes cooling the shed more difficult. A further level of sophistication is to have a refrigerated store with lights, which from the same trial kept the seed within 2.5–7.5°C for the whole period.

The time required to fill trays (three per 50 kg) is estimated at 120 person-minutes per tonne and the rate for filling the planter from trays can be up to 20 person-minutes per tonne in comparison with under 5 person-minutes per tonne from bulk bags (Bishop and Maunder, 1980). The filling time can be less than stated above if well mechanized.

10.5.3 Mini chitting

Although producing green sprouted seed gives the greatest yield and early harvest advantage, in many cases having the eyes open just prior to planting can suffice. Refrigerated storage and a good recirculation ventilation system to ensure potatoes are at a uniform temperature are required to achieve this. This avoids the high labour input of putting the seed into trays, the cost of the lights, and the problem of sprouts being broken off in the planter.

10.5.4 Alternatives to trays

Alternative, less labour-intensive systems than tray systems have been invented over the years. The Blackburn crate had a zigzag arrangement of mesh so that potatoes could be fed into the crate mechanically using a tractor front-end loader fitted with a box rotator. The spaces between the hanging mesh allowed light to get to most of the tubers. While these were reasonably successful, if sprouts grew too much the tubers became trapped between the mesh and would not come out. A similar system using plastic hanging nets suffered from the same problem. Every few years a new 'solution' to mechanized chitting systems is produced, but none but tray systems seems to have endured.

10.6 Chitting Store for 300 t of Seed

A spray foam-insulated chitting shed is shown in Fig. 10.19. The shed contains 300 t of seed held in trays 750 mm × 450 mm × 165 mm, with three trays holding 50 kg. There are 50 trays per pallet, so the loaded pallet has dimensions of 1.5 m × 1.9 m × 1.8 m. Pallets are stacked two high to give a stack of trays 3.6 m high. Gaps, 500 mm wide, are left on all four sides of each pallet to allow inspection of sprouts and to take the lights. Two 1.8-m-long fluorescent lights, each rated at 70 W, are suspended one above the other in the gaps to provide the light (Fig. 10.20). This provides well above the 65–70 W of light per tonne stated above. Lights are on continuously. The florescent tubes should be protected by perspex casings to prevent damage to tubes and the risk of injury to staff.

Fig. 10.19. Chitting shed for 300 t of seed. (Hamilton and Sons, East Lothian, UK.)

Fig. 10.20. Chitting light attached to the stack of trays. (Hamilton and Sons, East Lothian, UK.)

Fig. 10.21. Fridge unit for 300-t chitting shed. (Hamilton and Sons, East Lothian, UK.)

The tubers were put into trays on 17 January (H. Hamilton, West Lothian, Scotland, 2005, personal communication) and were planted on 7 April. They were therefore kept in the chitting shed for a period of 11.4 weeks. A refrigeration system (Fig. 10.21) is located at one end of the building, which serves to keep the temperature between 3 and 4°C.

10.7 Communication Between Seed Producer and Customer

Regardless of the efforts taken by growers to produce clean seed, there will always be some disease and blemishes present. Often the only evidence of this is in the rejects' skip, after the seed has been graded. If customers are informed of the history of the crop, this will help them decide how best to look after the seed when it arrives and where best to plant it. Information about the batch should be sent to the customer along with the seed.

10.7.1 Report on washed seed samples prior to grading

Some of the better seed suppliers produce reports on washed samples of seed, carried out prior to grading (Pseedco Ltd, Perth, Scotland, 2007, company information sheets). The benefit of inspecting pre-graded material is that this includes material which will be rejected on the inspection line, but which will give a true reflection of potential problems that could develop following planting, such as rotting or blackleg. An example of a washed and pre-graded report is shown in Box 10.2. As well as

Box 10.2. Example of a Washed and Pre-Graded Report (SAC Ltd, Edinburgh, UK)

SAC

Surface disease assessment (% in each category)

Severity of disease	Common scab	Powdery scab	Black scurf	Silver scurf	Black dot	Skin spot
0	92	62	100	74	100	100
1	2	16	0	6	0	0
2	6	16	0	12	0	0
3	0	4	0	4	0	0
4	0	2	0	4	0	0

Where 0 = nil, 1 = trace (1%), 2 = trace to 1/16th (1–6%), 3 = 1/16th to 1/8th (6–12%), 4 = >1/8th (>12%).

Eye-plug test (% of plugs)

Disease	% plugs
Black scurf	14
Black dot	0
Silver scurf	4
Skin spot	0

Rotting diseases and disorders (% of tubers)

Rotting disease	% tubers	Disorders	% tubers
Blight	0	Bruising	2
Dry rot	0	Growth cracks	0
Gangrene	0	Pest damage	0
Soft rot	0	Superficial rots	0
Other	0	Green	6

Mechanical damage (% of tubers)

Mechanical damage	% tubers
Loose skin	30
Cuts	2
Scuffs	16
Wounds	10
Thumbnails	0

details of variety, size, field name, producer and generation, the report indicates the percentage of:

- Disease (common scab, powdery scab, black scurf, silver scurf, black dot, skin spot).
- Damage-induced soft rotting.
- Blight-induced soft rotting.
- Damage-induced dry rot or gangrene.
- Blackleg-induced rots.
- Mechanical damage.
- Misshapes.

If the customer knows that there is a trace of blackleg in the seed, he knows not to grow it on land that could be subject to flooding or where irrigation application rates are uneven. Similarly, if there is some silver scurf present, he knows that he should harvest the subsequent crop early, ensure that it is dried and kept free from condensation during store loading and wound healing, and make certain the store is closed up within 3–4 days so that cooling can start as soon as possible.

10.7.1 Instructions delivered along with the seed

Documentation delivered along with the seed or with the invoice should advise customers to inspect the contents of boxes, bags or trucks on arrival and, if necessary, to dry the seed if it is moist. This should prevent bags being unloaded and left unopened in the back of a shed for weeks, with the associated risk of disease development should planting be delayed. The supplier should keep a sample of the seed dispatched so that he has evidence of the quality of the seed should a dispute arise (Ch12.15).

10.7.2 Temperature for planting seed

Experiments by Firman *et al.* (2004) have indicated that early planting of seed into cold soils produces more stems than would otherwise be expected. The effect is similar to that of storing seed at low temperature to encourage multiple sprouting and the production of small tubers. Optimally, seed should be planted into soils above 7°C so that growth is rapid and time to emergence is short.

10.8 Summary

Seed storage and preparation for dispatch have aspects that differ from the production of ware pre-pack or processing potatoes.

- Seed either needs boxes to separate out the numerous varieties and generations or a coordinated strategy so that one grower grows all the seed of one variety for a group.
- Seed storage requires ventilation with an evenly distributed stream of air following harvest to ensure that the heat generated from immature crops is removed, so that subsurface condensation is prevented and the crop is rapidly dried.

- Seed designed to maximize the production of large tubers or small tubers can now be produced to order using chronological age manipulation techniques.
- If access to seed stores for material for grading is required on a regular basis, door design or access management should aim to prevent warm humid air entering stores resulting in condensation on the stored crop.
- Warming of seed prior to grading is necessary if crops are stored at 3–4°C, both to minimize grader damage and to allow subsequent wound healing.
- As seed may spend many weeks after grading awaiting suitable planting conditions, wound healing along with ventilation to prevent subsurface condensation following grading are essential.
- Unless dispatched in refrigerated transport, seed should leave the grading area or store at a temperature above the likely dew-point temperature of the ambient air through which it will be travelling.
- Seed transported in refrigerated containers or lorries should be allowed to warm to above the dew-point temperature of the ambient air before opening doors, particularly in seed sent by sea to warm or tropical countries.
- Chitting of seed for ware production uses heat to initiate apical dominance, and light and cool storage to control subsequent sprout development.
- The use of pre-grading reports on washed samples, together with instructions stapled to seed deliveries, helps the customer with subsequent seed management, planting and crop husbandry.

11 Packhouse and Processing Facilities

Topics discussed in this chapter:

- Sourcing potatoes from field or store.
- Packing centrally or on-farm.
- Packhouse layout.
- Potato cleaning and washing equipment.
- Wash water treatment and disposal.
- Potato grading.
- Potato packing equipment.
- Bag stackers.
- Packaging alternatives.
- Partially processed products.
- Dispatch.

11.1 Cleaning and Sizing Prior to Packing

Potatoes arriving at a grading line, processor or packhouse may arrive from one or more sources. They may come from:

- The field in a tipping trailer, bulker lorry, boxes or bags.
- Bulk or box potato stores without being cleaned.
- Bulk stores, but via a cleaner/grader located on the farm.
- Box stores with on-farm cleaning/grading prior to dispatch using static, farm-owned equipment or mobile packhouse-owned systems.
- Dutch-type storage bins adjoining the grading area.
- Box or sack stores on site.

Potatoes for processing may in addition be washed on the farm to minimize the amount of soil for disposal as a waste from the factory.

Whether grading should be carried out on-farm, in a central grading facility, in a packhouse or at the processing factory depends on a number of considerations.

11.1.1 Minimizing contamination while maintaining bloom

When potatoes are harvested, adhering soil, haulm, damaged and diseased tubers (known as brock or cull potatoes) will be present along with marketable tubers. As tubers dry, increasing amounts of soil fall off the tubers whenever they are conveyed or handled. If tubers are stored for a period of months, a proportion will become diseased and unmarketable and tubers will shrink in diameter by a few millimetres due to evaporative moisture loss. Since packhouses want only marketable material, cleaning and grading on the farm avoids transporting soil and brock to the packhouse only for it to be returned to the farm for disposal. However, if on-farm grading equipment is old and worn, it can cause damage or loss of bloom to the crop. If this is the case the packhouse may prefer to accept the soil and brock along with ungraded potatoes, to ensure that the skins of the potatoes keep their bloom and suffer minimal damage. This waste material is likely to be considered as 'trade waste', which has then to be disposed of at a licensed waste treatment facility.

11.1.2 Washing potatoes on farm

Washing of potatoes on farm is only carried out if the crop is to be processed into French fries, crisps or for ready meals within 5–7 days, as disease can rapidly multiply on wet skins. Washing will remove any adhering soil and, if carried out in conjunction with grading, will allow simultaneous removal of brock and separation of tubers into different size grades. The wash water has then to be passed through a wastewater treatment plant to remove soil and reduce biological oxygen demand (BOD) to an acceptable level for either discharge to watercourses or reuse. Alternatively it can be applied to land using low-volume irrigation. The soil filters out inorganic soil particles and provides microorganisms which break down the sugars, starch and organic matter present. It is usually easier and cheaper to discharge dirty water to farmland than to treat it sufficiently to discharge it to sewer in built-up areas.

Brock potatoes can be fed to livestock on the farm, sold for stock feed or converted to starch if facilities are available.

11.1.3 Handling of cold potatoes

Potatoes removed directly from cold store will become wet with condensation if their temperature is below the dew-point temperature of the ambient air. Where potatoes are being washed prior to immediate use or sale, this is of no consequence.

In contrast, where potatoes are to be kept in bags or boxes for some weeks after removal from store, any wetting can result in disease initiation. Ware that is to be sold dry and which is put in Kraft bags some weeks prior to use is particularly vulnerable.

Potatoes removed from cold store at 3°C are likely to suffer damage or bruising during handling or transport. To prevent damage, crops should be warmed to at least 8°C before they are handled. To prevent condensation after they leave store, they should be warmed to a temperature above that of the dew-point temperature of the ambient air. A warming arrangement should preferably be designed into the store, as removal of cold crops from store to a warming area will result in the crop becoming wet before warming can begin (Ch9.12).

11.2 On-farm versus Off-farm Cleaning and Sizing

The first decision for grading and packing potatoes is whether this should be done on the farm or in a central facility. The central facility may be cooperatively owned, a private stand-alone company or part of a retail potato marketing group.

The advantages and disadvantages of on-farm or off-farm grading are considered in Table 11.1.

11.3 Packhouse Layout

Packhouses are divided into two areas: (i) a dirty area at intake where potatoes contaminated with soil and haulm enter the packhouse for cleaning; and (ii) a clean area where the cleaned material is sized, inspected and packed. Staff in the dirty area usually wear coveralls, while staff in the clean end wear white coats and white hats, and have to wash hands and change clothes before work. The sampling of material for quality control is carried out at reception, at critical control points during the process and prior to dispatch.

A packhouse may include some or all of the following:

- Reception for trailers, lorries, boxes and bags.
- Dry cleaning equipment to remove stones, soil and haulm.
- Washing plant followed by tuber drying equipment.
- Grading equipment to split material into different sizes.
- Inspection tables for examination of material prior to packing.
- Packaging machinery to put material into bags or boxes.
- Quality control laboratory and shelf-life assessment room.
- Pallet stacking equipment for stacking packs or bags on to pallets.
- Cool or chilled areas to hold produce prior to dispatch.
- Storage space for packaging material.
- Printers to produce labels with daily packing codes and display until dates.
- Offices, meeting rooms, staff canteens, changing accommodation and wash rooms.

Table 11.1. Consequences from on-farm or centralized grading and packing.

On-farm grading	Centralized grading or packing
Avoids transport of soil and brock to and from packhouse	Need either to transport soil and brock back to farms if permitted or to send them to landfill or a composting facility
Relatively easy to dispose of dirty wash water and soil if washing carried out	Will require wastewater treatment system, if washing carried out on site and plant is in a built-up area
Old or low-technology machinery may spoil bloom and damage potatoes	Can invest in high-technology, high-throughput, low-damage grading and packaging machinery
Provides jobs for farm staff during winter	If grading is seasonal, staff may be under-employed during periods when grading is not taking place. This can be avoided if casual labour is readily available
Use of family labour can keep costs down	Labour will mostly be hired, at or above minimum wage, with rest rooms and transport for staff required
Gain value from existing farm buildings if they meet supermarket standards	Costly investment in new buildings in possibly high rateable value area
Can take out over- and undersize material to sell locally for specialist high-value markets	Can establish specialist lines to use the over- and undersize material (e.g. punnets, bakers)
Growers with good reputation may get premium prices	Require intake monitoring system that ensures quality material is rewarded and poor material is penalized
If rural labour is scarce, seasonal staff may be difficult to obtain unless working conditions are pleasant	Town location may ease the getting of staff or be attractive to casual labour
Quality assurance and traceability may cost more per tonne due to small scale of operation	Can invest in high-technology computer-based quality assurance and traceability systems
Facility may be too small for supermarkets to deal with	Large facilities can supply 365 days per year and can buy in material to bridge gaps in crop availability and supply Christmas rush
Farmer has to market produce as well as grow it	Large facility can have its own marketing manager and agronomists

Conveyors or boxes are used to move the produce and waste between these, with skips or boxes placed by the machinery to accept soil, discarded potatoes, stones and haulm for disposal.

The packhouse lines illustrated in Fig. 11.1a and b are designed for packing potatoes into polythene bags, over-wrapped 600 mm × 400 mm × 300 mm free-flow trays, or boxes for bakers. The free-flow trays hold produce loose, allowing customers to select the potatoes they want. In the supermarket, rolls of polythene bags are placed by the free-flow trays, so that customers can fill the bags with the potatoes they select. In Fig. 11.1a and b, all produce movement is from top to bottom.

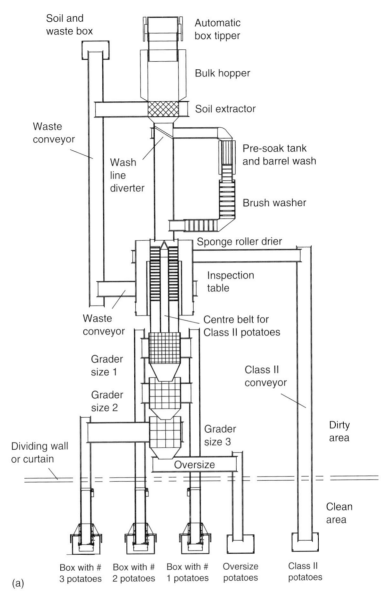

Soil and waste box

Automatic box tipper

Bulk hopper

Soil extractor

Waste conveyor

Wash line diverter

Pre-soak tank and barrel wash

Brush washer

Sponge roller drier

Inspection table

Waste conveyor

Centre belt for Class II potatoes

Grader size 1

Class II conveyor

Grader size 2

Grader size 3

Dirty area

Dividing wall or curtain

Oversize

Clean area

(a)

| Box with # 3 potatoes | Box with # 2 potatoes | Box with # 1 potatoes | Oversize potatoes | Class II potatoes |

Fig. 11 1. (a) Packhouse cleaning and grading line; (b) packhouse packing line. (Redrawn from plan supplied by R. Balls, consultant, Bedford, UK.)

11.3.1 Cleaning and grading line

The first stage in the packhouse system (Fig. 11.1a), the cleaning and grading line, has potatoes delivered to the packhouse in 1-t boxes, automatically tipped into a bulk hopper, then passed over a dry cleaning system to take out soil. The potatoes can be either delivered by conveyor to the inspection tables directly, or diverted via a pre-soak tank, barrel washer, brush washer and sponge drier, if potatoes are to be washed.

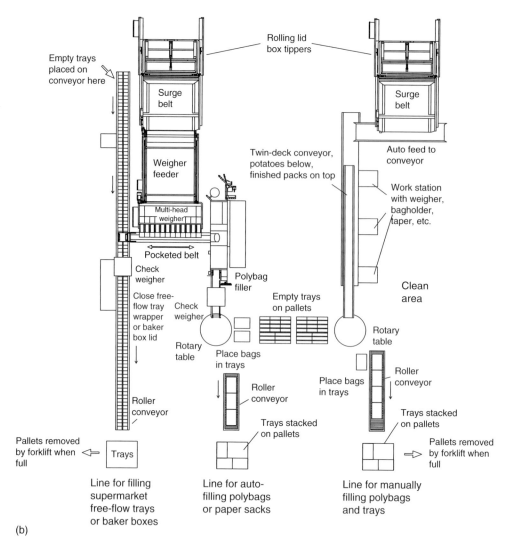

Fig. 11.1. *Continued.*

The potatoes go over an inspection table, where inspection staff remove rots, mis-shapes, blemished or green potatoes and transfer lesser-quality Class II potatoes to the centre belt. The Class I potatoes pass over three grader modules with square mesh screens, which grade the tubers from small to large. The size grades are diverted into three sizes (denoted by # on Fig. 11.1a) of grades, to fill separate 1-t interim holding boxes. The Class II potatoes avoid the grading system and are conveyed to their own box. A fourth oversize grade may go for catering or for processing.

The waste conveyor carrying soil, stones and reject potatoes delivers the waste to a box or skip situated beside the intake. A dividing wall, or curtain, separates the dirty operations of soil extraction, inspection and grading from the clean area where clean, sized crop is delivered into separate boxes. Once past the curtain, there is only clean, marketable produce.

11.3.2   Packing line

The second stage of the packhouse system (Fig. 11.1b) is all in the clean area. The clean, sized produce from the cleaning and grading line is delivered from its interim holding box to box tippers, fitted with rolling lids to minimize the height of drop when the 1-t boxes are being tipped. The surge belts ensure an even flow of potatoes is delivered from the tippers.

In the left-hand line for filling supermarket free-flow trays or baker boxes, the weigher feeder supplies the multi-head weigher, typically having 12 heads, which delivers accurately weighed lots into the pockets of the pocketed belt cross conveyor. The cross conveyor is timed so that each preset pocket lot is delivered into a lined supermarket free-flow tray passing below on its roller conveyor. As the free-flow tray capacity is bigger than the pocket lots, the system is set to stop the tray conveyor to receive a number of pockets per tray; e.g. a 10-kg tray would receive 5 × 2-kg pockets. The lining sheet on filled free-flow trays, or lid flaps on the baker box, is closed over the top, check weighed, possibly checked for metal contamination, labelled, and stacked on to pallets. The tray conveyor has a free-rolling end section, which allows the boxes or trays to accumulate and make palletizing independent of conveyor speed.

In the right-hand section of the left-hand line (Fig. 11.b), shown for filling polythene bags or paper sacks, the pocketed conveyor is reversed and discharges into plastic or paper bags. There are many variants of bag filling systems, ranging from bags manually placed on to the holders, to equipment which forms the bags from flat rolls of pre-printed plastic. A labeller automatically prints or adds a stick-on label to the full bags. These show contents, weight, pack and sell by date and barcode. The discharged full bags run along a flat belt conveyor and through a check weigher and metal detector. Any underweight bags or ones containing metal are removed. The bags drop on to a rotary table, where they are packed into trays and palletized. In very large installations the filling of bags into trays and tray palletizing is fully automated.

The right-hand box tipper and packing line (Fig. 11.b) depicts a very basic system used for manually filling polythene or paper bags and trays, or for filling bakers into various packs. This line is very flexible and can deal with almost any style or weight of pack but does not have the automatic filling that the left-hand line provides. As with the left-hand packing line, it starts with a box tipper followed by a surge belt feeding a cross conveyor. This delivers a steady stream of tubers one layer deep on to the main belt conveyor from which operators pick tubers and bag or box them, the bag or box being placed on a weigh scale. After labelling the rotary table conveys the polythene bags of potatoes for packing into customers' plastic trays, usually 600 mm × 400 mm × 300 mm if the packs are small, or directly on to the pallet if the packs are large.

For all packhouses, whatever the level of sophistication, there are a number of questions that need to be addressed. These are summarized in Table 11.2.

Table 11.2. Basic packhouse considerations.

Packhouse operation	Questions to ask
Reception	Is the crop always received in the same way (i.e. bulk, boxes, similar sized containers, etc.)?
	Is incoming material to be sampled routinely?
	Are facilities needed for incoming drivers to rest, wash and eat?
Cleaning	Is soil removed in a dry state before tubers are washed, essential if the wash water is to be kept clean?
	How is the reject material taken away?
	How is soil in the wash water kept in suspension?
	How is the contaminated wash water treated?
	Can the throughput be altered so that pickers on the roller tables or belts are not overburdened?
Grading	What quality control is required?
	Where does the second grade product go?
	Is individual operator accountability required?
Packing	How are the packing materials supplied to the operators and machines?
	Will many different packing configurations be required?
Dispatch	Does the end product need to be stored in more than one way?
	Does the area need to be chilled as part of a cool chain?
	Are all dispatch vehicles the same height and size?
	Is a final quality check required?

In the UK, packhouse operation comes under regulation (EC) No. 852/2004, Hygiene of Foodstuffs (EC, 2004), which is monitored by the local authority's department of environmental health.

The legislation for packhouses covers aspects of hygiene, the Hazard Analysis and Critical Control Point (HACCP) safe food production system and operator safety from machines, noise, dust and excessive working hours. While HACCP is not mandatory for packing potatoes, operators of such primary production facilities are encouraged to follow HACCP procedures as far as is possible (EC, 2004). Other legislation covers wastewater discharge consents, disposal of solid wastes and any environmental impact on the surrounding area. The clients of the packhouse may also have their own additional requirements, which should be addressed.

11.4 Components of the Packhouse or Processor Intake

Some of the equipment used in packhouses or the intake of a processing plant has been discussed in Ch2, Harvesting and store loading, and Ch10, Seed grading and preparation for planting. Further details of equipment may be found in these chapters. Quality assurance aspects are dealt with in Ch12.

11.4.1 Intake

The primary objective at intake is to ensure that the cleanest sample of potatoes possible enters the packhouse or processing plant. Soil, stones and waste organic matter can be put back to the land if separated from the potatoes at the farm. Once this material is brought to the packhouse or plant, in the UK they come under commercial waste regulations and have either to be processed into a useful product or treated (e.g. separated, composted, etc.) prior to disposal at a licensed waste disposal site. The potatoes being delivered to the cleaning and grading line will arrive in bulker lorries with conveyor discharge, tipping lorries or trailers, boxes or bags. In the UK prepack potatoes will normally arrive in boxes or bags to minimize damage.

11.4.2 Bulkers and tipping trailers and lorries

The bulker lorry, fitted with a bottom conveyor, discharges potatoes at a rate that the conveyor supplying the cleaner can accept. In contrast, a tipping lorry or trailer can rapidly discharge its load into a bulk reception hopper and get on the road again to collect more crop. The reception hopper acts as buffer storage. While this minimizes the waiting time for the tipping lorry or trailer, it can lead to potato damage if the drop to the hopper floor is too high. The size and dimensions of the hopper must match the trailer or lorry height when tipped and provide a reserve of tubers so that the cleaning line can still work even with an intermittent supply. Typical capacities are between 3 and 15 t of crop (Figs 2.5 and 2.7). The hopper is boat-shaped, having a long and preferably wide moving belt in the base. This belt can either be at a slight angle to the horizontal along its full length to feed on to the line or be in two parts, with a section of horizontal belt followed by another inclined upwards.

The supply from the bulk hopper must be readily adjustable so as to provide an optimum rate of supply to the sorting, sizing and packing operation. The rate of discharge can be adjusted using a manually operated variable speed motor but often proximity sensors or electronic eyes are used at buffer points (Fig. 11.2) to start and stop conveyors instead.

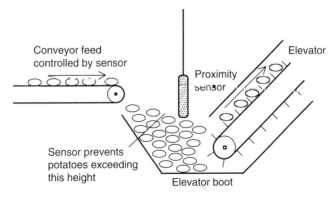

Fig. 11.2. Use of a proximity sensor to control flow of potatoes into the intake boot of an elevator.

A further type of hopper or modification is where the trailer can be backed partly along the hopper with wheel tracks on either side and can empty or finish emptying by drawing forward.

11.4.3 Box tipper unloaders

There are two types of box tipping systems: one which rapidly tips the potatoes into a hopper with a slow-moving discharge, the other which slowly tips the contents of the box on to a moving conveyor belt (Fig. 11.3). In the first, the reception hopper acts as a buffer, so that there is time for the forklift truck to fetch another box and place it in the tipper before the first box is empty.

In the second, either there will be a gap in supply as boxes are changed over, or two tippers are used. While one is discharging potatoes, the empty box in the other can be replaced with a full box. In the UK, these are gradually replacing the buffer hopper type, as they have been found to do less damage to the crop. Their tipping speed is controlled using a proximity switch to ensure a uniform discharge rate to the cleaning and washing equipment. The forklift driver is made aware that there is an empty box by a flashing light or claxon.

Where possible box tippers discharge at a high level, so that the crop does not need to be elevated to discharge on to the grader or inspection tables. The result is a flat grading system, with a minimum of drops that could otherwise do damage.

Some box unloading systems now have the facility to read a barcode or a RFID (radio-frequency identity device), which is discussed further in Ch12.

Fig. 11.3. Box tipper discharging potatoes delivered from a farm store. (Taypack, Inchture, Dundee, UK.)

11.4.4 Other unloading systems

Box tippers or rotating buckets fitted to a forklift truck or tractor avoid the need for a stationary tipper. Some need the base of the pallet box to be closed to prevent the pallet falling off the forks when rotated. Others have an arm above the top of the box to prevent it falling. The discharge height will be less controlled and if a bucket is used its edge will cut a small but significant percentage of tubers, which will have to be removed during inspection. In addition, equipment mounted on tractor front loaders is awkward to manoeuvre.

11.5 Dry Cleaning Equipment

If the packhouse requires that potatoes should be delivered in boxes 'as dug' (i.e. harvested directly into boxes and not handled thereafter), potatoes arriving straight from the field may have large amounts of adhering soil present. This may vary from moist and sticky to dry and crumbly, depending on the ground conditions at lifting. Any soil on crops arriving from store should be dry, assuming the store is well-ventilated. Potatoes removed from bulk stores by bucket can also contain soil. If, in contrast, crop has been pre-graded on farm or removed from a bulk store using an elevator fitted with a soil extractor, the amount of adhering soil will be minimal.

Where a considerable quantity of soil is present on delivered potatoes, a dry cleaning system should be used (Fig. 11.4). The separated soil can then be put into boxes or skips for return to the farm of origin (Fig. 11.5). This avoids the wash water becoming too contaminated with soil.

Fig. 11.4. Star wheel dry cleaner followed by endless screen grader. (Courtesy of RJ Herbert Engineering Ltd, Cambridgeshire, UK.)

Fig. 11.5. Discharge of undersize tubers (left) and soil (right) from dry cleaning equipment.

There are a number of ways to remove soil and trash from potatoes. In all cases a compromise has to be reached between the level of cleaning and the possibility of inflicting bruises or damage to the tubers.

11.5.1 Automatic stone and clod separators

Most modern harvesting systems will remove the majority of stones and clods lifted with the crop but some may still remain. Stones and clods, especially those with sharp edges, damage potatoes during transport and when the crop passes over cleaners and conveying equipment. Stones can also become lodged in cleaning equipment and graders, causing damage to equipment.

One separation system, which is also used on some harvesters, conveys the tubers and stones on a rubber pintle (or hedgehog) conveyor belt (Fig. 11.6), which is inclined so that one side is higher than the other. The denser, smaller stones settle between the pintles and are carried to the end of the conveyor, while the lower-density larger-diameter potatoes stay on top of the pintles and roll sideways over the conveyor on to a flat belt conveyor running parallel to the pintle conveyor. This system is simple and provides a reasonable degree of separation at intake volumes as high as 60 t of dirty product per hour.

An alternative is a rubber-coated rotating drum used to bounce potatoes and stones into different trajectories, so that they land on two separate belts. As potatoes and stones have different rebound characteristics, good separation can occur below

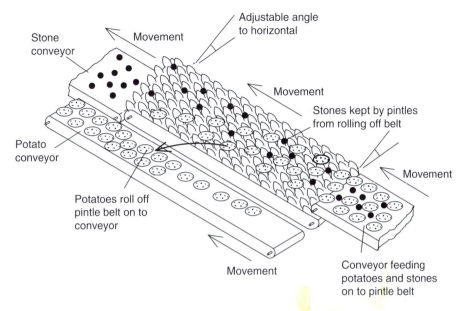

Stone conveyor

Movement

Adjustable angle to horizontal

Movement

Stones kept by pintles from rolling off belt

Potato conveyor

Movement

Potatoes roll off pintle belt on to conveyor

Movement

Conveyor feeding potatoes and stones on to pintle belt

Fig. 11.6. Sloping rubber pintle belt for separating stones from tubers.

the damage threshold (Feller *et al.*, 1987). When used in commercial situations and on some varieties, these machines have sometimes been found to cause damage so are now less popular than they once were.

In the past, X-ray separators were used to separate stones and clods from potatoes by sensing their relative permeability to X-rays, but these were complex, had health and safety implications when servicing radioactive components, and the fingers used to deflect potatoes and stones into the two separated streams tended to damage the potatoes.

A recent development uses a passive capacitance sensor, which acts similarly to the X-ray system in sensing the density difference between stones and potatoes but without the radioactive components that posed the health risk. It causes stones and clods to drop through the floor of the conveying system while leaving the potatoes to pass unchecked.

If potatoes are to be washed, the use of an upward flow of water ensures that all potatoes, regardless of their dry matter, will rise and the stones sink (Fig. 11.7). The potatoes can then be removed from the water free of stones.

11.5.2   Soil removal

Cleaners can be:

* Continental webs covered with plastic.
* A series of rotating steel spirals.
* A series of pairs of star wheels.

These are described more fully in Ch2.2 (Fig. 2.9a, b and c).

Fig. 11.7. De-stoner. (Courtesy of Haith Tickhill Group, Doncaster, UK.)

Brushes are sometimes used in wet cleaning systems after a soak tank to assist in reducing soil adhesion to the tuber surface.

11.6 Potato Washing Equipment

11.6.1 Shelf-life of washed potatoes

Washing potatoes increases the attractiveness and potential sale price of blemish-free potatoes but shortens their shelf-life. Fingers and Fontes (1999) claimed that the average shelf-life of potatoes is 30 to 40 days in perfect conditions but reduces to 7 to 15 days once they have been washed. Washing is therefore carried out just before sale or processing to minimize the development of disease blemishes or rotting. Were it not for the consumer preference for washed product, potatoes would be better sold unwashed.

11.6.2   Cooling of wash water

A heat exchanger may be fitted in a soak tank to act as a hydro-cooler to cool the potatoes to 3–9°C, to act as the start of the cool chain between packhouse and supermarket shelf.

Barrel washers

There are a variety of machines available on the market suitable for washing and drying potatoes (Clarke, 1996), but the most common is the barrel washer in which the tubers are rolled around inside a cylindrical barrel or drum (Fig. 11.8). The amount of cleaning depends on the speed of rotation of the barrel, the nature of its inside surface, and the residence time of the tubers within the washer. The residence time is adjusted by either changing the angle to the horizontal of the barrel axis or adjusting the discharge door opening to hold them back. The barrel is usually about 15% under water so that the potatoes keep falling back into the water. The barrel provides the agitation, while the water does the cleaning. Tubers can experience a high level of damage if they are 'thrown' against the sides of the barrel but the insertion of a 'brake sail' made of a thick plastic material can minimize the damage by slowing the fall of the cascading potatoes. This both reduces the potential damage to the crop and reduces the variations in mechanical load imposed on the barrel by the cascading potatoes (Geyer and Oberbarnscheidt, 1998). It is possible to have a dry barrel cleaner with sides made of bars to allow the dry soil to fall through.

Fig. 11.8. Barrel washer fed from a bulk hopper at rear and discharging to an inspection table in the foreground. (Courtesy of Potato Council, Oxford, UK.)

11.6.3 Management and treatment of wash water

The management and treatment of wash water in packhouses is often overlooked when new facilities are being planned. The importance of how best to treat wash water cannot be overstated and should be among the primary considerations when choosing a site for the packhouse. Common problems that arise are:

- Barrel washers filling up with soil, requiring them to be dug out by hand daily.
- Settlement of soil in piping between washing plant and water treatment plant and settling ponds.
- Insufficient land surrounding packhouses for the installation of treatment plant or settling ponds.
- Fine particles in wash water discharged to settling ponds that will not settle without the addition of chemical flocculants.
- Failure of treatment plant to meet suspended solids, BOD and chemical oxygen demand (COD) levels suitable for discharging into watercourses.

To conserve water, the dirty potatoes are usually conveyed through a tank of water, rather than being washed by a continuous flow of water from a clean water supply. Fresh top-up water is supplied to replace the water removed by the wet tubers and to prevent a build-up of disease in the wash water. The top-up water is added via spray jets, which rinse the washed tubers as they are conveyed from the washer (Fig. 11.9). The rinse water helps to remove any disease organisms residing on the tubers and reduces the chance of disease developing on the washed produce. Chlorine dioxide at 50 ppm is often put in the wash water to reduce the amount of

Fig. 11.9. Sprayed make-up water rinses wash water from clean tubers. (Courtesy of RJ Herbert Engineering Ltd, Cambridgeshire, UK.)

bacteria. All the equipment used to hold the wash water should therefore be either made from stainless steel or coated with a material which will resist attack by any bactericides added.

A decision has to be made on whether to let the soil settle below the barrel washer or to try to keep it in suspension by recirculating the wash water using high-volume water pumps. The tanks below barrel washers are fitted with removable sealed doors so that, if settlement does occur, the water in the tank can be drained and the wet soil removed using a shovel.

Plant to treat the water prior to reuse as top-up water or discharge to watercourses will include (Geyer, 1996):

- Sieves to remove large items of organic matter such as haulm and small tubers.
- A series of gravity settling tanks or hydro-cyclone soil/water separation equipment to remove the larger soil particles.
- Lagoons for removing very fine clay particles if these are present in the areas from which potatoes are sourced.
- Monitoring equipment to ensure water to be discharged meets the water discharge requirements (DETR, 1997).

If the water is to be discharged to a watercourse it will need to be treated in a biological treatment plant to reduce its BOD and COD to levels agreed by the environment agency concerned. In some cases, where there is the possibility of diseases such as brown rot in the water, it is heated to 70°C for 10 min prior to being placed in the settlement tank.

One way of reducing the heaviest and most easily settled soil and large pieces of organic matter is to use a trailer fitted with a sieve with a settling tank below (Fig. 11.10). This allows the soil and organic material to be returned to the field of origin, where it is tipped, spread out and ploughed in.

If there is plenty of farmland surrounding the packhouse, the final wastewater discharge can be through a low-volume irrigation rain gun, pulled back and forth across a field. The soil surface acts as size filter to prevent fine silt entering streams

Fig. 11.10. Dirty wash water screen mounted on a trailer to allow return of settled soil to the field. (Courtesy of Haith Tickhill Group, Doncaster, UK.)

and as a biological filter to reduce the BOD and COD of the wash water. This option is not available where packhouses are located in built-up areas or in industrial estates. In any new development, the requirements of the local water authority as to acceptable wastewater BOD/COD and suspended solids levels should be discussed before the final site for the packhouse or plant is chosen.

11.6.4 Drying potatoes after washing

Potatoes should be dried after washing to remove surface moisture. This is usually achieved using sponge rollers (Fig. 11.11), although more rarely air knives, which blow the water off the potatoes, are used.

11.7 Cleanliness of Packhouse or Processor Intake

Cleanliness of a packhouse or processor intake is vital if contracts with supermarkets are to be maintained. Mud and loose potatoes on the floor can quickly overwhelm staff and give a bad impression to visitors or supermarket inspectors. There should be easy access under and around all bulk hoppers, washers, graders and conveyors, so that these areas can be easily cleaned with shovels and brushes. Accidental discharges of potatoes to the floor should be cleared up right away and cleaning programmes should be routine.

11.8 Grader Design for Minimum Damage

Modern grading and packaging systems are designed to be as flat as possible. This avoids the large drops that occur when elevators discharge on to belts and reduces the need for buffer hoppers where potatoes, clods and stones can rumble about and abrade each other. A consequence of this approach is that potatoes must be

Fig. 11.11. Sponge rollers for removing surface water from washed potatoes. (Courtesy of RJ Herbert Engineering Ltd, Cambridgeshire, UK.)

discharged from boxes or trailers at high level (Fig. 11.12), so that there is room below grader screens for potatoes to fall on to belts for feeding to bag fillers without the need for secondary elevators.

Although all drops in potato handling should be less than 200 mm, the size of drop usually increases as the buffer hopper empties. Work carried out by O'Brien *et al.* (1980) on fruit and vegetable crop filling systems found that fitting a height sensor to an elevator with adjustable height control was the most effective way to limit damage. Where the use of a proximity sensor is not possible, a telescopic 'zig-zag' fall breaker should be considered. By constantly changing the direction of the potatoes, their downward velocity is slowed.

While rubber cushioning materials do reduce impacts on tubers as they land, any damp soil from the tubers tends to build up on the rubber, dry to a sandpaper-like texture and then abrade the skins of the tubers that follow. Static solid areas of any material over which potatoes have to roll or slide should therefore be kept to a minimum.

When cushioning is used, it must combine high surface wear resistance with retention of resilience over a long period (McRae, 1990). Armstrong *et al.* (1995) suggest that cushioning material should absorb at least 60% of the impact energy to minimize product rebound, it must be durable and easy to clean, and the uniformity between different production lots should be high. Bollen and Dela Rue (1995) evaluated a number of cushioning materials and stated that 'closed cell PVC foam was the best material with polyethylene foam and neoprene rubber exhibiting adequate characteristics over the energy range of the tests'. An instrumented sphere

Fig. 11.12. High-level tipper to allow grading/packing lines to be level, with a minimum of drops which can do damage. (Courtesy of Haith Tickhill Group, Doncaster, UK.)

(Zapp *et al.*, 1989) mounted on a pendulum was used to test the various padding materials at six energy levels between 0.3 and 1.8 J.

Significant damage arises in high flow rate systems from tuber to tuber collisions. While cushioning cannot reduce these, minimizing churning in hoppers or in the boots of elevators, or preventing changes in direction of conveyors, can.

Frictional or scuff damage to potatoes during handling is an area which has received only limited interest from researchers. The dynamic coefficient of friction has been investigated by Schaper and Yaeger (1992) for 25-kg lots of tubers and by Bishop (2007). Both teams showed significant differences in tuber damage between a range of surfaces for clean and dirty, and dry and wet tubers. Wet potatoes will slide more easily than dry and so receive less frictional damage. If tubers are both wet and dirty, particularly with soil having high clay content, there will be even more lubrication between the tuber and the handling surface and subsequently less scuff damage. The potential for scuffing is increased when tubers are transferred from one conveyor to another if tubers are rotating in the opposite direction to that of the second conveyor. On landing, the frictional force on the skin is increased compared with tubers landing without spin (Bishop, 1990).

As most potatoes for the ware or processing sector are washed, the problem of dust occurs only between trailer or box discharge and the washer. For more information on dust see Ch10.4.9, which discusses the problem in relation to seed potatoes that are graded dry.

11.9 Graders or Sizers

The size of a potato can be measured by its dimensions or by its weight. Some sizing can be done by 'vision' grading methods using camera technology and this is considered under Ch11.10.

11.9.1 Size graders

Size grading has been, and still is, the most common method of grading potatoes, even though some dimension methods such as screen mesh sizers have an accuracy of only ±40% as opposed to an optical weight grading system with an accuracy of ±9% (Glasbey *et al.*, 1988).

The most common form of size grading for ware potatoes is the endless screen, where the tubers pass over a series of conveyor belt-type square meshes. The size of aperture in the screens increases with each screen, so that the smallest tubers fall through first on to flat belt, cross conveyors. There is normally gentle agitation, but the size of aperture through which the tuber falls very much depends on the orientation of the potato if the variety is elliptical rather than round. If multiple size grades are required, this type of grader can become very long.

A second form of size grading is the riddle system, where the sieves or riddles are stacked one above the other with the largest mesh riddle uppermost (Fig. 10.5). This system is more suited to multiple grades than the endless screen, as it takes up less horizontal space. The tubers are thrown forward over the riddles by the oscillating action of the drive mechanism, and fall from one riddle to the next until their size prevents them falling further. The potential for damage has meant that riddle graders are losing popularity for ware potatoes and are used predominantly for seed where the small tubers are less prone to damage.

11.9.2   Diverging and expanding roller graders

Diverging roller graders, while expensive relative to endless screen graders, minimize damage to the potatoes. They consist of a series of rollers, rotating like those on a roller inspection table, but with the gap between the rollers increasing as the rollers move laterally from one end of the grader to the other. Potatoes being conveyed on the moving rollers fall through the gaps when these exceed the size of the potatoes. Flat conveyor belts below the grader, running at right angles to the flow of potatoes on the grader, remove the different size grades of potatoes for packing.

Expanding roller graders are similar in principle to diverging roller graders, but they incorporate diablo section rollers, similar in shape to the diablo roller used on a potato harvester. These better mimic the square hole of a square mesh riddle and expand in steps rather than the gradual expansion of the diverging roller grader. Simply winding a handle can alter the step increase.

11.9.3   Weight graders

An alternative method of size grading is by weight, where the potatoes are fed on to a conveyor made up of plastic cups, with each tuber having its own cup (Fig. 11.13).

Fig. 11.13. Cup weight grader. (Courtesy of RJ Herbert Engineering Ltd, Cambridgeshire, UK.)

The tubers and cups are passed over a weighing mechanism and routed depending on weight. This method has been popular in the apple industry for many years. Although a more costly method than traditional size grading, this system has become common for specific markets such as for bakers. Weight sizing has the potential to be more precise as the weight increases as the cube of the radius of a tuber.

11.10   Inspection

Almost all potato inspection to remove damaged, diseased or blemished tubers is still done manually. The inspection usually takes place on a roller table, where the tubers are slowly rotated under fluorescent lighting, to allow the operator to see their complete surface (Fig. 11.14). If a flat belt conveyor (Fig. 11.15) is used instead, a turner flap finger or roller should be located between each inspector to turn the tubers over so that both sides of tubers are checked.

Tubers are normally inspected at a rate of 3–6 tubers per second. This high rate of tuber lateral movement causes the number of defective tubers removed as a percentage of defective tubers present, termed the inspection efficiency, to vary considerably. Elstob *et al.* (1988) showed in a survey of inspection systems on UK farms that operator inspection efficiencies varied from 60% to as low as 8%.

Comfort is essential for staff to perform at their best. Seating should be available for those who find standing tiring. For those who prefer to stand, 100-mm step boards, like small pallets, should be available to allow inspection staff to raise or lower their stance to suit their height.

Some work has been done on the investigation of the speed of the tubers passing the operator in relation to the rotation speed of the tuber and the quantity of

Fig. 11.14 Roller-type inspection table to allow all sides of the tubers to be inspected. (Courtesy of RJ Herbert Engineering Ltd, Cambridgeshire, UK.)

Fig. 11.15. Tuber inspection using flat belts. (Courtesy of RJ Herbert Engineering Ltd, Cambridgeshire, UK.)

defects to be removed. Initial work, carried out by Malcolm and De Garmo (1953) with artificial wooden potatoes, suggested that each operator could inspect a maximum of 4.2–5.0 tubers per second at an optimum speed past the operator of 0.1–0.15 m/s and a rotation speed of 6–12 revolutions per metre of travel. Hunter and Yaeger (1970) carried out field trials with real tubers that had been artificially blemished, concluding that the feed rate could be adjusted to give 0.8 t of defects to be removed per hour per operator. In tests done by McRae (1985), there was 90% efficiency of defect removal with a total flow rate of 4.3 t/h when there was a 20% defect level, which can be expressed as the removal of 0.8 t/h.

Work by Bishop and Mortimer (1999) also showed that the forward velocity of the tubers can be 20% less at the edge as opposed to the centre of the roller table because of the drag effect of the table's side walls. Their work also found that the speed of rotation is influenced by the loading level and whether size grading has occurred prior to the roller table, as tubers of different sizes are more likely to 'mesh' and so not rotate. Overall the investigations indicate that the most efficient manual removal of defects will be when a roller table is working below maximum capacity, where less than 20% of the total tubers are expected to be removed and some size grading has been previously carried out.

11.10.1 Illumination for inspection

Both intensity of illumination and colour of the light are important if maximum inspection efficiency is to be achieved. The level of illumination should be no less than 500 lux (Zegers and van den Berg, 1988). While increased levels of lighting to

2000 lux showed no benefit, it is probably worth installing higher levels than 500 lux to compensate for subsequent deterioration in lighting level due to dirt and age of tubes or bulbs. In practice the illumination level is often below 500 lux. In work carried out by Hyde (1991), the levels varied in practice between 350 and 700 lux.

Zegers and van der Berg (1988) also suggested that for effective inspection the colour-rendering index of the light should be 85 or over, where natural light has a colour-rendering index of 100. A low colour-rendering index indicates a preponderance of a particular colour, which makes it more difficult to see certain defects on the tuber surface. The most common colour-rendering index found by Hyde (1991) was 62, for a cool-white fluorescent tube. Details of light output and rendering index are available from lighting data sheets.

The illumination source should be located so that inspection is never carried out in a shadow or in a light source of varying intensity. Where the fluorescent tubes run parallel to the direction of crop flow, the resultant light intensity tends to vary across the table. Balls (1986) describes a system where a 4.8-m-long fluorescent tube bank fitted with reflector above a 3.3-m-long table provided 1300 lux at the centre and 1000 lux near the sides of the roller tables. Balls also recommended that maintenance staff should replace the lights after every 6000 to 7000 h, rather than wait until they fail.

11.11   Combined Grading and Automatic Inspection

Considerable work has been carried out on automatic grading systems at the research level, where diseases, bruises and damage have been identified using near infrared reflectance (NIR) techniques (Tao *et al.*, 1990; Gall *et al.*, 1998). Current commercial systems almost universally use camera technology to grade potatoes. There are two common methods of presenting a flow of potatoes to camera systems. The first is a cable transport system, developed in the USA, which uses a system of parallel plastic cables to support a flow of potatoes. These pass through a camera unit which can grade and size individual tubers using image analysis techniques. The system rejects or grades using solenoid-driven fingers which knock tubers off the cable system on to an appropriate cross conveyor.

The other is a roller table transport system, which presents tubers to a camera and lighting system mounted overhead. The tubers are separated in the valleys between adjacent rollers using a singulator (Fig. 11.16) to provide separated tubers. The rollers rotate under the camera's active field of view (Fig. 11.17) to allow multiple images of rotating tubers to be captured. This permits a full inspection of each tuber surface.

Almost all camera-based vision grading systems use proprietary software to detect the periphery of each tuber and hence its size and shape characteristics. If colour cameras are employed, they also use the colour information to assess the tuber surface for damage or disease. This is achieved by relating the red, green and blue components of the sequential captured images to identify damage or disease based on pre-programmed, algorithm-based thresholds.

Rates of throughput have increased over recent years with high-end machine throughputs in excess of 30 t/h, depending on the size spectrum of the potatoes flowing through. One machine quotes 16.5 t/h for potatoes sized 45+ mm, and the same system will run in excess of 24 t/h with material sized 55+ mm.

Fig. 11.16. Singulator for separating flow into discrete tubers, as required for visual grading. (Courtesy of RJ Herbert Engineering Ltd, Cambridgeshire, UK.)

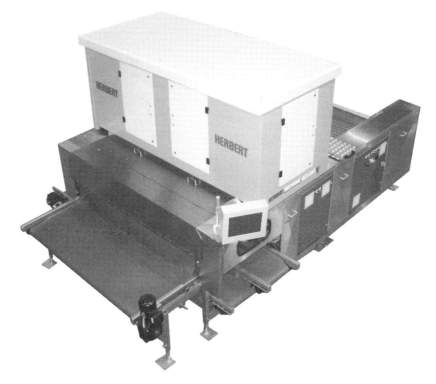

Fig. 11.17. Vision grader. (Courtesy of RJ Herbert Engineering Ltd, Cambridgeshire, UK.)

Although there has been much interest in these systems and a number have gone into commercial packhouses, the uptake for potato grading has been limited. They tend to be more popular in the grading and packing of fruit, where its higher value compared with potatoes can justify the high cost of vision grading.

11.12 Packaging

11.12.1 Bags and sacks

Potatoes for the ware market are sold in bags containing 1 kg, 2.5 kg, 5 kg, 10 kg or 25 kg. All bags are pre-printed with the farm, packhouse or supermarket name on the label. The 1-, 2.5- and 5-kg bags are made of low-density polyethylene (LDPE) film, with small holes to allow some ventilation. Alternatively nets made from polypropylene can be used. These are more expensive, but give better ventilation, reduce the likelihood of rotting if the product is washed, but allow more evaporative weight loss. The larger bags are made of natural Kraft, which is a strong brown paper made from wood pulp. Semi-bleached and fully bleached Kraft, off-white and white in colour respectively, are less commonly used for fresh produce. Hessian, sisal or woven polypropylene bags can also be used.

In the UK, the potatoes packed in LDPE plastic bags are usually washed. Potatoes put into Kraft bags are unwashed and are often marketed as 'Value Pack', when skin quality is not good enough to display the contents as a washed sample in see-through plastic bags. Double-walled paper bags may be used for strength, and can be filled directly on the harvester or in the packhouse for supply to wholesalers or retail outlets.

To protect the potatoes in plastic bags from damage, and to facilitate handling, the bags are put into either reusable polypropylene crates or trays for regular customers or 20- or 25-kg Kraft bags where there is no system for getting the crates back.

11.12.2 Bag filling and weighing systems

The equipment commonly used for filling the larger 20–25-kg sacks is the cleated-belt, elevator bagger (Fig. 11.18). To satisfy the Weights and Measures (Packaging Goods) Regulations 2006 (DTI, 2006), the local authority environmental service's trading standards department offers a service for calibrating weighing equipment and can visit packhouses to check weigher accuracy, should they get a complaint from commercial purchasers or the public. The most recent regulations specify that the average weight of bags should be as stated, superseding the previous requirement that every bag should be at or above the stated weight.

There are two approaches to ensure that the customer receives the stated weight in a bag. The first uses a low-cost cleated-belt elevator bagger, which slightly overfills the bag. The operator then places the bag on a calibrated platform weigher for check weighing (Fig. 11.19) to ensure that the bag is at or above the weight stated on the bag.

The second type of bagger fills a hopper suspended on calibrated load cells, which measure the weight of potatoes in the hopper prior to discharging the weighed batch of potatoes into the bag. More sophisticated designs have a second elevator for small tubers, which is used to top up the bag to the correct weight. This ensures

Fig. 11.18. Sack filler. (Courtesy of W.J. Morray Engineering Ltd, Braintree, UK.)

Fig. 11.19. Check weigher. (Courtesy of RJ Herbert Engineering Ltd, Cambridgeshire, UK.)

that the bag holds the required weight and is not overfilled. Once the target weight is obtained the sack is sealed or stitched.

Systems for filling plastic bags or punnet boxes are even more sophisticated. If a small bag of 1 kg is being filled and it weighs 990 g, another tuber is needed which may weigh 80 g, giving a 7% overweight for which the packer is not paid. To reduce this 'giveaway', a multi-head bagger is used. Between 20 and 25% of the required pack weight is fed into one of a bank of weigher heads, typically nine, and the weight of tubers in each head is measured. A computer decides the optimum combination of heads to give a weight just greater than the final desired pack weight, and these are emptied into the bag or punnet (Fig. 11.20). The automatic bagger will record the number of bags/punnets filled and can be interfaced with a personal computer to monitor production data.

All bags for the supermarket trade, and for many of the other retail outlets, have to be labelled post packing with contemporary data, including 'display until', 'use by' or 'sell by' date and a barcode which accords with the store till reader. These data cannot be pre-printed on the bag and have to be added post filling, either by printing directly on the bag or as a stick-on label; the latter have to be prepared and printed daily.

In a recent survey carried out by the British Potato Council the preferred way of buying fresh potatoes was the 2.5-kg bag (BPC, 2004b). The plastic bags are first filled with 2.5 kg of washed potatoes. These are then put into reusable plastic trays (Fig. 11.21) or Kraft paper sacks, commonly known as 'outers', to ease handling and provide cushioning from damage. A third machine stacks the Kraft sacks on a pallet (Fig. 11.22) while a fourth wraps the stack of bags with a film wrapper to stabilize the bags on the pallet.

Fig. 11.20. Multi-head weigher for filling pre-packs. (Courtesy of Newtec, 5230 Odense M, Denmark.)

Fig. 11.21. Pre-packs being packed into returnable plastic trays. (Taypack, Inchture, Dundee, UK.)

Fig. 11.22. Automated system for stacking Kraft bags on pallets.

11.12.3 Tailoring packing to specific selling lines

Increasingly potatoes are being packaged for specific selling lines, such as boxes with a known number of baking potatoes (Fig. 11.23) for a catering outlet or 750-g punnets of salad potatoes for a supermarket customer. The covering films are usually permeable and sometimes contain a light filter, as punnets may be displayed in a supermarket under illumination as high as 1000 lux. Potatoes have to compete with simple-to-prepare rice or pasta, so any packaging system which identifies an idea for a meal and simplifies meal preparation, thereby increasing sales, may be attractive to supermarkets (Fig. 11.24). Another example, which requires food processing equipment, is partially cooked roasting potatoes in an aluminium tray, where one has only to remove the over-wrap before placing in the oven (Fig. 11.25). Packhouses have to invest in the required packaging or cooking machinery, which should only be done after discussion with potential buyers.

11.12.4 Modified atmosphere packaging

The use of modified atmosphere packaging (MAP), where the gas composition in the packaging is altered to extend shelf-life or inhibit disease, is rarely used for fresh potatoes. When it is used, it is primarily for punnets where the tubers have unset

Fig. 11.23. Bakers being packed into boxes.

Fig. 11.24. Microwave-ready tray of salad potatoes, herbs and pats of butter.

Fig. 11.25. Partially cooked, oven-ready roast potatoes.

skins. Where potatoes are sold peeled for sale to catering outlets, MAP is the normal mode of packaging.

11.12.5 Biodegradable packaging

A limited amount of biodegradable packaging is now on the market that will break down completely, or almost completely, into carbon dioxide and water over a period of months. This has considerable potential, as waste packaging can be put into the kitchen receptacle for food waste for subsequent collection by the local authority for centralized composting or for composting at home. One of the alternative materials used for biodegradable packaging is potato starch, which provides an outlet for potatoes rejected for the fresh market. At the time of writing, corn-starch seems the preferred feedstock but there is some commercial work with a combination of starch types. The present higher cost of this material compared with oil-based LDPE results in it being used primarily for higher-value organic product lines.

11.13   Packhouse Automation

To satisfy the increasingly demanding quality assurance requirements for packed produce, to reduce labour costs, to improve packhouse efficiency and to provide effective feedback to growers, ever greater automation is being introduced. Table 11.3 summarizes some of the reasons for this development.

11.14   Dispatch Areas and Chills

The packed potatoes are held in a dispatch area until enough produce is available to fill a lorry. The product is most likely to be washed tubers, held in 2.5-kg bags, in free-flow trays or in Kraft bags stacked on a pallet. As the potatoes are likely to be damp from being washed, blemish disease or rots may develop, particularly if the holding area is not kept cool. Free-flow trays (also called reusable plastic crates, RPCs) have many advantages in being reusable, easily cleaned, strong and water-resistant. However, because of their highly permeable sides, they can allow the product to warm if exposed to warm ambient air.

Potatoes cooled by passing them through a hydro-cooler (i.e. cooled water tank) prior to packing can be kept at low temperature by wrapping the crates or bags on the pallets with plastic film. This both stabilizes the crates on the pallet and restricts warm air from penetrating the crates or bags during their stay in the dispatch area. In trials carried out on 2.5-kg bags of potatoes at 4.0°C, packed six to an RPC and exposed to an airflow 1–2 m/s in a room at 20°C, the flesh temperature of the tubers warmed up by 3.5°C in 1 h without a film wrap and by only 1.7°C when encased in film wrap. In the former case a 0.5–1.0°C temperature difference occurred across the width of the pallet.

Table 11.3. Reasons for packhouse automation.

Primary aim	Secondary benefits
Increased productivity	Higher product flow rate
	Better coordination of different operations to keep the system consistently at near full capacity
Improve quality of products	More accurate sizing and quality grading
	Potatoes can be packed into a greater range of lines
	Less manual handling of products, giving reduced mechanical damage
Reduce cost	More efficient use of hand labour
	Fewer conveying lines due to faster grading, resulting in a more compact system
	Less out-of-grade material as a result of more accurate sizing and sorting
Facilitate accountability	Identify incoming material by producer's barcode or RFID[a]
	Can log number of bags or trays, weight, size, maturity, grade, number and type of packages
	Record any chemical used
	Record sampling of stored tubers
	Record buying and selling prices and hours and wages of workers
	Can calculate packing efficiency on a daily basis
Ability to control operations	Complete control of volume and lines packed
	Can link supermarket tills' barcode readers with packhouse to indicate volume and type of sales
	Instant feedback helps to match supply to demand

[a]RFID, radio-frequency identity device.

Although potatoes are best dispatched at above 8–10°C to prevent handling damage, they tend to be treated as part of the cool chain for vegetables for supply to supermarkets, so are usually cooled to the cool chain temperature. Minimizing rough handling at this temperature is vital.

11.15 Reject Potatoes

Unmarketable potatoes are usually sold for feeding to livestock. To be acceptable to farmers, they should be free from soil, sprout-free and have no contaminants like plastic film that could choke animals. Alternatively, misshapen or blemished tubers can be used for potato flour, potato starch or as a feedstock for the manufacture of biodegradable packaging. Effective agronomy can maximize the number of tubers in the desired size range, which minimizes pack out loss. Pack out rates of 80% are common (A. Clarkson, Greenvale, Norwich, UK

2006, personal communication), so reject potatoes may constitute 20% of the total received at intake.

11.16 Summary

This chapter has focused on packhouse systems, their layout and the equipment used. From the discussion above, a number of conclusions can be drawn.

- The packhouse must be able to receive potatoes from either field or store.
- On-farm cleaning and sizing allows soil, stones and organic matter to be returned to either the field where they were grown or elsewhere on the farm.
- On-farm packing allows the grower to market over- and undersize potatoes as bakers or salad potatoes, which might otherwise be disposed of as out-grades by the packhouse.
- Into-store cleaning and sizing at harvest help to ensure a clean sample can be provided at packhouse intake.
- In UK law, any soil, stones and organic matter that enters an off-farm pack-house is immediately classified as a waste. Subsequent waste disposal should be accompanied by waste transfer notes for delivery to licensed waste treatment facilities.
- The use of old on-farm cleaning and sizing equipment may cause damage and loss of bloom if used for cleaning or grading material prior to delivery to the packhouse.
- Centralized packhouses can afford more sophisticated cleaning and grading equipment and have economies of scale which are difficult to achieve with on-farm packing.
- In the UK most packhouses are supplied with potatoes in boxes or bags.
- Packhouses have a dirty reception area to receive product and a clean area after washing to grade, inspect, pack, store and dispatch product. The two areas are separated by a wall or curtain.
- The intake area will have a covered area to protect delivered boxes from rain, to allow a buffer stock of boxes to stored; it may have a cleaning and grading line, and box tippers to discharge material on to the cleaning and packing lines.
- If potatoes are accompanied by soil, this should be removed using a dry clean-ing system prior to washing.
- After dry cleaning, the potatoes are delivered into a soak tank, passed through a barrel washer and then excess water removed by a sponge drier.
- A hydro-cooler may be fitted in the washing system to cool the potatoes to between 3 and 9°C, to act as the start of the cool chain between packhouse and supermarket shelf.
- Sizing is carried out using a continuous mesh screen grader followed by inspec-tion and packing into 2.5-kg or 1-kg polythene bags.
- In larger packhouses, weight or optical graders may be used.
- Automation of packhouse systems simplifies traceability and quality control and reduces staff costs.

- Bags are loaded into trays, which are stacked on pallets and stabilized using a plastic film wrapper.
- Pallets may be stored in a chill prior to dispatch.
- A number of different selling lines, targeted at the convenience buyer, are being tried to add value to potato sales.
- MAP is used for sales of peeled potatoes or new potatoes with unset skins.
- Biodegradable packaging enables it to be disposed of along with food waste but at present is used mainly for higher-value organically grown product.
- Reject potatoes are best fed to livestock and should be free from plastic, soil and stones.

12 Quality Assurance

Topics discussed in this chapter:

- Quality assurance defined.
- Alternative quality assurance schemes.
- Components of a quality assurance scheme.
- Quality control.
- HACCP.
- Traceability.
- Standards to set quality.
- Standards for futures trading.
- Sampling methodology.
- Sampling at intake.
- Quality control on a potato packing line.
- Paper- and computer-based quality assurance systems.
- Sharing information within the company.

12.1 Introduction

This chapter covers quality assurance, alternative quality assurance schemes, quality control, the HACCP (Hazard Analysis and Critical Control Point) safe food production system and statutory requirements for the sale of potatoes for the futures market. Since quality assurance is easier to understand if its use is linked to a particular situation, the descriptions will focus primarily on a quality assurance system for a packhouse producing potatoes packed in plastic bags for dispatch to a supermarket distribution centre. Since the grower's traceability system must link with that of the packhouse, this too is explained. As there are differences for systems for processing potatoes, reference is made to these from time to time.

The chapter concludes with a discussion on the relative merits of paper trail versus computer-based quality control systems and emphasizes how important it is to maximize useful information collected while minimizing form-filling paperwork.

12.2 Quality

The packhouse's quality control system starts with the packhouse procurement manager assessing potatoes while they are still in the field or store. Quality control then follows the crop through the packhouse and finishes by monitoring samples of potatoes dispatched to the supermarket distribution centre or merchant.

The most appropriate definition for the quality of potatoes is 'suitability for a particular use' (Abbott, 1999) or more fully 'quality is the extent to which the product comes up to the users expectations'. It is important to define quality so that customers know precisely what they are buying. A customer buying 'value-pack' potatoes in 10-kg Kraft bags may not be concerned about skin blemishes, which in a pre-pack could result in the product being left on the shelf unsold.

There are emotional as well as analytical considerations. A bright shiny skin or 'bloom' may attract the purchaser to buy, but is of little real benefit if the potatoes are to be peeled. Such qualities may require sophisticated instrumentation to measure. The relative importance of different quality attributes can change from harvesting and handling, to purchase from a grower to purchase by the final consumer. The harvesting team may concentrate on minimizing damage and picking off any rots, while a buyer may be looking for a clean, soil-free sample with a good bloom. The final customer may search the supermarket shelves for a bag of small sized, good-tasting potatoes, which look attractive on the plate.

The easiest quality standards to use are those like size, which is simple to control and easy for quantifying out-grades. Temperature of product is normally another, but in packing potatoes compared with cooking food, is of lesser importance. Disease and damage are much harder to define, as removal of excessively damaged or diseased material on a conveying system usually involves a rapid visual assessment by pickers on the inspection line. What is often required is a simple accept/reject approach, which the 'would I buy it?' approach solves (Ch12.11). This is by definition subjective, and has to be backed up by disease and damage assessments so that staff on one shift can apply the same standard as that on the next. Such a standard may alter depending on the general quality of the potatoes coming on to the market at the time. Shoppers may select a blemish-free sample in preference to one covered in silver scurf, but if all items on sale have silver scurf they may just select the samples that look best.

Research on consumers' approach to taste quality is limited. However, BPC and other trade associations do run taste panels on new varieties and are actively funding work on what biological and genetic issues contribute to taste (Winfield *et al.*, 2005). There are increasing numbers of studies of consumers worldwide to identify what influences customers to choose a particular variety, size grade and pack (Krueger, 1994; Moskowitz, 1994). The combination of consumer panels together with an increased understanding of the science behind taste, texture, colour and cooking performance is helping to identify the aspects of potato quality that are required for the different markets (Connor, 1994).

12.3 Quality Assurance

Quality assurance covers a number of management procedures that combine to form an efficient, well-organized business that safeguards its customers and the environment. These procedures are defined as follows:

- Quality assurance is the procedure by which quality is maintained or assured (e.g. ISO 9001 (2000), EFSIS-FABBL Farm Assurance Scheme, UK Assured Produce 'Red Tractor' scheme, Tesco supermarket's 'Nature's Choice' scheme).
- Quality control is the process used in a quality assurance scheme to check that the produce quality meets the standards quoted at sale.
- Due diligence indicates that all reasonable precautions have been taken to remove or reduce risks and hazards to production staff and potential consumers.
- HACCP schemes monitor production and divert product from the processing line should it fail a measurement at any stage in the production chain.
- Traceability is the ability to trace the history of any produce; the seed from which it came, the field where it was grown, the chemicals that were applied to it, the lifting, storage and grading procedures, and the dispatch date and destination.

12.4 Quality Assurance Schemes

Processors and retailers have responded to consumer concerns on food quality by implementing food safety and quality assurance programmes. The UK Assured Produce Scheme, developed by the UK National Farmers Union in consultation with retailers, offers potato growers a standardized approach to on-farm and pack-house quality assurance (CMi certification; CMi plc, Oxford, UK). Nature's Choice, produced independently by the UK supermarket chain Tesco, is similar to the Assured Produce Scheme. The DEFRA/Scottish Executive Seed Certification Scheme, for the production of seed, assures the quality of seed potatoes (Ch10.4). Examples of quality assurance schemes not specific to agricultural crops include ISO 9001:2000 (ISO, 2000), which succeeds ISO 9000 and before that BS 5750.

In addition to these schemes, purchasing managers or retailers or supermarkets will supply the packhouse with a manual specifying their own specific quality assurance requirements, the details they require to be put on labels such as 'display until' dates, 'use by' dates and dispatch information.

The quality assurance scheme has a number of requirements:

- A complete record of any product sold should be held on file so that if a problem with the product occurs, its cause can be established.
- The quality of any product, be it high, average or low, is defined and all of that product sold should meet the quality specified.
- Routine pesticide use on seed, growing crops or at crop intake, and the application of specific pesticides in response to disease, weeds or insects should all be justified on the quality control paperwork.

- Crop production practices should not adversely affect soils, watercourses, biodiversity or the wider environment.
- Waste materials, pesticide runoff and reject potatoes with associated botanical matter should be minimized and their disposal should not harm the environment.
- Operator health and safety practices should be documented and operators trained and tested in their operation.

The quality assurance scheme therefore encourages good crop management and the production of product to a defined quality standard while discouraging practices like the routine prophylactic use of chemicals, the disposal of chemicals to watercourses, the burning or burying of plastics and the use of excessive and unnecessary packaging.

12.5 Quality Control

Central to the quality control department is their sampling room or laboratory (Fig. 12.1). This consists of the usual laboratory benches and is likely to contain the following pieces of equipment:

- Rotating barrel sample washer.
- Bench weighing machines.
- Mechanical and hand potato peelers.
- French fry cutter.
- Crisp slicer.
- Crisp/chip fryer.

Fig. 12.1. Tuber samples from intake material are washed, peeled and cut in half to assess any disease, damage, bruising or hollow heart. (Taypack, Inchture, Dundee, UK.)

Fig. 12.2. Pre-packs stored under lights for 8 days to replicate a supermarket display cabinet. (Taypack, Inchture, Dundee, UK.)

On the walls will be charts for identifying disease, tables showing maximum levels for blemishes and rots, and charts for identification of the fry colour of French fries and crisps.

Adjoining the laboratory is the quality control office, where the quality assurance manuals are kept. It is here that the most up-to-date reference manual is kept. This is where the paper records, if paper is still used, of previous samples of product are kept and the computers holding the data of material processed over the preceding year are located. In the laboratory or office, there will be a series of shelves (Fig. 12.2), holding bagged samples of produce kept at room temperature until their 'use by' date expires. At the room temperature of 20°C, these samples give early warning of potential rotting or development of blemish diseases in produce sold, so that the retailers can be warned early of potential problems with their produce on display.

Samples will be brought to the quality control laboratory by fieldsmen, the potato procurement manager, or by agronomists and quality control staff who are constantly on the road visiting farms and stores.

12.6 HACCP Procedures

HACCP is entirely concerned with food safety and is a legal requirement within the EU (EC, 2004). If potatoes are sold to a packer rather than to a merchant, agent or broker (Fig. 12.3), the produce will be packed using HACCP procedures.

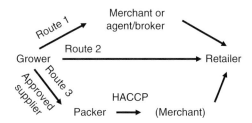

Fig. 12.3. Alternative routes from grower to retailer.

The packhouse will be subject to regular inspection by the supermarkets they supply to ensure that these procedures are being maintained.

The main objective of any HACCP system is to ensure that, in a continuous-flow operation with a series of stages, produce cannot move on to the next stage unless it has met the requirements for the previous stage or group of stages. The requirements, or critical limits, form critical control points. Should the produce fail a critical limit, there is an action programme designed into the system to divert the product from the production line. HACCP is therefore a systematic approach to the identification, evaluation and control of food hazards.

There are seven principles, or steps, in this systematic approach (CCFH, 1997).

1. Conduct a hazard analysis.
2. Determine the critical control points.
3. Establish critical limits.
4. Establish monitoring procedures.
5. Establish corrective action.
6. Establish verification procedures.
7. Establish record-keeping and documentation procedures.

A key requirement for a critical control point is that the measurement must be able to be carried out quickly, to ensure that unacceptable product is immediately diverted away from the normal production line. This is simple where produce has to achieve a specified temperature but is more difficult if, for example, produce has to achieve <2% bruising.

A detailed description of how to implement HACCP procedures in any specific situation is beyond the remit of this book, but it is important to stress that any HACCP procedure must be fully supported by the senior management. There will be occasions when material fails to reach the standard required, which will incur loss of income. It is, however, better to lose some income than to lose a reputation for quality produce. The stages in a HACCP study are summarized in Table 12.1. An example of a flow diagram for pre-packing potatoes is shown in Fig. 12.4.

One of the weaknesses of some HACCP plans is that once they are developed they are not updated. So it is important to build into the plan a verification schedule to validate and review the procedures.

Various organizations provide training in HACCP procedures and certification is available from organizations such as the Royal Institute of Public Health and Hygiene (London, UK).

Table 12.1. Development of a Hazard Analysis and Critical Control Point (HACCP) plan.

Action	Detail
Assemble the HACCP team	Assemble a group which has specific knowledge and expertise to develop the plan. This should include people from within the company as well as an outsider with 'fresh eyes'
Describe the product and its distribution	General description of the product (e.g. bakers, salad potatoes, etc.) and the processing method
Describe the intended use and the consumers of the food	Intended consumer (e.g. processor, supermarket, etc.)
Develop a flow diagram that describes the process	Outline all the steps in the process under the control of the packhouse
Verify the flow diagram	Verify the accuracy of the flow diagram with modifications as necessary

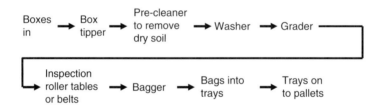

Fig. 12.4. Flow chart of a pre-pack grading line.

12.7 Traceability

Growers supplying packhouses or processors will have their own traceability systems. These document:

- Seed source, variety and generation used when planting a field.
- Field name or number used to grow the crop.
- Seed treatment, herbicides and other pesticides applied to the growing crop.
- Irrigation applications.
- Harvest date.
- Final destination or location in farm store.

Packhouse or processor staff purchasing crop from, or packing crop for, the grower require access to these data should a problem occur, but will not necessarily store the information itself. Packhouse or processor staff will start monitoring the crop as it is delivered for packing or processing, or will visit stores filled with their source material.

12.8   Standard for Selling Crops on to the Futures Market

In the UK the BPC Ware Prescription is the statutory standard for quality and size and its quality regulation states that the total of damage, blemish, misshapen or diseased tubers must not exceed 5% of the sample by weight (Gall *et al.*, 1998). While the quality of potatoes should agree with the BPC Ware Standard, the seller may deliver up to a maximum of 10% faults. According to the London International Financial Futures and Option Exchange (LIFFE), where the proportion of faults exceeds 5%, the seller 'shall make an allowance to the buyer of 1% of the settlement price for each 1% or part thereof of the faults over 5%. Dry matter shall be a minimum of 18%.'

12.9   Labelling Lifted Crops

Bulk crops have to be identified by the field or the store they came from. Unlike potatoes in boxes they cannot be identified once they have been moved, so strict recording of loads is necessary.

Boxes are always labelled (Fig. 12.5). Information required for stacking in store should be printed large enough for the forklift driver to read from his seat. In seed crops the label will usually include the:

- Variety.
- Generation (e.g. Super Elite generation 2, i.e. SE2).
- Name or number of field.
- Name of grower.

Fig. 12.5. Label to identify box contents.

The label is either stapled to boxes in the field as they are loaded or as the boxes enter the store or packhouse. The loss of a label can render the box contents worthless.

Boxes can be identified by:

- Label with information printed on it.
- Barcode label plus some printed information.
- RFID fitted to the box.

12.9.1   Label system

The label system is cheap, simple, easy to read and can be in a colour to differentiate varieties or generations, but the amount of information is limited. Individual boxes are rarely identified. Most of the information has to be kept elsewhere.

12.9.2   Barcode label

Barcode labels can now be produced on a portable computer fitted with a printer as the boxes leave the field or enter the packhouse or store. The label can have some typed information on it so that farm staff can identify the boxes visually. The barcodes allow individual boxes to be identified, detailed data of the contents of each box to be kept on computer and rapid reading of box number prior to it being tipped to empty.

12.9.3   Identification by radio-frequency identity device

An RFID fixed to each box allows the box to carry its own information. It acts as a transponder, being able both to transmit and to respond to an external electronic receiver/transmitter. It therefore needs no battery. Details of the box's contents, weight and history can therefore be stored on the RFID itself and the information subsequently read by another receiver/transmitter fitted on the box tipper at intake to the packhouse or by an operator. The RFID can also be used to identify ownership and history of the empty boxes. Though more expensive than the label and barcode systems, it has greater capacity for storing and conveying information. However, staff without a receiver cannot identify the contents of the box. The use of RFIDs is widespread in the retail trade where it is a proven technology. Their size can be as small as the full stop on this page, but normally with potatoes they are larger and have a bigger aerial so that the transmission range is greater.

12.10   Sampling Protocols

Quality control requires critical limits to be set and crop to be diverted away from the product stream should these limits be exceeded. These limits may be <2% area of tuber skin to be covered with silver scurf lesions, or <10% common scab, or <1% greening. Because the location of each growing tuber in the ridge is different (Fig. 1.14), each tuber experiences different environmental conditions. To get an average with a low standard deviation, a large number of tubers have to be sampled. Sampling predicts the average quality of a batch and possibly the quality distribution (Abbott, 1999), but it does not identify that a few tubers in the sample may be totally rotten or severely deformed while some may be of very high quality. Averages may need verbal statements added.

The number of tubers required for detecting a disease or defect is shown in Table 12.2. For example, if the number of rots should not exceed 2% of the product, 150 tubers should be assessed. This would best be sampled in three separate sub-samples of 50 each to act as replicates. The results will be in the 95% confidence level.

12.11 Sampling of Incoming Product

Initial sampling of potatoes establishes whether or not a particular batch of potatoes is suitable for the proposed end market. Tests may assess:

- Size distribution.
- Damage index.
- Bruising and blackspot.
- Growth deformities.
- Presence of sprouts.
- Pest damage.
- Disease.
- Bloom.
- Potential to rot.
- Would I buy it?

Additional tests for the processing market are:

- Dry matter.
- Sugar content.

Table 12.2. Number of randomly selected tubers required to find defective tubers in a load. (Adapted from EC, 2002; 95% confidence level.)

Maximum level of defect permitted (%)	Minimum sample size (no. of tubers) required to detect presence of defect	Minimum sample size (no. of tubers) required to estimate % level of defect
20	15	45
10	30	90
5	60	180
2	150	450
1	300	900

Fig. 12.6. Size gauge to check samples are within specification. (Taypack, Inchture, Dundee, UK.)

12.11.1 Size distribution

Random samples of tubers are checked for size (Fig. 12.6) and the various grades weighed to find out the percentage in each grade. The actual sizes used are matched to the intended market.

12.11.2 Damage index

The traditional damage index is based on the amount of tuber flesh or pulp that will be lost when a tuber is peeled to remove the damaged area (Robertson, 1970).

A sample of 100 tubers taken at random is divided into four categories:

- Undamaged.
- Scuff damage.
- Peeler damage (<1.5 mm deep).
- Severe damage (>1.5 mm deep).

The damage index of the sample is calculated as in Ch2.3.1.

While the above damage index is simple to understand and determine in the field, store or grading area, it fails to identify how the damage is being caused and omits any mention of bruising. A more rigorous approach is given in Table 12.3 (Bouman, 1995, 1996).

Table 12.3. Damage assessment system proposed by the European Association of Potato Research.

Damage	Severity	Surface measurement	Depth (mm)	Definition
		% of tuber surface		
Scuffing	None	0	0	The skin (epiderm) is damaged
	Slight	0–10	0	and removed partly or totally
	Moderate	10–20	0	
	Severe	>20	0	
		Area of tuber (cm²)		
Cuts	None	0	0	Part(s) of the tuber is (are)
	Slight	0–1.7	0–1	sheared, or shearing may
	Moderate	1.7–5.1	1–2.5	be accompanied by loss of
	Severe	>5.1	>2.5	tissue
Crushing	None	0	0	Rupture of tuber tissue caused
	Slight	0–1.7	0–1	by compressive forces
	Moderate	1.7–5.1	1–2.5	
	Severe	>5.1	>2.5	
Splitting	None	0	0	The tuber has splits in the
	Slight	0–1.7	0–1	flesh, propagated from the
	Moderate	1.7–5.1	1–2.5	surface
	Severe	>5.1	>2.5	
Bruising	None	0	0	Subsurface damage to
	Slight	0–1.7	0–2	the tuber tissue, which
	Moderate	1.7–5.1	2–5	subsequently causes blue/
	Severe	>5.1	>5	grey to black discoloration of the flesh

Washington State University has developed a bruise classification system which puts the tuber into one of seven categories (Baritelle *et al.*, 1999):

- No bruise.
- Blackspot.
- Crush.
- White spot/white knot.
- Internal shatter.
- External shatter.
- External cracking.

Evans and McRae (1996) describe a system which counts the number of peeler strokes needed to remove all damage and bruised areas to arrive at a damage index.

12.11.3 Bruising and blackspot

When tubers are hit by stones or other tubers, dropped on to a hard surface or exposed to pressure damage from the weight of crop above, bruising damage (blackspot) to cells below the skin can occur. Blackheart in contrast is caused by depleted oxygen in the atmosphere surrounding the tubers, causing cells in the centre of the tuber to die off. Both are invisible to inspection staff unless the tuber is sliced open. Bruised tubers are therefore very difficult to remove on a packhouse inspection line; so whole crops are often rejected even if only a few bruised tubers are present.

Bruising may take 3–4 days to develop into its black or greyish appearance and can have a serious impact on crisp quality (Fig. 12.7). To identify potential bruising from mechanical damage, samples can be put into a hot box (Fig. 12.8) and kept at high RH by circulating air over a water bath. Bruises show up within 12–14 h. Samples of 25 tubers are placed in the hot box overnight. By morning the bruises should be visible. The hot box (Fig. 12.8) is maintained at a temperature of 34–36°C and RH of 95–98%. It has shelves to hold the sample trays, a heater element to keep the box warm, a wet wick to produce the high RH and a circulating fan to reduce temperature differentials within the box.

To assess bruise damage, the tubers should be peeled to reveal bruising. The categories are:

nil bruising = no bruise
slight bruising = removed in less than two peels (<3 mm)
severe bruising = needs more than two peels to remove (>3 mm).

The hot box method is simple, effective and can also be used to determine the risk of crop breakdown in store.

Another way of increasing the speed of bruise blackening is to hold potatoes in oxygen maintained at 150 kPa (1.5 bar) pressure at 37°C. Bruising will become apparent within 7 h (Duncan, 1973). When the oxygen was humidified the time was reduced to 5 h (Melrose and McRae, 1987). However, owing to safety and cost

Fig. 12.7. Crisps with blemishes due to bruise damage. (Courtesy of Potato Council, Oxford, UK.)

Fig. 12.8. By placing samples of potatoes at intake in a hot box overnight, bruising or rotting becomes visible within 12–14 h. (Taypack, Inchture, Dundee, UK.)

considerations, compressed air at 300 kPa (3 bar) was used instead. This showed up bruising as rapidly as when oxygen was used (McRae and Melrose, 1990). However, due to the complexity of these methods and the relative simplicity of the hot box, growers and packhouse staff routinely use the hot box for rapid bruise detection.

12.11.4 Growth deformities and pest damage

Growth deformities and pest damage are normally estimated from visual inspections of samples, with the occurrence of each being calculated as percentages.

12.11.5 Presence of sprouts

The presence of small sprouts may be unacceptable in a pre-pack sample. Longer sprouts can be broken off during grading, but the presence of odd sprouts in a pre-pack sample is undesirable. Sampling would identify percentage incidence of sprouting and average length of sprouts.

12.11.6 Disease

Since the growing and storage conditions of the individual tubers in a crop are never completely uniform, disease and blemishes will occur in some tubers but not in others. To assess the severity of a single disease, the presence or absence of the disease on a number of individual tubers is recorded and expressed as a percentage incidence. Where the tubers have a blemish disease, such as silver scurf, the percentage of surface area affected by the disease on each tuber is recorded and used to calculate an average percentage area affected.

12.11.7 Bloom

Bloom is the ability of the skin to reflect light and so give a bright, shiny appearance rather than a dull, matt look (Ch1.3). While this is purely cosmetic, supermarkets want a bright bloom as consumer research suggests that customers prefer potatoes with a good bloom. Bloom meters (Bowen *et al.*, 1996) are available (Fig. 12.9) only on an experimental basis at present, but these may become standard pieces of equipment in future.

12.11.8 Potential to rot

Potatoes that rot on the supermarket shelf are one of scenarios that packhouse staff fear most. Organisms that cause rot can be present within the flesh of tubers at harvest, but may not show up as rots until days or weeks later. When the possibility of rot is present, the same hot boxes used to accelerate bruising, but kept at 20°C rather than 36°C, can be used to accelerate rotting. Samples are usually kept in the hot box for 12–24 h.

Fig. 12.9. Bloom meter. (Courtesy of SAC Ltd, Edinburgh, UK.)

12.11.9   Would I buy it?

With all these tests, it is possible to lose sight of the critical test: would I, as a supermarket customer, buy these potatoes? One packhouse uses this as their standard. Potatoes are given three qualities:

    0 = rots present
    1 = poor grade
    2 = buy grade.

This provides a simple accept/reject decision, which is easy for staff to operate. It needs to be backed up by the previous tests, as growers will not like material they supply to be rejected on such a system without more detailed explanation.

12.11.10   Sugar content

Sugar content is crucial in potatoes destined for the crisping or French fry markets. Assessments are made prior to harvest as levels reduce with time and so dictate harvest date. Levels increase in store with time or if held at too low a temperature, resulting in darker fry colours the longer they are stored.

12.11.12   Dry matter

Dry matter is especially important in processing crops as it dictates how much cooking oil is absorbed into the crisp or chip. While in the past the aim was to minimize the amount of costly cooking oil used, dietary concerns of excessive oil in crisps and French fries is now making oil absorption an even more important factor.

   Dry matter is measured using a hydrometer. An electronic version (Fig. 12.10) allows a sample of potatoes to be suspended in water and the percentage dry matter to be read directly from the meter. Alternatively a small core can be removed from a tuber, trimmed and weighed so that the weight is known of a precise volume of tuber. From this weight reading and the use of charts, the dry matter can be found (Fig. 12.11).

12.12   Crop Assessment Prior to Dispatch

Packhouse managers usually develop a special relationship with a group of core growers, who supply the bulk of produce packed. With one organization (Nick Winmill, Greenvale AP, Norwich, UK, 2008, personal communication), these core, or committed, growers are required to carry out a quality assessment of each load prior to it being dispatched. The results are recorded in triplicate on a Good Collection Note (GCN). These minimize rejections of material at the packhouse intake and avoid the associated transport costs that such rejections incur. The

Fig. 12.10. Electronic hydrometer for tuber dry matter determination. (Courtesy of Martin Lishman, Lincolnshire, UK.)

assessment provides valuable feedback to the grower, which can lead to changes in husbandry in future seasons.

With certain high-risk crops such as new potatoes, lightly set salad potatoes or loose skin bakers, growers are required to identify specific risks. These have a red, amber and green classification so that packhouse staff can deal appropriately with the crop.

Growers not in the core group should adopt this approach, as this practice may increase the chance of their being asked to join the core group of suppliers. It also informs other buyers of the quality of the material being supplied and reduces the likelihood of unwarranted rejection of loads in time of excessive supply.

12.13 Monitoring Crop In and Out and Waste Produced

Monitoring crop in, soil and trash removed, over- and undersize material, discarded potatoes and final product dispatched is a mass balance exercise (Fig. 12.12). While not strictly necessary for quality assurance, it is a vital management exercise. It allows growers to be rewarded or penalized on the basis of percentage of saleable product of the potatoes they have supplied. It also allows targets for improvements in crop quality to be set and feedback as to whether these targets have been met. 'If you don't measure you can't manage.'

(a)

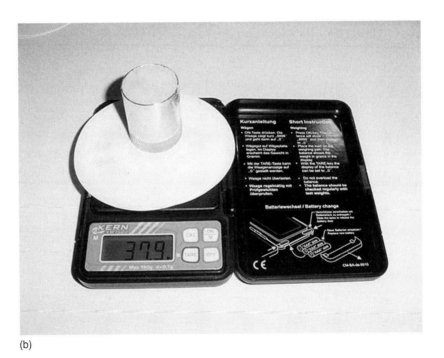

(b)

Fig. 12.11. Tuber dry matter determination: (a) core sampler; (b) balance for weighing core sample. (Courtesy of Martin Lishman, Lincolnshire, UK.)

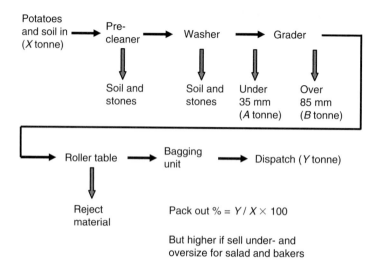

Fig. 12.12. Mass balance for a packing line.

12.14   Critical Control Points on Inspection, Grading and Packing Line

Key to any HACCP system on a continuous-flow production line are sensors that can rapidly measure deviation from an acceptable critical value and so divert product that fails to meet the critical value from the production line. The system can therefore be used with processed crisps or French fries, where the product can be passed under fast-operating optical sensors that monitor bruise damage or other imperfections and discard unsuitable material.

With potatoes however, fast-acting automatic sensors are in their infancy. The traditional approach on a grading line is for the staff inspecting and removing unsuitable tubers to slow the flow of produce so that they can keep pace, to ensure the quality of the saleable product is maintained. If the rejection rate becomes excessive, the grader is stopped and the batch of potatoes discarded. Thorough testing of potatoes at intake is designed to avoid this wasteful scenario.

Developments in automatic monitoring are underway that make continuous and rapid monitoring of potatoes possible.

12.14.1   Grading by sieve, weight or tuber size

Grading by sieve has been the traditional method of sizing potatoes and suits continuous-flow production. Weight grading is also suitable but is more expensive and tends to be used for the more valuable grades of potatoes such as bakers. Optical grading, using cameras to measure tuber size, offers the potential to include disease monitoring on the continuous flow of produce. This meets the HACCP requirements of being able to automatically divert sub-standard tubers from the main flow of produce. Disease recognition is, however, still in the development stage.

Table 12.4. Potato blemish disease assessment chart.

Variety	Grower	Date	Time
Saturna	A.G. Monks & Sons	08/05/08	09.30 hours

Measure	Sample 1	Sample 2	Sample 3
Rots	Nil	Nil	Nil
Blemish disease	Silver scurf	Silver scurf	Silver scurf
Incidence (%)	5	3	25
Area (%)	15	9	30
Bloom	Medium	Medium	Medium
Deformed tubers (%)	2	0	0
Would I buy it?	Yes	Yes	No

12.14.2 Visual inspection

Until optical grading systems become more common, sampling will continue to be carried out visually by inspection staff using tables like Table 12.4 above. This sampling has to be carried out at intake on sample batches, as it is not possible to do as a HACCP critical control point.

12.15   Final Product Sampling

Most packhouses will use a quality checklist similar to that shown in Table 12.5. The end user may have their own format with the range of factors reduced to those likely in that particular season.

   The list is extensive but in most cases a zero can be put in most of the boxes. The system is designed to ensure that the same result would be obtained by whoever carries out the assessment, so that it is objective rather than subjective.

12.16   Accessibility to Quality Assurance System's Information

Information on the product passing through the packhouse is collected from a number of different sources. These include:

- Grower's own data, available on request.
- Test diggings of crops by fieldsmen and agronomists.
- Damage testing during harvesting.
- Data logging of store climate, including stored crop sampling and inspection.
- Procurement manager's inspection reports of crops in store.
- Sampling at intake.
- Sampling and monitoring of product during cleaning, grading and packing.
- Final product sampling.
- Shelf-life testing.

Table 12.5. Quality check sheet.

Date 06/04/08	Variety Maris Piper	Destination AJC Ltd	Identity no. AJM 2398	Grower A.J. Monks & Sons
Factor	**Sample 1**	**Sample 2**	**Sample 3**	**Sample 4**
Sample number	2398/1	2398/2	Etc.	
Time (hours)	10.30	10.40		
Correct variety	✓	✓		
Correct label	✓	✓		
Correct label position	✓	No		
Display until	12/04/08	12/04/08		
Intact seal	✓	✓		
Stated weight (g)	1000	1000		
Actual weight (g)	1045	1095		
Number of tubers	19	17		
Size range (mm)	35–45	35–45		
Oversize	0	1		
Undersize	0	0		
Temperature	8°C	7°C		
Skinned	8	7		
Misshape	0	0		
Damage	3	1		
Pressure bruising	0	0		
Bruising	0	0		
Green	0	0		
Slugs	0	0		
Soft	0	0		
Soil adhesion	0	0		
Sprouting	0	0		
Growth cracks	0	0		
Silver scurf	0	5		
Black scurf	0	0		
Black dot	0	0		
Skin spot	0	0		
Scab	0	2		
Checker's initials	RSJ	RSJ		

A great deal of information is therefore collected on a routine basis. There is, however, a tendency for data to be collected by individual staff members but not shared. The procurement manager may have a clear assessment of the quality and type of crop available to him but this information may not be available to the sales force. Opportunities for sale of specific crops may therefore be lost. The quality control staff may be recording increased amount of damage on crops at intake, but the procurement manager or growers may be unaware of this until it is too late to tackle the problem.

Much information that could be useful in identifying problems of disease development or adverse sugar levels, such as store climate records and crop inspections, may never be used. The advent of store data loggers often results in store climate data being logged automatically but rarely examined. The results of inspections of stored crop by store managers may remain in their pocket notebooks, but never accessed unless a major crisis occurs.

A well-set-up quality assurance system will aim to provide maximum information to all staff with the minimum of unnecessary form-filling. It should:

• Provide staff with a method of making collected data immediately available to other staff within the organization.
• Identify the origin or previous store location of any particular batch of potatoes.
• Allow quality control staff to find where and why problems occurred.

12.16.1 Paper-based systems

Where possible the system should not require information collected to be retyped into a database, as this can result in delays in making the information available. It is usually better to have paper forms, which can be instantly inserted into a file.

The problem with even the best paper-based system is that the information is likely to be held in different locations, requiring the troubleshooter to visit numerous offices. Information may be distributed among the procurement manager, the quality control department, salesmen and the grower's farm office. This is where computers can help.

12.16.2 Computer-based systems

With computers, all permitted staff can access the collected data in the comfort of their own office. If the system is set up well, troubleshooting can be easy.

Successful computer systems require staff to enter information into the system as it is collected, preferably on to a hand-held computer, which may also provide a prompt for the information required. The computers have to be robust to cope with being dropped in mud or rained upon. The computer shown in Fig. 12.13 can transmit data to the packhouse via a telephone link, can be used as a mobile phone, can have a Global Positioning System installed, and can be fitted with a barcode or RFID reader. With such a system the information fed in can become immediately available to other staff. In contrast, paper-based systems are unavailable to others until entered on to the computer network. With busy people, instilling a regime that updating records is more important than solving the latest crisis is difficult. A combination of sufficient, but not excessive, data recording, backed up by office staff available to assist record-keeping, should be sufficient to ensure material is entered into the computer system on a regular basis.

With any record-keeping system, it must be simple, it should only record information that will be used and it must have the backing of staff. The ability to upgrade the system in response to suggestions by staff is vital.

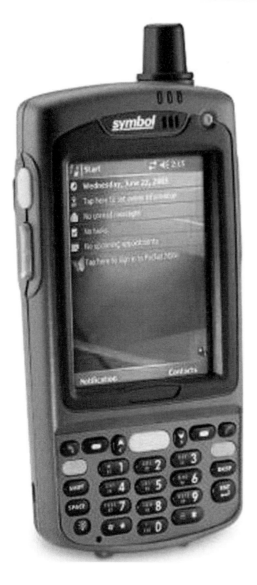

Fig. 12.13. Hand-held computer. (Courtesy of Motorola, Basingstoke, Hampshire, UK.)

12.17 Summary

Quality assurance, due diligence and traceability have become key components of any food packaging organization. With particular reference to a packhouse for pre-pack potatoes, this chapter has discussed the following:

- Quality assurance, the scheme adopted by the packhouse to ensure product quality is maintained.
- A traceability system between field and product dispatch to facilitate trouble-shooting and reassure the public that due diligence is being practised.

- Quality control, the sampling and monitoring system used to ensure product lies within the quality quoted.
- HAACP, the system whereby material on a continuous-flow packing line can be diverted from the line should it fail a critical control point test.
- The use of labels on boxes, or documentation with bulk loads, to track product once it leaves the field.
- Sampling procedures for monitoring potatoes dispatched by the grower and taken in by the packhouse.
- The benefit of monitoring packhouse intake, output, discards, soil and trash, to establish a mass balance, useful for improving management.
- Sampling of the final product together with shelf-life monitoring.
- Improving accessibility by staff troubleshooters to information collected and recorded by other packhouse staff.

13 Marketing and Costs

Topics discussed in this chapter:

- Cost and returns from storage and grading.
- Potato price at harvest and after storage.
- World potato production and trade data.
- Calculation of storage operating costs.
- Storage costs compared.
- Comparison of home stored crops with imports.
- Risks of investing in storage.

13.1 Cost and Returns from Storage and Grading

The cost of growing and harvesting a crop represents approximately 70–75% of the total cost of supplying potatoes to a packer (J. Wootton, Andersons Midlands, Bury St Edmunds, UK, 2008, personal communication). The remaining 25–30% of the cost involves cleaning, separation into different size grades, loading into store, the depreciation cost of store buildings and equipment, store running costs, and dispatch. The packer would then inspect, wash and grade, pack and dispatch to a supermarket distribution centre. The extra income over and above what could be obtained from selling the crop straight off the field should pay for these postharvest operations and provide a margin to pay for future investment. If this is not the case, the crop should be sold directly from the field at harvest.

13.1.1 Sale of potatoes directly from the field

The sale of potatoes directly from the field has advantages and disadvantages. The advantages are:

©CAB International 2009. *Potatoes Postharvest* (R. Pringle, C. Bishop and R. Clayton)

- No investment is required for grading equipment and storage.
- Reduces labour demand at harvest.
- Can store crop in the ground until required so long as the climate allows.
- Leaves grading and storage to a merchant, cooperative or packhouse, which is likely to have sophisticated grading and storage.
- Economies of scale may favour a large central storage and grading facility rather than small scattered stores with their own grading lines.

The disadvantages are:

- Growers have to sell immediately after lifting so are vulnerable to low prices in glut years.
- Cannot remove out-grades (e.g. bakers or salad potatoes), which can sell for a premium.
- May have to pay higher cost to merchant, cooperative or packhouse for grading and storage than if these are carried out on the farm.
- Loss of winter employment for farm staff.
- Crop left in the ground can delay or prevent the planting of the next crop if this is autumn sown or if the climate allows two or more successive crops a year.

13.1.2 Provision of storage

The onset of cold winter weather may force growers to provide storage, or storage may be a marketing decision. Storage may be built to:

- Protect the crop from frost over the winter period.
- Prevent the crop sprouting prior to sale.
- Take advantage of the rise in potato price as time from harvest increases.
- Allow sale of crop at the optimum price over the storage period.
- Satisfy processors' and packhouses' requirements for a continuous supply of material 365 days per year.

If storage is in field clamps or low-cost structures, the decision to store will cost little and so is easily made. If it involves the building of a sealed, insulated, environmentally controlled store, this is a major investment decision. The annual cost of a store built specifically to supply potatoes at the end of a storage season, when prices are high, may become financially uncompetitive if imports from countries in other climatic zones undercut the price of the stored material.

Prices of potatoes can rise or fall by large amounts in years of shortage or glut; the crop has an 'inelastic' response to demand in that a small shortfall in supply can result in a disproportionate rise in price. However, high prices cannot be relied upon in years of shortage as shoppers may simply change to rice, pasta or flour-based foods. In the UK in 2004 the price of ware potatoes ex-farm was only £6/t more in May than at harvest the previous October. Growers would have obtained only £6/t for storing potatoes (Fig. 13.1). In 2003 and in 2005, however, the difference was £58 and £42/t, respectively.

There are few alternative uses for redundant potato stores, so they should be built or modernized only if they fit into the supply infrastructure. In the UK the main

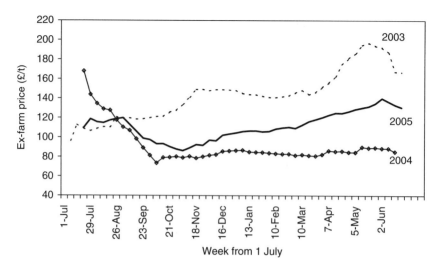

Fig. 13.1. Ex-farm UK ware potato prices (2003–2005).

influence over the last 20 years on farm storage has been the increasing size and specialization of potato and packing enterprises. Economy of scale is the driving force.

13.2   Global Potato Production

While world potato production appears static, this is due to the 1.8% per annum decrease in production within industrialized countries counteracting the 3.8% per annum rise in production in developing countries. Table 13.1 shows the main potato-producing countries, in decreasing order of importance, for the different regions of the world. Data of imports and exports of primary commodity show where competition from other countries can threaten the profitability of producers who store potatoes. In the industrialized countries 12% of production is exported, while in developing countries the quantity is less than 2%. In developing countries potatoes are largely consumed where they are grown. What the table does not show is competition from growers within their own country, where different climatic regions allow potatoes direct from the field to compete with material coming out of store.

13.3   Effect of Climatic Area on Production

Near the equator, two or sometimes even three crops of potatoes can be grown each year. This reduces the length of time that harvested potatoes need to be stored. With two harvests per year, with some being early potatoes and some left in the ground after they are ready, storage need be for only a couple of months. This allows simple, low-cost, rustic stores to be used (Ch5.1). Sample weather data for different climatic areas are included in Ch3.6 and in more detail in Appendix 2 to assist strategic decisions as to whether the climate allows storage in the field or when storage should be provided.

Table 13.1. Potato production, growth rate, relative importance and trade for the main potato-producing countries of the world. (From FAO, 2006.)

Country	Production (×10³ t) 2005	Production (×10³ t) 1996	Annual growth (%) 1996–2005	Yield (t/ha) 2005	Trade (×10³ t) Imports, 2005	Trade (×10³ t) Exports, 2005
Asia (n=36)	132,848	95,910	3.9	15.1	1,070	1,165
China	73,462	53,079	3.8	15.0	34	243
India	23,631	18,843	4.1	17.0	3	80
Bangladesh	4,855	1,492	22.5	14.9	4	5
Iran	4,830	3,140	5.4	25.5	2	166
Turkey	4,090	4,950	−1.7	26.5	13	72
Japan	2,752	3,087	−1.1	31.7	1	0
Pakistan	2,025	1,064	9.0	18.1	11	21
Nepal	1,739	898	9.4	11.8	45	0
Kyrgyzstan	1,142	562	10.3	15.0	0	1
Indonesia	924	1,110	−1.7	14.7	6	15
Republic of Korea	894	731	2.2	27.3	17	0
Syrian Arab Republic	608	439	3.8	20.6	21	10
Africa (n=44)	16,200	10,732	5.1	10.8	450	469
Egypt	2,500	2,626	−0.5	20.2	74	359
Algeria	2,157	1,150	8.8	21.6	65	0
Malawi	1,800	703	15.6	12.0	0	0
South Africa	1,768	1,552	1.4	33.4	1	49
Morocco	1,479	1,250	1.8	23.8	35	41
Uganda	585	318	8.4	6.8	0	0
Latin America (n=29)	15,864	14,157	1.2	16.5	333	128
Peru	3,290	2,309	4.2	12.4	0	0
Brazil	3,130	2,406	3.0	22.0	18	0
Argentina	2,432	2,275	0.7	29.5	1	33
Colombia	1,754	2,801	−3.7	17.5	0	23
Mexico	1,635	1,282	2.8	26.0	58	2
Chile	1,116	828	3.5	20.1	17	1
Bolivia	762	626	2.2	5.7	0	0
Ecuador	339	454	−2.5	7.0	0	0
Cuba	313	365	−1.4	25.4	33	0
Europe (n=37)	130,795	162,040	−1.9	17.2	6,477	6,350
Russian Federation	37,280	38,652	−0.4	12.1	383	25
Ukraine	19,462	18,410	0.6	12.8	3	0
Germany	11,624	13,558	−1.4	42.0	492	1,302
Poland	10,369	27,217	−6.2	17.6	76	25
Netherlands	6,777	8,081	−1.6	43.4	1,646	1,583
France	6,681	6,249	−0.7	42.2	262	1,585
UK	5,961	7,228	−1.8	43.4	424	208
North America (n=2)	23,478	26,703	−1.2	39.5	548	722
USA	19,091	22,618	−1.6	43.5	357	289
Canada	4,386	4,085	0.7	28.2	190	433
Australasia (n=2)	1,788	1,764	0.1	36.9	0	73
Australia	1,288	1,308	−0.2	34.4	0	46
New Zealand	500	456	0.9	45.3	0	27
World (n=150)	320,978	311,310	0.3	16.7	8,902	8,907
Industrialized countries	158,813	193,594	−1.8	19.1	7,025	7,145
Developing countries	162,164	117,716	3.8	15.3	1,877	1,762

n=number of countries in the region.

13.4   Continuous Supply to the Market

Crops which were once seasonal are increasingly available all the year round. Strategies to provide a continuous supply of potatoes include:

- Storage for 6 to 9 months during which sprout growth is suppressed and disease development minimized by ambient-air cooling, refrigeration or sprout suppressants to control sprout growth.
- Storage for 1 to 2 months using non-environmentally controlled rustic stores or cellars to protect crops from insects and animals.
- Transportation of produce from: a growing area at different latitude or a growing area at different altitude.
- Combinations of the above.

If the growing costs in the home country and exporting country are similar, the cost of transport compared with the annual operating cost of storage will determine which is the more competitive. If the producer country has lower production costs, due to low mechanization and labour costs (Fig. 13.2), this will favour imports against storage at home.

Since crops lose moisture during storage, which tends to reduce bloom and firmness, the transported crop is likely to have superior skin finish to the stored crop. It may therefore attract a premium. This too has to be taken into account when evaluating the costs and benefits from building new storage.

Fig. 13.2. Low production costs and seasonal difference allow Egyptian exports to compete with UK-produced potatoes. (Courtesy of D.C. Inglis, Velcourt Ltd, Suffolk, UK.)

13.5 Calculating Storage and Grading Costs

To calculate what storage and grading will add in cost to every tonne of potatoes sold, the capital fixed cost of the building and equipment, and the variable costs of energy, equipment maintenance and labour should be added. This total sum is then divided by the total tonnage that passes through the facility to find a cost per tonne stored and graded.

13.5.1 Fixed costs

Fixed costs are those which are borne by the producer regardless of whether or not a crop is stored or graded. Fixed costs for a store and its associated grading equipment include:

- Capital cost of the building and ventilation equipment.
- Capital cost of the grading equipment.
- Interest on the money borrowed to pay for the building and grading equipment.
- Insurance cover for the building and equipment.
- Any rates or fixed payments to government or local authority.

All the above costs are converted to an annual charge to the business, as it would be unrealistic for the business to pay the whole cost of an investment in a single financial year. The annual charges are calculated as shown below.

13.5.2 Calculation of fixed costs

Depreciation

$$\text{Depreciation per annum} = \frac{\text{Capital cost of building or machine}}{\text{Expected life (years)}}.$$

For example, for a £100,000 building with a life of 20 years

$$\text{Depreciation per annum} = \frac{£100,000}{20} = £5,000$$

Interest charge

In the year of purchase a sum equal to the whole capital cost of the building or machine has to be borrowed. If the machine is written off over a period of years, the amount of the loan outstanding in the last year of write-off is zero. The average interest charge per annum is therefore the average of the capital sum borrowed in the first year and last year multiplied by the interest rate paid to the bank for the loan.

For example, for a loan of £100,000 at an interest rate of 7%:

$$\text{Interest per annum} = \frac{\text{Initial loan} + \text{Final loan}}{2} \times \frac{\text{Interest rate}}{100}$$

$$= \frac{£100,000 + £0}{2} \times \frac{7}{100}$$

$$= £50,000 \times 0.07$$

$$= £3,500.$$

Box 13.1 presents an alternative to this straight-line depreciation scenario.

Insurance

The insurance is proportional to the capital cost and type of building or machine. An approximate cost of insurance is 1% of the capital cost of the building or machine, though a more precise figure should be sought for detailed costings.

Employer's liability insurance may have to be included if this is a completely new facility.

Rates or payment to local authority

In the UK, business rates are not paid on farm buildings. Rates are paid on commercial buildings such as packhouses. This may differ in other countries.

13.5.3 Calculation of variable costs

Repairs and maintenance

These are proportional to the capital cost of the machines used and their hours of use per annum. For machinery, figures can be taken from Table 13.2.

For example, if a grader costs £30,000 and operates for 300h per annum:

$$\text{Repair cost} = £30,000 \times \frac{3.0 + 0.5}{100} = £1050 \text{ per annum}$$

For buildings, repairs and maintenance costs are approximately 1% of capital cost.

Electricity or fuel costs

Electricity costs are dependent on the electrical power rating of the equipment being used. If an electric motor is rated at 1 kW on the aluminium plate which has the motor's serial number, it will consume approximately 1 kWh of electricity every hour it is run. This figure assumes that electric motors are usually about 1.25 times oversized for the load they drive and that they are about 80% efficient.

For equipment like forklifts, the fuel consumption should be assessed from fuel purchase records or from the forklift supplier. The fuel consumption will be very much less than the fuel used at full power, as power use during transport and when stationary will be well below maximum. A very approximate figure can be obtained using the figure of 0.25 l of diesel per kilowatt-hour of power produced, with

Box 13.1. Use of amortization table instead of straight-line depreciation and interest

Straight-line depreciation can be criticized for the annual charge not following the curved drop in value of an investment that occurs in practice. Instead, the following formula can be used:

$$\text{Annual charge} = C\left(\frac{r(1+r)^n}{(1+r)^n - 1}\right),$$

where

C = capital investment
r = rate of interest (e.g. 0.08)
n = years of repayment.

This annual charge combines both depreciation and the interest paid on borrowed capital, so a single figure replaces the two used in the calculation in Ch13.5.2 above. Table B.13.1 (SAC, 2005/6) gives annual charges to service the capital and interest per £1000 borrowed for rates of interest between 5 and 20% and depreciation life of 1–40 years.

Table B.13.1. Amortization table.

	Percentage rate of interest										
Year	5	6	7	8	9	10	12	14	16	18	20
1	1050	1060	1070	1080	1090	1100	1120	1140	1160	1180	1200
2	538	545	553	562	569	576	592	607	623	639	655
3	367	374	381	388	395	403	417	431	445	460	475
4	282	289	296	302	309	316	330	343	357	372	386
5	231	237	244	251	257	264	278	291	305	320	334
6	197	203	210	216	223	230	243	257	271	286	301
7	173	179	186	192	199	206	219	233	248	262	277
8	155	161	168	174	181	188	202	216	230	245	261
9	141	147	154	160	167	174	188	202	217	232	248
10	130	136	142	149	156	163	177	192	207	223	239
11	120	127	134	140	147	154	169	183	199	215	231
12	113	119	126	133	140	147	162	177	192	209	225
13	106	113	120	127	134	141	156	171	187	204	221
14	101	108	114	121	129	136	151	167	183	200	217
15	96	103	110	117	124	132	147	163	179	196	214
20	80	87	94	102	110	117	134	151	169	187	205
25	71	78	86	94	102	110	128	146	164	183	202
30	65	73	81	89	97	106	124	143	162	181	201
40	58	66	75	84	93	102	121	141	160	180	200

For example, the annual charge to service the interest and capital repayments on £8000 repayable over 20 years at 7% would be £94 × 8 = £752.

Table 13.2. Estimated annual repair cost (as percentage of purchase price). (From Nix, 2008.)

Machine	Approximate annual use (h)				Additional use per 100 h
	500	750	1000	1500	ADD
Tractors and forklifts	5	6.7	8.0	10.5	0.5
	Approximate annual use (h)				Additional use per 100 h
	50	100	150	200	ADD
Potato harvesters	3.0	5.0	6.0	7.0	2.0
Stationary potato grading equipment and spray applicators	1.5	2.0	2.5	3.0	0.5

average power used as 40% of maximum. A typical forklift for 1-t boxes would have a maximum power of 40–45 kW.

For example, for a 40 kW forklift truck and with diesel costing £0.45/l:

$$\text{Fuel cost} = 40 \times \frac{40}{100} \times 0.25 \times £0.45$$

$$= £1.80 \text{ per hour.}$$

Consumables

Various consumables like potato bags, sprays, lubricating oil, etc. will be required during storage and grading. These have to be included in the operating costs.

Labour cost

Labour cost can be either a variable or a fixed cost. For seasonal work like potato grading it is usual to employ labour that would otherwise not be employed. Some staff will be employed whether or not grading is practised. Since this labour could be doing useful work elsewhere on the farm, it is best to cost labour as a variable cost. After carrying out the calculation to find the total operating cost of the store and grading area, the implications of employing fixed or seasonal labour should be considered before making the final decision to invest in a grading system.

13.5.4 Total operating cost

Operating cost per annum for a building or equipment
= Fixed costs + Variable costs,

where

Fixed costs = annual depreciation + interest + insurance + building rates if applicable

Variable costs = annual repair and maintenance + energy + consumables + labour.

Operating cost per tonne stored or graded $(\pounds)$

$$= \frac{\text{Operating cost per annum } (\pounds)}{\text{Tonnes stored or graded per annum}}.$$

The operating cost per tonne of potatoes produced or sold is a single figure which is used to determine whether storage or grading is likely to be profitable.

13.6 Storage Capital Costs

The capital costs for three stores are presented in this section, with a sketch of the store layouts given in Fig. 13.3. So that the cost of bulk, airspace-ventilated box and positively ventilated box stores can be compared, the same umbrella building was used for all three systems. As these prices will go out of date in time, the present-day price of potatoes and some key elements that go into building and storage costs are listed in Table 13.3.

13.6.1 Ambient-air cooled bulk store

The costings in Table 13.4 relate to the 1950-t bulk store shown in Fig. 13.3a. This is a bulk store fitted with ambient-air cooling, but no refrigeration, with potatoes stored 4 m deep. The store fabric is plasticized steel sprayed with 60-mm foam insulation in the walls and 80-mm in the roof. Airflow is 0.02 m³/s/t, with air cooling being intermittent and not humidified. Maximum air speeds in the main

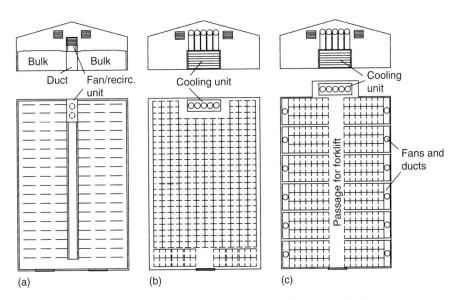

Fig. 13.3. Plan and elevations of: (a) a 1950-t processing potato bulk store; (b) a 1860-t box pre-pack store; and (c) a 1512-t box seed store.

Table 13.3. Prices in the UK of potatoes, building materials and energy (2007).

Item	Details	Unit	Price
Potato prices	From BPC weekly market figures, 5 October 2007		
	Average price	£/t	£108
	cv. Maris Piper/Romano	£/t	£155
	cv. Estima	£/t	£120
Building costs	From SAC Farm Buildings Cost Guide, 2005/6		
Cement	Ordinary Portland	£/t	112.60
Universal steel beams	305 × 165 × 40 kg/m	£/t	655.30
Timber	75 mm × 150 mm sawn white wood	£/m	2.70
Insulation	Expanded polystyrene (SD), 50 mm	£/m^2	4.10
Plasticized steel sheeting	Colour coated, box profile, 0.7 mm	£/m^2	7.20
Building labour	Cost of labour to the employer	£/h	15.20

Note: £1 = US$2.00 or €1.40 (October 2007).

Table 13.4. Cost of an ambient-air cooled bulk store holding 1950 t.

Structural item	Cost (£)	Percentage of cost	Cost per tonne (£)
Foundations, floor and ducts	83,868.93	32.4	
Portal frames and load-bearing walls	51,301.73	19.8	
Cladding	34,274.44	13.2	
Insulation	38,393.14	14.8	
Doors and guard rails	3,364.00	1.3	
Main duct	9,445.08	3.6	
Fans and controller	8,971.80	3.5	
Ventilation louvres	12,720.48	4.9	
Wiring	16,772.93	6.5	
Total	259,112.52	100.0	132.88

and lateral ducts are restricted to 6 m/s and the laterals are tapered, all to ensure uniform air distribution from ducts.

The cost of the foundations, floor and lateral ducts make up 32% of the cost of the store. The relatively low building cost per tonne of potatoes stored compared with the other two stores is due in part to the maximum utilization of space that bulk storage offers and that expensive storage boxes are not required.

Table 13.5. Cost of a space-ventilated ambient/fridge box store holding 1860 t.

Structural item	Cost (£)	Percentage of cost	Cost per tonne (£)
Foundations and floor	48,689.52	21.2	
Portal frames	33,744.92	14.7	
Cladding	34,274.44	14.9	
Insulation	38,393.14	16.7	
Doors and guard rails	3,364.00	1.5	
Ventilators	5,424.48	2.4	
Wiring	16,772.93	7.3	
Ventilation and refrigeration unit	49,000.00	21.3	
Total excluding boxes	229,663.43	100.0	123.47
Cost of boxes	87,420.00		47.00
Total including boxes	317,083.43		170.47

13.6.2 Ambient-air/fridge-cooled space-ventilated box store

The costings in Table 13.5 relate to the 1860-t box store shown in Fig. 13.3b. This type of store is typically used for ware potatoes for the pre-pack market. This is a spray foam-insulated, portal frame building, with an ambient-air/fridge cooling unit located at the back of the store. Incoming ventilation air can be mixed with store air to ensure the incoming air is not too much cooler than the crop. Alternatively, the refrigeration system can be switched on with the ventilation system switched to recirculation. Air distribution is over the top of the boxes, with the return air encouraged to return to the intake of the air/fridge cooling unit via the 'ducts' formed by the pallet apertures of boxes making up the stack.

The boxes add a considerable additional cost of storage over bulk storage, and are justified in part by the improved skin finish and reduction in damage that boxes provide.

13.6.3 Ambient-air/fridge-cooled positively ventilated box store

The costings in Table 13.6 relate to the 1512-t positively ventilated box store shown in Fig. 13.3c. This type of store suits seed potato production, where tubers and voids are small; with positive ventilation a significant advantage to ensure air flows uniformly through the whole crop. Since seed is usually grown in poorer soils, in cool climates and often harvested from wet soils, good drying capability is required. The increasing adoption of earlier harvesting of seed to minimize disease development on growing tubers requires well-ventilated storage to remove the high respiration heat associated with early lifting. The wall ducts and the access passage up the middle of the store do however occupy expensive building space. Since seed is sold throughout the year, ease of access to rows of different varieties is convenient. The relatively low tonnage of crop that can be housed, the

Table 13.6. Cost of a positively ventilated ambient-air/fridge box store holding 1512 t.

Structural item	Cost (£)	Percentage of cost	Cost per tonne (£)
Foundations and floor	48,689.52	18.4	
Portal frames	33,744.92	12.8	
Extension to accept ambient/fridge unit	8,703.72	3.3	
Cladding	34,274.44	13.0	
Insulation	38,393.14	14.5	
Doors and guard rails	3,364.00	1.3	
Posi-vent™ ducts (12 off)	25,838.07	9.8	
Ventilators	5,424.48	2.1	
Wiring	16,772.93	6.3	
Ventilation and refrigeration unit	49,000.00	18.5	
Total excluding boxes	264,205.23	100.0	174.74
Cost of boxes	71,064.00		47.00
Total including boxes	335,269.23		221.74

additional cost of the positive ventilation and the cost of boxes all make this the most costly form of storage.

13.6.4 Store insurance costs

In addition to the capital cost of the store is the annual insurance premium. As with all fixed costs this is paid whether or not the store is used. Insurance companies have concerns relating to fire danger from foam-filled composite panels or foam-sprayed building cladding. Prior to a building being erected, it is a wise precaution to contact the grower's insurance company to get an estimate of the likely premium. Insurance companies may have preferred constructions or materials that will be cheaper to insure. Where an actual figure for annual insurance premium cannot be obtained, a figure of 0.5–1% of the capital cost can be used.

13.7 Calculation of Store Variable Costs

Store variable costs include:

- Electricity for ventilation and refrigeration.
- Fuel or electricity for heating if required.
- Repair and maintenance of the building fabric.
- Repair and maintenance of the cooling plant.
- Forklift operating cost for loading and unloading the store.
- Store management including cleaning, sampling and chemical application.

13.7.1   Electricity for ventilation and refrigeration

An example of the approximate energy use for a 1000-t refrigerated store, with insulation U-value of 0.32 W/m²°C, is given in Table 13.7. This was calculated using a mathematical model developed by FEC Services Ltd, Kennilworth, UK for the *BPC Store Managers' Guide* (BPC, 2001a). The store was not fitted with ambient-air cooling, which would have reduced cooling costs. A unit cost of energy, including standing charges, was assumed to be £0.10/kWh. Total energy cost for the year from October until July was £9.32/t.

13.7.2   Fuel for heating

In the UK and similar mild weather maritime areas, heat is rarely required to keep the crop from cooling below the set-point temperature. The heat generated from the potatoes themselves, provided the store is sufficiently well insulated, is usually enough to keep the crop at the desired level.

Heaters are commonly installed in the roof space of stores to maintain roof-space temperatures during cold nights, to prevent surface cooling of the crop, as this can result in subsurface condensation. Such heating in the UK is for short periods only so its energy cost can be largely ignored.

13.7.3   Repair and maintenance of the building fabric

The annual repair and maintenance of the building fabric is usually assumed to be approximately 1% of the building cost (Nix, 2008).

Table 13.7. Approximate energy costs for a 1000-t refrigerated store, Lincolnshire, UK. (From BPC, 2001a.)

Month	Energy consumption (kWh/t)	Unit cost of electricity (£/kWh)	Cost (£/t)
October	16.385	0.100	1.64
November	7.532		0.75
December	6.180		0.62
January	5.592		0.56
February	5.534		0.55
March	7.249		0.72
April	8.514		0.85
May	10.614		1.06
June	12.427		1.24
July	13.340		1.33
Total			9.32

13.7.4 Repair and maintenance of the cooling plant

The repair and maintenance cost of an ambient-air cooling/fridge unit can be found by pricing a maintenance contract with the supplier. With refrigeration equipment using CFC refrigerants this is now mandatory (Ch7.1). While the cost will vary with distance from the supplier, a figure of 1.5% of the capital cost of the plant and equipment provides an approximate value (W. Leslie, Farm Electronics, Grantham, UK, 2007, personal communication).

13.7.5 Forklift operating cost for loading and unloading the store

The decision to select box storage forces a grower to purchase an industrial-type, small-wheeled, forklift truck. The whole cost of the forklift may have to be borne by the potato enterprise or, if there are other tasks for it to do, its cost may be shared over a number of enterprises. The operating cost of the forklift and operator is calculated in the same way as that for storage, by determining the fixed and variable costs and dividing these by the hours used. This provides an operating cost that can be included in the overall cost of storage.

13.7.6 Store management including cleaning, sampling and chemical application

The management of a 1500- to 2000-t potato store will vary on the level of inspection carried out, whether temperature monitoring is manual or automatic and whether problems occur. The main inputs will be at loading and unloading and if applying CIPC to the crop. For a pre-pack store the management input could be about 1.5 h per weekday. If the manager's annual average cost including National Insurance, Employers' Liability Insurance, a premium over the minimum wage and some overtime is £20,460 (SAC, 2005/6), the hourly rate would be £11.22.

13.8 Calculation of Store Operating Costs

Having calculated the fixed and variable costs, these can now be put together to determine the store operating costs. The store chosen for this example is the pre-pack store in Ch13.6 (Table 13.5) for 1860 t, with an ambient-air/fridge cooling system and potatoes stored in boxes. We shall use the single amortization figure method rather than the straight-line depreciation plus interest on half-capital method.

The period chosen to depreciate or amortize the store over is based partly on the actual life of the building, which may be 30–50 years, and partly on the risk that growing and storing potatoes may become unprofitable. Where long-term profitability appears secure a life of 20 years may be chosen. This was the period chosen for Table 13.8. Were there more doubt about the long-term viability of potato growing on the farm, a shorter life of 10–15 years should be taken.

Table 13.8. Calculation of store operating costs for 1860-t pre-pack store.

Item	Original sum	Annual cost (£)	Cost per tonne (£)
Fixed costs			
Amortized cost of store over 20-year life, 7% interest	£94/1000 × £229,663.43[a]	21,588.36	11.60
Amortized cost of boxes at £47.00 each	£94/£1000 × £87,420	8,217.48	4.42
Building insurance, 1.5%		2,296.63	1.85
Total fixed costs			17.25
Variable costs			
Electricity for ventilation and cooling			9.32
Repair and maintenance at 1% of building cost (excl. plant and equipment)		1,584.70	0.85
Repair and maintenance at 1.5% of cooling plant and equipment capital cost		1,067.96	0.57
Forklift operating cost		6,092	3.28
Store management cost		2,861.10	1.54
Total variable costs			15.56
Total store operating cost			32.81

[a]Capital cost of 1860-t store taken from Table 13.5.

The total store operating cost per tonne stored, assuming a depreciation period of 20 years, is therefore £32.81. This assumes that 100% of the potatoes stored are sold. If only 95% are actually sold, the cost would increase proportionally to £35.19/t. To justify the cost of putting up a store, the sale price in June/July must therefore be £34.54/t more than the grower would get for selling straight from the harvester. From the prices in Fig. 13.1, the average increase in price over the 3 years was £29.30, so storage until July did not pay. A better approach would be to store until May only, when prices were somewhat higher and the electricity and management costs would be less.

The store operating cost therefore produces a single figure which allows capital investments to be compared with potato prices and transport costs, and so aids investment planning.

13.9 Calculation of the Cost of Grading

The operating cost of grading is carried out in the same way as for storage. The building, the plant and the equipment are depreciated as before, with the building assumed to have a life of 20 years and plant a shorter life of about 7–10 years. The electricity costs can be obtained by adding up the power ratings of the electric motors and lights. This will give the approximate consumption in kilowatt-hours for the energy used every hour grading is taking place. The cost of labour is added. Insurance, repairs and maintenance, and forklift costs are added and a final operating cost per hour obtained. If the potato throughput is known, the cost per tonne can be calculated by dividing the grading plant operating cost per hour by the throughput of potatoes.

13.10 Summary

Growers should erect storage buildings only if there is clear financial benefit from doing so or the crop would be at risk from the weather. The risk of having to accept low prices from selling crop straight off the field may be justification enough, but other factors should be considered.

- The difference between the selling price of potatoes at the end of storage and at harvest should cover the operating costs of storage.
- This calculation must take into account that a proportion of the stored crop will be rejected due to blemishes, misshapes and disease.
- The decision to invest in new storage should consider the changes in potato growing and export worldwide as it may be cheaper to ship potatoes from another climate zone than to grow and store at home.
- In higher latitudes only one crop a year is possible, so storage is necessary to supply potatoes when growth does not occur. Nearer the equator, crops may be planted sequentially, and at different altitudes, to deliver potatoes to market all the year round.
- To obtain a single figure per tonne for the cost of storage, the store operating cost should be calculated.
- The store operating cost combines both building and equipment fixed and variable costs.
- The fixed costs include annual depreciation, interest payments on borrowings and insurance.
- The variable costs include annual energy use, repairs and maintenance, consumables and labour.
- Depreciation and interest charges can be calculated separately or a single amortization charge, which combines both depreciation and interest payments, can be used.
- The store operating cost is lowest for bulk storage of processing potatoes, and highest for box storage of seed. The cost of pre-pack potatoes lies between the two.

Appendix 1: Metric–US Imperial Conversion Tables

Imperial–Metric Conversions

Imperial unit $\times$ A = Metric unit
Metric unit $\times$ B = Imperial unit

	Imperial	A	Metric	B
Length	in	25.4	mm	0.03937
	ft	0.3048	m	3.281
	yd	0.9144	m	1.094
Area	acre	0.4047	ha	2.471
	ft^2	0.0929	m^2	10.76
	yd^2	0.8361	m^2	1.196
Volume	ft^3	0.02832	m^3	35.31
	yd^3	0.7646	m^3	1.308
	gal (UK)	4.546	l	0.220
	gal (UK)	0.004546	m^3	219.969
	gal (US)	3.785	l	0.264
	gal (US)	0.003785	m^3	264.172
	bu (US)	35.2390	litre	0.028378
	bu (UK)	35.3687	litre	0.027496
Weight	oz	28.35	g	0.03527
	lb	0.4536	kg	2.205
	cwt (UK)	50.8	kg	0.01968
	cwt (US)	45.36	kg	0.02205
	ton	1.016	t	0.9842
Velocity	ft/min	0.00508	m/s	196.85
	ft/min	0.01829	km/h	54.68
Yield	cwt/acre	0.1255	t/ha	7.968

Continued

	Imperial	A	Metric	B
	ton/acre	2.511	t/ha	0.398
Airflow	ft³/min	1.699	m³/h	0.5886
	ft³/min	0.0004719	m³/s	2118.9
	ft³/s	0.02832	m³/s	35.314
Airflow per unit mass of potatoes	ft³/min/cwt (US)	37.46	m³/h/t	0.0268
	ft³/min/ton	1.726	m³/h/t	0.5794
Density	lb/ft³	16.018	kg/m³	0.0624
Specific volume	ft³/lb	0.0624	m³/kg	16.018
	ft³/ton	0.0279	m³/t	35.84
Pressure	lb/ft²	47.880	N/m² (Pa)	0.02089
	lb/in²	68.948	mbar	0.01450
	inH₂O w.g.	249.1	N/m²	0.004014
Energy	therm	105.5	MJ	0.009478
	Btu	1.055	kJ	0.9478
Power	hp	0.746	kW	1.341
Temperature	°F	(°F−32)×0.5556	°C	(°C×1.8)+32
Heat transfer	Btu/h	0.2928	W	3.4152
	Btu/s	1.0541	kW	0.9487
Thermal conductivity (λ)	Btu ft/ft² h°F	1.7307	W/m °C	0.5778
Thermal resistance (R)	ft² °F h/Btu	0.1761	m² °C/W	5.6732
Thermal conductance or transmittance (U-value)	Btu/ft² °F h	5.678	W/m² °C	0.176

NB: Imperial thermal resistance $(R) = \frac{1}{\lambda} \times 5.6732$, where λ is in W/m °C and R is in ft³ °F h/Btu.

Metric–Metric and Imperial–Imperial Conversions

Column 1 × A = Column 1 to column 2
Column 2 × B = Column 2 to column 1

	Column 1	A	Column 2	B
Weight	cwt (UK)	112	lb	0.0089286
	cwt (US)	100	lb	0.0100
Pressure	mm w.g.	100	mbar	0.0100
	mm w.g.	9.81	Pa	0.1019
	N/m²	1	Pa	1
	mbar	0.0100	N/m²	100
Energy	kWh	3.6	MJ	0.2778

Physical Properties

	Metric		Imperial	
Density of potatoes	650–700	kg/m^3	41–43	lb/ft^3
Specific volume of potatoes	1.43–1.54	m^3/t	51–55	ft^3/t
Density of air at 20°C	1.23	kg/m^3	0.0768	lb/ft^3
Specific heat of air	1,005	J/kg °C	0.240	Btu/lb °F
Specific heat of potatoes	3.80	kJ/kg	1.63	Btu/lb
Angle of repose of potatoes	35	degrees	35	degrees
Respiration rate at 10°C (50°F)	10	W/t	34	Btu/h/ton
Calorific value of LPG	50	MJ/kg	21,495	Btu/lb
	25.5	MJ/l	684,477	Btu/ft^3
Calorific value of oil (35 s)	45.5	MJ/kg	19,560	Btu/lb
	39.6	MJ/l	1,062,953	Btu/ft^3

Notation

Unit/prefix	Name
°C	degree Celsius
°F	degree Fahrenheit
bar	= 0.9869 atmospheres
Btu	British thermal unit
bu (US or UK)	bushel
cwt (US or UK)*	hundredweight (20/ton)
ft	foot
gal	gallon
ha	hectare
hp	horsepower
in	inch
J	Joule
k	kilo ($\times 10^3$)
km	kilometre
l	litre
lb	pound
m	metre
M	mega ($\times 10^6$)
m	milli ($\times 10^{-3}$)
mm	millimetre
μ	micro ($\times 10^{-6}$)
oz	ounce
Pa	Pascal (N/m^2)
t	tonne
therm	unit of heat = 100,000 Btu
ton	ton
W	Watt
w.g.	water gauge
yd	yard

*Long ton = 2240 lb; short ton = 2000 lb.

Appendix 2: Weather Data for World Climatic Zones

Region/country	Location	Latitude	Longitude	Elevation (m)	Coldest month	Coldest temperatures (°C)				Ave. RH (%)	Warmest month	Warmest temperatures (°C)				Ave. RH (%)
						Average		Record				Average		Record		
						Min.	Max.	Min.	Max.			Min.	Max.	Min.	Max.	
Asia																
Bangladesh	Dhaka	25°N	90°E	10	Jan	12	25	7	31	46	Jun	26	32	22	36	72
China	Beijing	40°N	117°E	44	Jan	-10	1	-23	14	50	Jul	21	31	15	41	72
	Wuzhou	23°N	111°E	120	Jan	8	16	1	28	72	Jul	26	32	22	38	78
India	Hyderabad	17°N	78°E	38	Dec	15	28	8	33	57	May	27	40	19	44	44
Indonesia	Medan	4°N	99°E	25	Jan	22	29	18	34	80	May	23	32	18	36	78
Iran	Isfahan	33°N	52°E	1590	Jan	-4	8	-19	18	64	Jul	19	37	9	42	28
Israel	Haifa	32°N	35°E	125	Jan	9	18	-2	26	61	Aug	24	32	18	37	70
Japan	Hakodate	42°N	141°E	34	Jan	-7	0	-22	13	76	Aug	18	26	9	33	85
Kazakhstan	Kazalinsk	49°N	62°E	68	Jan	-15	-9	-33	5	84	Jul	18	32	10	41	47
Republic of Korea	Seoul	38°N	127°E	70	Jan	-9	0	-22	12	65	Aug	22	31	14	37	76
Nepal	Kathmandu	28°N	85°E	2230	Jan	2	18	-2	25	80	Jul	20	29	18	33	84
Pakistan	Islamabad	34°N	73°E	508	Jan	2	16	-4	24	44	Jun	25	40	14	48	23
Syrian Arab Republic	Damascus	34°N	36°E	609	Jan	2	12	-6	21	69	Aug	18	37	13	45	34
Turkey	Ankara	40°N	33°E	850	Jan	-4	4	-25	15	78	Aug	15	31	4	38	40
Africa																
Algeria	Algiers	37°N	3°E	25	Jan	9	15	1	24	71	Aug	22	29	18	42	65
Egypt	Alexandria	31°N	30°E	7	Jan	11	18	3	28	66	Aug	23	31	18	41	70
Malawi (Mozambique)	Zumbo	16°S	13°E	1154	Jun	13	28	6	36	54	Nov	23	37	17	49	48
Morocco	Marrakech	31°N	8°W	466	Jan	4	18	-2	28	77	Aug	20	38	14	47	53
South Africa	Pretoria	26°S	28°E	1534	Jul	3	19	-4	24	52	Jan	16	27	9	35	59
Uganda	Kampala	0°N	33°E	1190	Aug	16	25	12	29	78	Jan	18	28	12	33	66
Latin America																
Argentina	Bahia Blanca	39°S	62°W	75	Jul	4	14	-7	26	68	Jan	17	31	6	42	52

continued

Region/country	Location	Latitude	Longitude	Elevation (m)	Coldest month	Coldest temperatures (°C)				Ave. RH (%)	Warmest month	Warmest temperatures (°C)				Ave. RH (%)
						Average		Record				Average		Record		
						Min.	Max.	Min.	Max.			Min.	Max.	Min.	Max.	
Bolivia	La Paz	16°S	68°W	4014	Jun	1	17	-3	21	-	Jan	6	17	1	25	-
Brazil	Parana	13°S	48°W	74	Jul	9	33	3	36	70	Jan	14	32	9	38	81
Chile	Santiago	34°S	71°W	476	Jul	3	15	-4	27	76	Jan	12	29	6	36	54
Columbia	Bogota	5°N	74°W	2640	Dec	9	19	4	23	71	May	11	19	7	23	72
Cuba	Havana	23°N	82°W	59	Jan	18	26	10	32	75	Jul	24	32	19	34	76
Ecuador	Quito	0°S	79°W	2800	Jul	7	22	1	26	62	Jan	8	22	3	26	74
Mexico	Mexico City	19°N	99°W	2240	Jan	6	19	-3	23	57	Jun	13	24	9	31	65
Peru	Cajamarca	7°S	78°W	2750	Jul	5	21	-2	25	60	Dec	8	22	-4	26	62
Europe																
France	Paris	49°N	2°E	109	Jan	1	6	-12	15	84	Jul	15	25	9	40	70
Germany	Hanover	52°N	10°E	56	Jan	-3	3	-29	14	84	Jul	13	23	5	35	75
Lithuania	Vilnius	55°N	25°E	156	Jan	-11	-5	-28	4	86	Jul	12	23	5	35	68
Netherlands	Utrecht (De Bilt)	52°N	5°E	13	Jan	-1	4	-25	13	86	Jul	13	22	4	34	72
Poland	Warsaw	52°N	21°E	100	Jan	-6	0	-29	11	87	Jul	15	24	7	35	75
Russian Federation	Moscow	56°N	38°E	209	Jan	-16	-9	-32	2	80	Jul	13	23	5	32	61
Sweden	Stockholm	60°N	18°E	20	Jan	-5	-1	-28	10	84	Jul	14	22	8	35	67
UK	York	54°N	1°W	?	Jan	1	6	-14	15	89	Jul	12	21	5	31	74
Ukraine	Kiev	51°N	31°E	179	Jan	-10	-4	-25	8	84	Jul	15	25	9	34	63
North America																
Canada	Halifax	45°N	64°W	145	Jan	-9	0	-27	14	76	Jul	13	23	4	37	73
USA	Boise	37°N	103°W	866	Jan	-6	3	-33	17	76	Jul	14	32	4	45	38
Australasia																
Australia	Adelaide	35°S	139°E	40	Jul	7	15	0	23	70	Dec	15	28	6	46	36
New Zealand	Napier	39°S	177°E	2	Jul	5	13	-3	22	77	Jan	14	24	5	34	62

Source: BBC (2007).
RH, relative humidity.

Appendix 3: Thermal Conductivity (λ Values) for Various Building Structural Components

Thermal Conductivity (λ) of Some Materials

Material	λ (W/m °C)
Aluminium	160
Asbestos sheet	0.23–0.40
Bitumen	0.16
Brick	0.81
Clay roof tiles	0.85
Clay soil	1.50
Concrete, aerated	0.12–0.20
Concrete, no fines	0.60–0.90
Concrete slabs	1.44–2.0
Concrete tiles	1.10
Cork, granulated	0.04
Cork slab	0.05
Expanded polystyrene	0.037
Extruded polystyrene	0.029
Glass	1.05
Glass fibre	0.036–0.04
Granite	2.50
Gravel	0.30
Hardboard	0.08
Limestone	1.50
Loam soil	1.20
Mineral wool felt	0.039
Mineral wool slab	0.045
Plasterboard	0.16
Plywood	0.14
Polyethylene	0.50

Continued

Material	λ (W/m °C)
Polyurethane foam	0.026
PVC	0.16
Roofing felt	0.19
Sand cement	0.55
Sandstone	1.30
Sawdust	0.08
Steel	50
Straw bales	0.072
Straw slab	0.11
Timber, hardwood	0.15
Timber, softwood	0.13
Urea formaldehyde foam	0.036
Vermiculite (loose)	0.065
Wood chipboard	0.15

Source: CIBSE (2006).
Increasing the moisture content, temperature or density of any material increases the λ value.

Thermal Resistance of Airspaces

Airspace	Thermal resistance (m² °C/W)
Cavity, 20 mm or more	0.18
Cavity, 5 mm	0.11
Cavity, one face clad with reflective foil facing into cavity	0.35

Surface Resistance

The surface resistance depends on the surface colour and air speed over the surface. However, typical surface resistances (sum of inside and outside) follow.

Component	Surface resistance (m² °C/W)
Roof	0.15
Walls	0.176

Appendix 4: Theoretical Derivation of the Dimensions of Lateral Ducts used in Bulk Storage

In a grain store the airflow through the grain creates sufficient backpressure to nearly equalize the airflow coming from both ends of the main duct. Potatoes create much less backpressure, so uniform airflow must be achieved through more precise design of the ductwork. The aim is to obtain a reasonably uniform flow, even though potatoes are not present or are only part loaded. The calculation below (Rastovski and van Es, 1981) is designed to set the parameters for good duct design. Its conclusions are used in Ch6.4.1.

Equal Airflow from Ducts Along Their Length

The aim is to have the same airflow flowing to each lateral from the main duct and for each lateral to supply a uniform airflow along its length. For air to flow through the duct outlet there must be a pressure difference between the inside of the duct and outside. This initial theory looks at how static pressure within the duct causes the air to flow in the laterals and how the duct should be designed to provide uniform air distribution.

The relationship between the static pressure in the duct and the air speed through a hole in the duct is given by:

$$\Delta P_s = \frac{1}{2} k \rho v^2 \tag{A1}$$

where

ΔP_s = static pressure difference between inside and outside of the duct (Pa)
k = constant friction coefficient for the hole
ρ = density of air ($1.23 \, \text{kg/m}^3$)
v = velocity of air flowing though hole (m/s).

Rearranging Eqn (A1) in terms of velocity gives:

$$v = \sqrt{\frac{2\Delta P_s}{k\rho}} \tag{A2}$$

The flow of air through an outlet is given by:

$$Q = Av, \tag{A3}$$

where

Q = airflow (m³/s)
A = area of outlet (m²).

Combining Eqns (A2) and (A3) gives:

$$Q = A\sqrt{\frac{2\Delta P_s}{k\rho}} \tag{A4}$$

which allows the flow of air through the outlet to be found from the static pressure difference and the size of the outlet.

Tapered ducts

For the duct shown in Fig. A.1, changing from A_1 to A_2 in cross-section, the total pressure (P_{t1}) at outlet 1 is:

$$P_{t1} = P_{s1} + P_{v1},$$

where

P_{s1} = static pressure in the duct (Pa)
P_{v1} = velocity pressure in the duct (Pa).

The total pressure in the duct declines from outlet 1 to outlet 2 due to friction between the duct wall and the air. If this loss is given by the value ΔP, then:

$$P_{t2} = P_{t1} - \Delta P. \tag{A5}$$

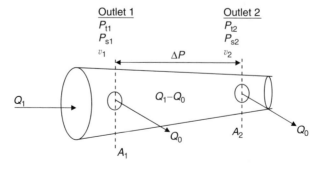

Fig. A.1. Section of duct between two outlets showing static pressure at each outlet.

The total pressures at outlet 1 and 2 are:

$$P_{t1} = P_{s1} + \frac{1}{2}\rho v_1^2.$$

and

$$P_{t2} = P_{s2} + \frac{1}{2}\rho v_2^2.$$

Eqn (A5) can therefore be rewritten as:

$$P_{s2} + \frac{1}{2}\rho v_2^2 = P_{s1} + \frac{1}{2}\rho v_1^2 - \Delta P.$$

The difference in static pressure between outlet 1 and 2 is then:

$$P_{s1} - P_{s2} = \Delta P - \frac{1}{2}\rho\left(v_1^2 - v_2^2\right). \tag{A6}$$

Since pressure losses always arise when the air velocity changes, the entire velocity pressure will not be converted into static pressure but only a part (η) of it.
Eqn (A6) then becomes:

$$P_{s1} - P_{s2} = \Delta P - \frac{1}{2}\eta\rho\left(v_1^2 - v_2^2\right). \tag{A7}$$

In air ducts η will usually vary between 80 and 90%, i.e. η = 0.8 to 0.9.
By expressing air velocity in terms of airflow/area of duct (Eqn (A3)), Eqn (A7) can be expressed as:

$$P_{s1} - P_{s2} = \Delta P - \frac{1}{2}\eta\rho\left[\left(\frac{Q_1}{A_1}\right)^2 - \left(\frac{Q_1 - Q_0}{A_2}\right)^2\right]. \tag{A8}$$

The static pressure in the duct will remain unchanged, i.e. $P_{s1} - P_{s2} = 0$, if the velocity of air in the duct declines in such a way that the resulting gain in *static pressure* equals the loss in *total pressure* along the duct, i.e.

$$\Delta P = \frac{1}{2}\eta\rho\left[\left(\frac{Q_1}{A_1}\right)^2 - \left(\frac{Q_1 - Q_0}{A_2}\right)^2\right]. \tag{A9}$$

A duct which conforms to Eqn (A9) is called an equal pressure duct and will distribute air uniformly along its length. Its cross-section declines steadily in the direction of the airflow, so it is tapered, being widest at the fan end and narrowest at the far end.
Eqn (A9) can be rearranged to determine the cross-sectional area A_2 of the duct at the second outlet:

$$A_2 = \frac{Q_1 - Q_0}{\sqrt{\left(Q_1/A_1\right)^2 - 2\Delta P/\eta\rho}}. \tag{A10}$$

Table A.1. Inlet/end cross-sectional area of distribution ducts of different lengths.

Duct length (m)	Duct inlet/duct end cross-sectional ratio
5	4:1
10	7:1
15	10:1
20	13:1

To achieve equal airflow from each outlet, Eqn (A10) would have to be used to calculate the taper between each outlet. Manufacturing a duct with the taper varying along its length is too complicated for potato storage, so a duct with a uniform taper is all that is required. While not stated in their text, Rastovski and van Es (1981) appear to have used a value of 10 Pa for the resistance of the duct (ΔP). This enabled them to recommend the following tapers (Table A.1).

The ratios in Table A.1 can be used for lateral duct design, or where a fan blows into a single tapered duct (Ch6.4.1). If the lateral of the single fan duct is an underground duct, its width will be constant and its base will taper vertically upwards, being a maximum at the inlet and a minimum at the far end of the duct. Its cross-sectional area should accord with the values in Table A.1.

Ducts of Constant Cross-section

In a constant cross-section duct, such as the main duct in a bulk store (Fig. A.2) having lateral duct outlets along its length, the velocity of the air flowing along the main duct will keep decreasing as air flows out from the laterals, until the velocity will be zero at the end of the duct furthest from the fan.

The total pressure at the beginning of the duct is:

$$P_{ti} = P_{si} + \frac{1}{2}\rho v_i^2. \tag{A11}$$

The total pressure at the end of the duct is:

$$P_{se} = P_{ti} - \Delta P. \tag{A12}$$

As there is no flow at that end of the duct, there is no velocity pressure. The velocity pressure at the inlet has transformed into static pressure with the same loss η that was discussed in Eqn (A7).

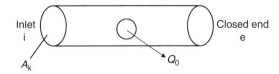

Inlet
i

A_k

Q_0

Closed end
e

Fig. A.2. Duct of constant cross-section.

Combining Eqns (A11) and (A12) gives:

$$P_{sc} - P_{si} = \frac{1}{2}\eta\rho v_i^2 - \Delta P$$

$$= \frac{1}{2}\eta\rho\left(\frac{Q_1}{A_k}\right) - \Delta P, \tag{A13}$$

where

A_k = constant cross-sectional area of duct (m²).

The static pressure difference between inside and outside of duct ΔP_s determines the velocity of air v_o discharged through an opening. This pressure rises as the hole size decreases, i.e.

$$\Delta P_s = \frac{1}{2}k\rho v_o^2,$$

where

v_o = velocity of air through outlet to laterals.

The static pressure at the inlet is therefore determined by:

$$\Delta P_i = \frac{1}{2}k\rho v_0^2.$$

The static pressure at the end of the duct is then equal to the total pressure (i.e. static pressure plus velocity pressure) at the inlet minus the loss due to the friction between the airflow and the duct:

$$\Delta P_e = \frac{1}{2}k\rho v_0^2 + \frac{1}{2}\eta\rho v_i^2 - \Delta P. \tag{A14}$$

The ratio between the static pressure at the end and beginning of the duct becomes:

$$\frac{\Delta P_e}{\Delta P_i} = \frac{1/2\,k\rho v_0^2 + 1/2\,\eta\rho v_i^2 - \Delta P}{1/2\,k\rho v_0^2}, \tag{A15}$$

which simplifies to:

$$\frac{\Delta P_e}{\Delta P_i} = 1 - \frac{\Delta P_e}{\Delta P_i} + \frac{\eta v_i^2}{k v_0^2}.$$

Since the velocity is equal to Q/A, this ratio can be written as follows:

$$\frac{\Delta P_e}{\Delta P_i} = 1 - \frac{2\Delta P A_0^2}{k\rho Q^2} + \left(\frac{A_0}{A_k}\right) \times \frac{\eta}{k}, \tag{A16}$$

where

A_o = cross-sectional area of all discharge openings (m²)
A_k = cross-sectional area of duct (m²)

Q = volume airflow at duct inlet (m³/s)

ΔP = pressure loss in the duct (Pa).

The static pressures ΔP_e and ΔP_i determine the discharge velocities at the end and at the inlet of the duct respectively, i.e.

$$\Delta P_e = \left(\frac{Q_e}{A_0}\right)^2 \times \frac{k\rho}{2} \tag{A17}$$

and

$$\Delta P_i = \left(\frac{Q_i}{A_0}\right)^2 \times \frac{k\rho}{2}. \tag{A18}$$

Eqn (A16) then becomes:

$$\left(\frac{Q_e}{Q_i}\right)^2 = 1 + \left(\frac{A_0}{A_k}\right)^2 \times \frac{\eta}{k} - \frac{2\Delta P}{k\rho} \times \left(\frac{A_0}{Q_i}\right)^2, \tag{A19}$$

where

Q_e = volume of air discharged at the end of the duct (m³/s)

Q_i = volume of air discharged at the inlet (m³/s).

In a smooth-walled duct the pressure loss in the duct ΔP is usually relatively small, so the value

$$\frac{2\Delta P}{k\rho} \times \left(\frac{A_0}{Q_i}\right)^2$$

can be disregarded.

Eqn (A19) then is simplified to:

$$\left(\frac{Q_e}{Q_i}\right)^2 = 1 + \left(\frac{A_0}{A_k}\right)^2 \times \frac{\eta}{k}. \tag{A20}$$

Since resistance factor $k \approx 1$ for the discharge through an opening and $\eta = 0.8$–0.9, the ratio of airflow coming from the first and last outlet in the constant cross-section

Table A.2. Uniformity of airflow to laterals and backpressure loss at lateral outlets for various ratios of total lateral duct cross-sectional area to main duct cross-sectional area. (From Rastovski and van Es, 1981.)

Total lateral cross-sectional area/main duct cross-sectional area	Airflow in last lateral/ airflow in first lateral	Backpressure at lateral outlets assuming 6 m/s at entry to main duct (Pa)
0.5	1.13	89
1.0	1.35	22
2.0	2.00	6

duct is mainly determined by the ratio of the combined area of all the outlets in the duct, A_o, to the cross-section of the duct, A_k. Table A.2 shows the ratios of airflows in the last compared with the first duct for different ratios of A_o/A_k.

If the air velocity in the duct is limited to 6 m/s, the backpressure caused by the size of outlet used to feed the laterals will be as in Table A.2. These values were calculated using Eqn (A18) assuming that the inlet air speed to the main duct is 6 m/s, the ratio A_o/A_k is as shown in the table and $k = 1$.

From Table A.2 it is clear that there is a compromise between having a high static pressure in the duct due to the outlet holes being small, to obtain a relatively uniform airflow between first and last outlet, and the benefits of keeping the static pressure low so that low backpressure fans can be used. The compromise is to have the total lateral outlet duct area equal to the cross-sectional area of the main duct, which gives a static pressure of 22 Pa and an airflow ratio of 1.35.

References

Abbott, J.A. (1999) Quality measurement of fruits and vegetables. *Postharvest Biology and Technology* 15, 207–225.

Adams, M.J. (1975) Potato tuber lenticels: development and structure. *Annals of Applied Biology* 79, 265–273.

Adams, M.J. and Griffith, R.L. (1978) The effect of harvest date and duration of wound healing conditions on the susceptibility of damaged potato tubers to infection by *Phoma exigua* (gangrene). *Annals of Applied Biology* 88, 51–55.

Agrios, G.N. (1988) *Plant Pathology*, 3rd edn. Academic Press, London.

Armstrong, P.R., Brown, G.K. and Timm, E.J. (1995) Cushioning choices can avoid produce bruising during handling. In: Kushwaha, L., Serwatowski, R. and Brook, R. (eds) *Proceedings of Harvest and Postharvest Technologies for Fresh Fruits and Vegetables*. American Society of Agricultural Engineers, St Joseph, Michigan, pp. 183–190.

Balls, R. (1986) Make light work of grading. *Potato World* 3(4), 14–15.

Baritelle, A.L., Hyde, G.M., Thornton, R.E. and Bajema, R.W. (1999) A classification system for impact related defects in potato tubers. In: *Proceedings of the 14th Triennial Conference of the European Association for Potato Research*. Wageningen Pers, Wageningen, The Netherlands, pp. 655–656.

BBC (2007) *BBC Weather – Round the World*. British Broadcasting Corporation, London; available at http://www.bbc.co.uk/weather/world/features/roundtheworld.shtml (accessed November 2007).

BCPC (1996) *Survey of Pesticide Use on Seed and Ware*. British Crop Protection Council, Farnham, UK.

BCPC (1997) *Second Seed Treatment Usage Review 1996*. British Crop Protection Council, Farnham, UK.

Bishop, C.F.H. (1990) On-farm grading and cleaning of potatoes and onions – techniques and equipment to improve returns. *Agricultural Engineer* 45, 40–42.

Bishop, C.F.H. (1992) *Energy efficiency of cooling systems for potato storage*. Silsoe College, Cranfield University, Cranfield, UK.

Bishop, C.F.H. (1994) Drainage pipes for ventilation. *Agricultural Engineer* 49, 21–23.

Bishop, C.F.H. (2007) Effect of tuber condition and surface on chemical application and grading. Presented at *4th EAPR/FNK/UEITP Potato Processing Conference*, Grantham, UK, 17 January 2007.

Bishop, C.F.H. and Garlick, T.E. (1998) Effectiveness of roller table tuber application systems. In: *Aspects of Applied Biology*, Vol. 52. *Production and Protection of Sugar Beet and Potatoes*. Association of Applied Biologists, Warwick, UK, pp. 305–308.

Bishop, C.F.H. and Maunder, W.F. (1980) *Potato Mechanization and Storage.* Farming Press, Ipswich, UK.

Bishop, C.F.H. and Mortimer, D.R. (1999) Analysis of tuber motion on roller tables. In: *Proceedings of the 14th Triennial Conference of the European Association for Potato Research.* Wageningen Pers, Wageningen, The Netherlands, pp. 172–173.

Bishop, C.F.H. and Stenning, B.C. (1997) Solar-powered cooling for tropical potato storage. *Agricultural Mechanization in Asia, Africa and Latin America* 28(2), 57–60.

Bishop, C.F.H., Thorogood, A.J., Duran, T. and Devres, Y.O. (2000) Reduction of potato damage by radiant heating. *Potato Research* 43, 413–426.

Bissonnette, H.L. (1993) *Potato Production and Pest Management in North Dakota and Minnesota.* Extension Bulletin No. 26. NDSU Extension Service, Fargo, North Dakota.

Bollen, A.F. and Dela Rue, B.T. (1995) Padding materials for handling horticultural products – development of an evaluation procedure. In: Kushwaha, L., Serwatowski, R. and Brook, R. (eds) *Proceedings of Harvest and Postharvest Technologies for Fresh Fruits and Vegetables.* American Society of Agricultural Engineers, St Joseph, Michigan, pp. 129–135.

Bouman, A. (1995) *Mechanical Damage in Potato Tubers – Inquiry of the EAPR Engineering Section.* Institute of Agricultural and Environmental Engineering, Wageningen, The Netherlands.

Bouman, A. (1996) Damage systems for potatoes. In: *Proceedings of the 13th Triennial Conference of the European Association for Potato Research.* Wageningen Pers, Wageningen, The Netherlands, pp. 383–384.

Bowen, S.A., Muir, A.Y. and Dewar, C.T. (1996) Investigations into skin strength in potatoes: factors affecting skin adhesion strength. *Potato Research* 39, 313–321.

BPC (2001a) *BPC Store Managers' Guide.* British Potato Council, Sutton Bridge Experimental Unit, Spalding, UK.

BPC (2001b) *Potato Store Hygiene and Disinfection to Improve Seed Health and Ware Quality.* Project Report No. 2001/5. British Potato Council, Oxford, UK.

BPC (2002a) *Getting the Best from CIPC. Optimising CIPC Application and Distribution in Stored Potatoes.* Growers Advice Note. British Potato Council, Sutton Bridge Experimental Unit, Spalding, UK.

BPC (2002b) *Potato Sprout Suppressants.* Product Guide No. 2002/3. British Potato Council, Sutton Bridge Experimental Unit, Spalding, UK.

BPC (2004a) *National Bruising Survey.* British Potato Council, Oxford, UK.

BPC (2004b) *Potatoes, a Healthy Market.* British Potato Council, Oxford, UK.

BPC (2005a) *Health and Safety in Potato Stores.* British Potato Council, Oxford, UK.

BPC (2005b) *BPC Seed Storage Survey 2004.* British Potato Council, Oxford, UK.

BPC (2006a) *Independent Variety Trials 2005.* Project Report No. 2006/7. British Potato Council, Oxford, UK.

BPC (2006b) *Managing the Risk of Black Dot.* British Potato Council, Sutton Bridge Experimental Unit, Spalding, UK.

Briddon, A. (2007) *BPC Research Review: the Use of Ethylene for Sprout Control.* British Potato Council, Oxford, UK.

Briddon, A., Cunnington, A.C., Miller, P.C.H. and Duncan, H.J. (1999) The optimisation of application and distribution of chloropropham. In: *Proceedings of the 14th Triennial Conference of the European Association for Potato Research.* Wageningen Pers, Wageningen, The Netherlands, pp. 194–195.

BSI (1992) *BS 7611:1992. Specification for potato storage boxes for mechanical handling.* British Standards Institution, London.

BSI (1997a) *BS EN 12086:1997. Thermal insulating products for building applications. Determination of water vapour transmission properties.* British Standards Institution, London.

BSI (1997b) *BS 476:Part 7:1997. Fire tests on building materials and structures. Method of Test to Determine the Classification of the Surface Spread of Flame of Products.* British Standards Institution, London.

BSI (1999) *BS EN 12942:1999. Respiratory protective devices. Power assisted filtering devices incorporating full face masks, half masks or quarter masks. Requirements, testing, marking.* British Standards Institution, London.

BSI (2001) *BS EN 12667:2001. Thermal performance of building materials and products. Determination of thermal resistance by means of guarded hot plate and heat flow meter methods. Products of high and medium thermal resistance.* British Standards Institution, London.

Building Regulations (2007) *Approved Document B (Fire safety) – Volume 2 – Buildings other than dwellinghouses (2006 Edition, Amended 2007)*. Department for Communities and Local Government, London.

Burfoot, D. (1997) *Control of Heat and Mass Transfer Processes in Crop Stores*. Final Project Report, MAFF Project OC 9318. Ministry of Agriculture, Fisheries and Food, London.

Burke, J.J., O'Donavan, T. and Barry, P. (2005) Effect of seed source, presprouting and desiccation date on the processing quality of potato tubers for French fry production. *Potato Research* 48, 69–84.

Burton, W.G. (1966) *The Potato*, 2nd edn. Veenman en Zonen, Wageningen, The Netherlands, p. 382.

Burton, W.G. (1973) Physiological and biochemical changes in the tuber as affected by storage conditions. In: *Proceedings of the 5th Triennial Conference of the European Association for Potato Research*. Wageningen Pers, Wageningen, The Netherlands, pp. 63–81.

Burton, W.G. (1989) *The Potato*, 3rd edn. Longman Scientific and Technical, London.

Burton, W.G. and Hannan, R.S. (1957) Use of γ-radiation for preventing the sprouting of potatoes. *Journal of the Science of Food and Agriculture* 12, 707–715.

Burton, W.G. and Wiggington, M.J. (1970) The effect of a film of water upon the oxygen status of a potato tuber. *Potato Research* 13, 180–186.

Burton, W.G., Mann, G. and Wager, H.G. (1955) The storage of ware potatoes in permanent buildings II. The temperature of unventilated stacks of potatoes. *Journal of Agricultural Science, Cambridge* 46, 150–163.

Burton, W.G., van Es, A. and Hartmans, K.J. (1992) The physics and physiology of storage. In: Harris, P.M. (ed.) T*he Potato Crop*, 2nd edn. Chapman and Hall, London, pp. 608–727.

Cargill, B.F. (1976) *The Potato Storage – Design, Construction, Handling and Environmental Control*. Michigan State University, East Lancing, Michigan (endorsed by ASAE).

Cargill, B.F., Brook, R.C. and Forbush, T.D. (eds) (1989) *Potato Storage Technology and Practice – Proceedings of an International Symposium, June 27–29, 1985, Michigan State University, East Lansing, MI*. ASAE Publication No. 01-89. American Society of Agricultural Engineers, St Joseph, Michigan.

Carpenter, G.A. (1972) The design of permeable ducts and their application to the ventilation of livestock buildings. *Journal of Agricultural Engineering Research* 17, 219–230.

Cayley, G.R., Hide, G.A., Lewthwaite, R.J., Pye, B.J. and Vojvodic, P.J. (1987) Methods of applying fungicide sprays to potato tubers and description and use of a prototype electrostatic sprayer. *Potato Research* 30, 301–317.

CCFH (1997) Hazard Analysis and Critical Control Point (HACCP) Systems and Guidelines for its Application. Annex to CAC/RCP-1 (1969), Rev 3 (1997). Codex Committee on Food Hygiene. In: *Codex Alimentarius –Food Hygiene – Basic Texts*. Food and Agriculture Organization of the United Nations/World Health Organization, Rome.

CIBSE (2006) *Properties of Humid Air. Section C*. Chartered Institution of Building Services Engineers, London.

CIP (2007) *Potato /*. International Potato Center, Lima; available at http://www.cipotato.org/potato/ (accessed September 2008).

Clarke, B. (1996) Packhouse operations for fruit and vegetables. In: Thompson, A.K. (ed.) *Postharvest Technology of Fruit and Vegetables*. Blackwell Science, Oxford, UK, pp. 189–212.

Clayton, R. and Cunnington, A. (1996) Drying to deter disease. In: *Sutton Bridge Experimental Station Annual Review 1996*. Potato Marketing Board, Sutton Bridge Experimental Unit, Spalding, UK.

Clayton, R.C. and Blackwood, J. (2001) *Effect of Duration of Condensation and Temperature on Infection of Potatoes by Skin Spot, Dry Rot and Gangrene*. Report No. 2001/10. British Potato Council, Oxford, UK.

Clayton, R.C., Hardy, C. and Pringle, R.T. (1998) Development of *Helminthosporium solani* in response to temperature and condensation on stored potatoes. In: *Abstracts of the 7th International Congress of Plant Pathology*, Vol. 2, Paper 2.5.23.

Clayton, R.C., Wale, S.J. and Blackwood, J. (2000) *Potato Store Hygiene and Disinfection*. British Potato Council, Oxford, UK.

Connor, M.T. (1994) An individualized psychological approach to measuring influences on consumer preferences. In: Mac-Fie, H.J.H. and Thomson, D.M.H. (eds) *Measurement of Food Preferences*. Blackie Academic and Professional, London, pp. 167–201.

Cook, R.J. and Papendick, R.I. (1978) Role of water potential in microbial growth and development of plant disease, with special reference to post harvest pathology. *Horticultural Science* 13, 559–564.

Cunnington, A. and Dowd, G. (2003) *An Investigation into the Effects of CIPC Use on the Processing Quality of Stored Potatoes.* Report No. 2003/8. British Potato Council, Oxford, UK.

DEFRA (2000) *Fertiliser Recommendations for Agricultural and Horticultural Crops. (RB209)*, 7th edn. Department for Environment, Food and Rural Affairs, London; available at http://www.defra. gov.uk/farm/environment/land-manage/nutrient/index.htm (accessed December 2006).

DEFRA (2006) *The Seed Potatoes (England) Regulations 2006. 2006 No. 1161.* Department for the Environment, Food and Rural Affairs, London; available at http://www.opsi.gov.uk/si/si2006/ uksi_20061161_en.pdf (accessed 5 February 2007).

DETR (1997) The Urban Waste Water Treatment (England and Wales) Regulations 1994. Working document for dischargers and regulators. Department of the Environment, Transport and the Regions and the Welsh Office.

Devres, Y.O. and Bishop, C.F.H. (1995) The effect of potato store operation on energy usage and weight loss. *Potato Research* 38, 251–256.

Dickens, J.C., Harrap, F.E.G. and Holmes, M.R.J. (1962) Field experiments comparing the effects of muriate and sulphate of potash on potato yield and quality. *Journal of Agricultural Science* 59, 319–326.

Dossat, R.J. (1981) *Principles of Refrigeration*, 2nd edn. John Wiley & Sons, New York.

Dowley, L.J., Carnegie, S.F., Balandras-Chatot, C., Elliseche, D., Gans, P., Schober-Butin, B. and Wustman, R. (1999) Guidelines for evaluating disease resistance in potato cultivars. Foliage blight resistance (field test) *Phytophthora infestans* (Mont.) de Bary. *Potato Research* 42, 107.

DTI (2006) *Weights and Measures Act 1985 and secondary legislation including the Weights and Measures (Packaged Goods) Regulations 2006.* Department of Trade and Industry, London.

Duncan, H. (2006) *CIPC Application and Environmental Issues.* Final Report. British Potato Council, Oxford, UK.

Duncan, H.J. (1973) Rapid bruise development in potatoes with oxygen under pressure. *Potato Research* 16, 306–310.

Duncan, H.J. and Kraish, S. (1999) Thermal fogging of potato stores with sprout suppressant formulation. In: *Proceedings of the 14th Triennial Conference of the European Association for Potato Research.* Wageningen Pers, Wageningen, The Netherlands, pp. 92–93.

EC (2000) Regulation (EC) No 2037/2000 of the European Parliament and of the Council of 29 June 2000 on substances that deplete the ozone layer. *Official Journal of the European Union* 29.9.2000, L244/1–L244/24.

EC (2002) Commission Directive 2002/63/EC of 11 July 2002 establishing Community methods of sampling for the official control of pesticide residues in and on products of plant and animal origin and repealing Directive 79/700/EEC (Text with EEA relevance). *Official Journal of the European Union* 16.7.2002, L187/30–L187/43.

EC (2004) Regulation (EC) No 852/2004 of the European Parliament and of the Council of 29 April 2004 on the hygiene of foodstuffs. *Official Journal of the European Union* 25.6.2004, L266/ 3–L266/21.

Elphinstone, J., Stead, D. and Wale, S.J. (2004) *Non-water Control Measures for Potato Common Scab.* Report Ref. 248. British Potato Council, Oxford, UK.

Elstob, R.H., Kirkland J., McRae, D.C. and Fleming, J. (1988) *Investigation of Factors which Limit the Efficiency of Operation of Roller Tables.* Report for the Potato Marketing Board by the Scottish Centre for Agricultural Engineering. Scottish Agricultural College, Edinburgh.

EU (1995) Plant protection products regulations (PPPR) directive 91/414/EEC. Will become the overarching EU regulating body taking over from Plant Protection Products (Basic Conditions) Regulations 1997 and the Pesticides Safety Directorate (PSD). EU, Brussels.

Evans, S.D. and McRae, D.C. (1996) Peeler system of strokes. In: *Proceedings of the 13th Triennial Conference of the European Association for Potato Research.* Wageningen Pers, Wageningen, The Netherlands, pp. 377–378.

FAO (2006) *Summary of world potato supply statistics 1990–1998.* Food and Agriculture Organization of the United Nations, Rome; available at http://faostat.fao.org/site/601/default.aspx (accessed December 2007).

FEC (1985) *Potato Storage – A Farm Electric Centre Handbook.* Farm Electric Centre, Kenilworth, UK.

Feller, R., Margalin, E., Hetzoni, A. and Galilil, N. (1987) Impingement angle and product interference effects of clod separation. *Transactions of the American Society of Agricultural Engineers* 30, 357–360.

Fellows, J. (2004) *Factors Associated with Internal Damage and Bruising in Potato Tubers (BRUCE).* Report No. 2004/4. British Potato Council, Oxford, UK.

Fingers, F.L. and Fontes, P.C.R. (1999) Manejo pos-colheita da batata (Post-harvest management of the potato). *Informe Agropecuario, Belo Horizonte* 20, 105–111, 197.

Firman, D.M., Allison, M.F. and Fowler, J.H. (1999) Dormancy and determinacy experiment. In: *Cambridge University Potato Growers Research Association Annual Report 1998.* CUPGRA, Cambridge, UK, pp. 24–29.

Firman, D.M., Allen, E.J. and Shearman, V.J. (2004) *Production Practices, Storage and Sprouting Conditions Affecting Number of Stems per Seed Tuber and the Grading of Potato Crops.* Project Report No. 2004/14. British Potato Council, Oxford, UK.

Fuglie, K., Khatana, V., Ilangantileke S., Singh, J.P., Kumar, D. and Scott, G. (2000) Economics of potato storage in Northern India. *Quarterly Journal of International Agriculture* 39, 131–148.

Gall, H., Muir, A.Y., Fleming, J., Pohlmann, R., Göcke, L. and Hossack, W. (1998) A ring sensor system for the determination of volume and axis measurements of irregular objects. *Measurement Science and Technology* 9, 1809–1820.

Gaze, S.R., Stalham, M.A. Newbery, R.M. and Allen, E.J. (1998) *Manipulating Potato Tuber Dry Matter Concentration through Soil Water Status.* Progress Report for 1998, Project No. 807/182. British Potato Council, Oxford, UK.

Gaze, S.R., Stalham, M.A., Newbery, R.M. and Allen, E.J. (1999) *Manipulating Potato Tuber Dry Matter Concentration through Soil Water Status.* Final Report. British Potato Council, Oxford, UK.

Geyer, M. (1996) Waste water from potato washing. *Kartoffelbau* 47, 256–258.

Geyer, M. and Oberbarnscheidt, B. (1998) Washing root crops and potatoes – reducing the mechanical load in drum washers. *Landtechnik* 53, 98–99.

Glasbey, C.A., McRae, D.C. and Fleming, J. (1988) The size distribution of potato tubers and its application to grading schemes. *Annals of Applied Biology* 113, 579–587.

Grähs, L.-E., Hylmö, B. and Wickberg, C. (1977) Bulk storing of potatoes. A second condensation problem. *Acta Agriculturae Scandinavica* 27, 156–158.

Grähs, L.-E., Hylmö, B. and Johansson, A. (1978) The two point temperature measurement – a method to determine the rate of respiration of a potato pile. *Acta Agriculturae Scandinavica* 28, 231–236.

Gray and Robinson (1988) Effect of temperature on the rate of development of *Erwinia caratovora* bacteria. *Laboratory Practice* 37. (Precise title and pages unknown)

Gray, E.G. and Paterson, M.I. (1971) The effect of the temperature of potato tubers on the incidence of mechanical damage during grading and of gangrene (caused by *Phoma exigua*) during storage. *Potato Research* 14, 251–262.

Hardy, C.E., Burgess, P.J. and Pringle, R. (1997) The effect of condensation on sporulation on potato tubers naturally infected with *Helminthosporium solani* in simulated store conditions. *Potato Research* 40, 169–180.

Hawkes, J.G. (1979) History of the potato. In: Harris, P.M. (ed.) *The Scientific Basis for Improvement,* 2nd edn. Chapman and Hall, London, pp. 1–12.

Herbert (2006) *Herbert cleaners.* RJ Herbert Engineering Ltd, Wisbech, UK.

Heywood, E.W.K., Garthwaite, D.G. and Thomas, M.R. (2006) *Pesticide Usage Survey Report 215 – Potato Stores in Great Britain 2006.* Pesticide Usage Survey Team, Central Science Laboratory/Department for Environment Food and Rural Affairs/Scottish Executive & Rural Affairs Department, York, UK/London/Edinburgh; available at http://www.csl.gov.uk/newsAndResources/resourceLibrary/articles/puskm/potstores2006.pdf (accessed September 2008).

Hide, G.A., Hall, S.M. and Read, P.J. (1994) Control of skin spot and silver scurf on stored cv King Edward potatoes by chemical and non chemical means. *Annals of Applied Biology* 125, 87–96.

Hilton, A., Linton, S. and Less, A. (2001) *Resistance to Potato Blemish Diseases.* Report No. 2001/13. British Potato Council, Oxford, UK.

Holloway, S. (1990) Monitoring of temperature and related conditions in commercial potato stores. MPhil. thesis, Silsoe College, Cranfield University, Cranfield, UK.

Hooker, W.J. (ed.) (1981) *Compendium of Potato Diseases.* American Phytopathological Society, St Paul, Minnesota.

HSE (2005a) *Control of Vibration at Work Regulations 2005. SI 2005 No. 1093.* Health and Safety Executive, London; available at http://www.opsi.gov.uk/si/si2005/20051093.htm (accessed September 2008).

HSE (2005b) *The Work at Height Regulations 2005 (as amended). A brief guide. INDG401.* Health and Safety Executive, London; available at http://www.hse.gov.uk/pubns/indg401.pdf (accessed September 2008).

HSE (2005c) *Control of Substances Hazardous to Health (Fifth Edition). The Control of Substances Hazardous to Health Regulations 2002 (as amended). Approved Code of Practice and Guidance.* Series L5. HSE Books, London.

Huijsmans, J., Bouman, A. and Keen, A. (1990) Web rod covers to reduce potato bruising. In: *Proceedings of the 11th Triennial Conference of the European Association for Potato Research.* Wageningen Pers, Wageningen, The Netherlands, pp. 241–242.

Hunt, G.L. (1982) Cheap potato storage in the tropics. In: Nganga, S. and Shideler, F. (eds) *Cheap Potato Seed Production for Tropical Africa.* International Potato Center, Lima, pp. 155–170.

Hunt, G.L. (1990) Low cost ware potato storage in the hot humid tropics. In: *Proceedings of the 11th Triennial Conference of the European Association for Potato Research.* Wageningen Pers, Wageningen, The Netherlands, pp. 335–336.

Hunter, J.H. (1985) *Heat of Respiration and Weight Loss from Potatoes in Storage.* Proceedings of the ASAE, Paper No. 85-4035. American Society of Agricultural Engineers, St Joseph, Michigan.

Hunter, J.H. and Yaeger, E.C. (1970) *Use of a Float Roll Table in Potato Grading Operations.* Bulletin No. 690. University of Maine, Orono, Maine.

Hyde, G.M. (1991) *Lighting Environment for Manual Sorting of Potatoes and Onions.* Proceedings of the ASAE, Paper No. 91-3548. American Society of Agricultural Engineers, St Joseph, Michigan.

Hylmö, B., Persson, T., Wickberg, C. and Sparks, W.C. (1975a) The heat balance in a potato pile. I. The influence of the latent heat of the removed water. *Acta Agriculturae Scandinavica* 25, 81–87.

Hylmö, B., Persson, T., Wickberg, C. and Sparks, W.C. (1975b) The heat balance in a potato pile. II. The temperature distribution at intermittent ventilation. *Acta Agriculturae Scandinavica* 25, 88–91.

Hylmö, B., Persson, T. and Wickberg, C. (1976) Bulk storing of potatoes. *Acta Agriculturae Scandinavica* 26, 99–102.

Ingram, G.H. and Storey, R.M.J. (1997) *Guide to Potato Tuber Treatments in the UK.* British Crop Protection Council, Farnham, UK.

ISO (1993) *ISO 9705:1993. Fire tests – Full-scale room test for surface products.* International Organization for Standardization, Geneva, Switzerland.

ISO (2000) *ISO 9001:2000. Quality management systems – requirements.* International Organization for Standardization, Geneva, Switzerland.

Jackson, K. (2005) *BPC Seed Storage Survey 2004.* British Potato Council, Oxford.

Johansson, A. (1998) Continuous ventilation with humidified air: a method to maintain optimal storage conditions in box stores for (seed) potato. Presented at *EAPR Engineering Section Meeting*, Scottish Agricultural College, Aberdeen, UK, March 1998.

Kerr, S. and Parrish, H. (2005) *Independent Variety Trials 2002–2004.* Research Report No. 2005/11. British Potato Council, Oxford, UK.

Krijthe, N. (1948) *Influence of Temperature and Light on the Early Potato 'Eersteking' Under Storage Conditions.* Mededeling No. 73. Laboratorium Voor Plantenpysiologisch Onderzoek, Wageningen, The Netherlands.

Krueger, R.A. (1994) *Focus Groups: A Practical Guide for Applied Research*, 2nd edn. Sage Publications, Newbury Park, California.

Lerew, L.E. and Bakker-Arkema, F.W. (1976) Cooling and heating of potato piles. In: Cargill, B.F. (ed) *The Potato Storage – Design, Construction, Handling and Environmental Control.* Michigan State University, East Lancing, Michigan, pp. 379–401.

Lund, B.M. and Kelman, A. (1977) Determination of the potential for the development of bacterial soft rot of potatoes. *American Potato Journal* 54, 211–255.

McRae, D.C. (1985) A review of developments in potato handling and grading. *Journal of Agricultural Engineering Research* 31, 115–138.

McRae, D.C. (1990) Post harvest handling, sizing and inspection. In: *Proceedings of the 11th Triennial Conference of the European Association for Potato Research.* Wageningen Pers, Wageningen, The Netherlands, pp. 171–172.

McRae, D.C. and Melrose, H. (1990) Rapid development of latent bruising in potatoes using pressurised humidified air. In: *Proceedings of the 11th Triennial Conference of the European Association for Potato Research.* Wageningen Pers, Wageningen, The Netherlands, pp. 233–244.

McRae, D.C., Caruthers, C.J. and Porteous, R.L. (1975) The effect of drop height on potato damage. *6th Triennial Conference of the European Association for Potato Research.* Dordrecht, The Netherlands, pp. 108–109.

McRae, D.C., Melrose, H. and Hutchinson, P. (1990) The use of high tensile tubular web rods to reduce potato damage of harvesters. In: *Proceedings of the 11th Triennial Conference of the European Association for Potato Research.* Wageningen Pers, Wageningen, The Netherlands, p. 502.

MAFF (1982) *Avoiding Potato Damage – A Guide to Reducing Losses in the Potato Crop.* HMSO, London.

Malcolm, D.G. and DeGarmo, E.P. (1953) Visual inspection of products for surface characteristics in grading operations. Marketing Research Report No. 45, Production and Marketing Administration, US Department of Agriculture, Washington, DC.

Martino, F. and Gow, K. (1994) Recherche et developpement d'un transducteur pour l'estimation de la condensation sur les pommes de terres stockees. BSc. thesis, School of Electronic and Electrical Engineering, Robert Gordon University, Aberdeen, UK.

Mathews, G.A. (2000) *Pesticide Application Methods*, 3rd edn. Blackwell Science Ltd, Oxford.

Maunder, W.F., Pullen, D.E.R., Peirson, S. and Basford, W.D. (1990) Practical experience of the cause and prevention of potato damage during harvest, grading and packing. In: *Proceedings of the 11th Triennial Conference of the European Association for Potato Research.* Wageningen Pers, Wageningen, The Netherlands, pp. 173–174.

Mawson, K., Statham, O.J.H. and Cunnington, A.C. (1992) An observational evaluation of two cooling systems in new commercial stores. In: *Sutton Bridge Experimental Station Annual Review 1992.* Potato Marketing Board, Sutton Bridge Experimental Unit, Spalding, UK, pp. 36–38.

Mawson, K., Statham, O.J.H. and Cunnington, A.C. (1993) An observational evaluation of two cooling systems in new commercial stores. In: *Sutton Bridge Experimental Station Annual Review 1993.* Potato Marketing Board, Sutton Bridge Experimental Unit, Spalding, UK, pp. 38–39.

Meinl, G. (1972) Studies on the respiration intensity of potato tubers as an indicator of damage caused by different methods of harvesting and handling. *Archiv fur Acker- und Pflanzenbau und Bodenkunde* 16, 21–30.

Melrose, H. and McRae, D.C. (1987) Rapid development of bruises in potatoes by means of a humidified pressurised oxygen tank. Departmental Note SIN/491. Scottish Institute of Agricultural Engineering, Pencuik, UK.

Mitchell B. (2000) A year in the life of the potato. *Potato Newsletter* (September). Scottish Agricultural College, Aberdeen, UK, pp. 7–10.

Molema, G.J., Bouman, A., Verwijs, B.R., Van Den Berg, J.V. and Klooster, J.J. (2000) Subcutaneous tissue discoloration in ware potatoes. 1. A chain analysis in the Netherlands. *Potato Research* 43, 211–224.

Morris, D.A. (1966) Intersprout competition in the potato. 1. Effect of tuber size, sprout number and temperature on sprout growth during storage. *European Potato Journal* 9, 69–85.

Moskowitz, H.R. (1994) *Food Concepts and Products: Just-in-time Development.* Food and Nutrition Press, Trumbull, Connecticut.

Munters (2007) *Evaporative Humidifier/Cooler Humimax.* Munters Component AB, Sollentuna, Sweden.

Nix, J. (2008) *Farm Management Pocket Book*, 38th edn. The Pocketbook, Melton Mowbray, UK.

Nolan, S., Firman, D.M. and Allen, E.J. (2000) *Production and Use of Healthier Seed in order to Improve Ware Quality.* Report No. 2000/8. British Potato Council, Oxford, UK.

O'Brien, M., Poasch, R.K. and Garrett, R.E. (1980) Fillers for fruit and vegetable damage reduction. *Transactions of the American Society of Agricultural Engineers* 23, 71–73.

Percival, G. and Bain, R. (1999) Light induced bio-control of potato tuber blight. In: *Abstracts of the 14th Triennial Conference of the European Association for Potato Research.* Wageningen Pers, Wageningen, The Netherlands, pp. 176–177.

Peterson, C.L., Wyse, R. and Neuber, H. (1981) Evaluation of respiration as a tool in predicting internal quality and storability of potatoes. *American Potato Journal* 58, 245–256.

PMB (1991–1995) Reconditioning and processing quality of stored potatoes. *Sutton Bridge Experimental Station Annual Reviews, 1991–1995.* Potato Marketing Board, Sutton Bridge Experimental Unit, Spalding, UK.

Potter, K. (2000) An assessment of continuous, low-rate ventilation with humidified air for ambient air-cooling of potatoes in the UK climate. PhD thesis, University of Aberdeen, Aberdeen, UK.

Prange, R.K., Kalt, W., Daniels-Lake, B.J., Liew, C.L., Page, R.T., Walsh, J.R., Dean, P. and Coffin, R. (1998) Using ethylene as a sprout control agent in stored 'Russet Burbank' potatoes. *Journal of the American Society for Horticultural Science* 123, 463–469.

Preston, D. and Glynn, M. (1995) Bruising evaluation of different cultivars on machines. *Spudman* 33, 16–23.

Pringle, R.T. (1989) Positive ventilation for potatoes in boxes. *Farm Building Progress* 98 (October), 25–31. Scottish Agricultural College, Aberdeen, UK.

Pringle, R.T. (1993a) Potato boxes. *Farm Buildings Progress* 111 (January), 19–24. Scottish Agricultural College, Aberdeen, UK.

Pringle, R.T. (1993b) Potato store ventilation. *Farm Buildings Progress* 113 (July), 24–28. Scottish Agricultural College, Aberdeen, UK.

Pringle, R.T. and Robinson, K. (1996) Storage of seed in pallet boxes. 1. The role of tuber surface moisture on the population of *Erwinia* bacteria. *Potato Research* 39, 205–222.

Pringle, R.T. and Thompson, J. (1987) *Temperature of Potatoes During Transport by Lorry.* Report for the Scottish Seed Potato Development Council, Edinburgh, by the Engineering Division, North of Scotland College of Agriculture. Scottish Agricultural College, Aberdeen, UK.

Pringle, R.T., Potter, K., McGovern, R.E. and Hardy, C. (1997) *Control of Heat and Mass Transfer Processes in Crop Stores.* MAFF Contract Reference CSA 2597. Scottish Agricultural College, Aberdeen, UK.

Pseedco Ltd (2004) Pseedco Ltd Date-smart™ bold trial. *Potatoes in Practice.* Scottish Crop Research Institute & Scottish Agricultural College, Dundee, UK.

Randall, J.M. (1975) The prediction of airflow patterns in livestock buildings. *Journal of Agricultural Engineering Research* 20, 119–215.

Randall, J.M. and Battams, V.A. (1979) Stability criteria for airflow patterns in livestock buildings. *Journal of Agricultural Engineering Research* 24, 361–374.

Rasmussen, R. (1989) Technical design of potato stores. In: Cargill, B.F., Brook, R.C. and Forbush, T.D. (eds) *Potato Storage Technology and Practice – Proceedings of an International Symposium, June 27–29,*

1985, Michigan State University, East Lansing, MI. ASAE Publication No. 01-89. American Society of Agricultural Engineers, St Joseph, Michigan, pp. 127–139.

Rastovski, A. and van Es, A. (1981) *Storage of Potatoes: Post Harvest Behaviour, Store Design, Storage Practice, Handling.* Centre for Agricultural Publishing and Documentation, Wageningen, The Netherlands.

Read, P.J. and Hide, G.A. (1984) Effects of silver scurf (*Helminthosporium solani*) on seed potatoes. *Potato Research* 27, 145–154.

Read, P.J., Storey, R.M.J. and Hudson, D.R. (1995) A survey of black dot and other fungal tuber blemishing diseases in British potato crops at harvest. *Annals of Applied Biology* 126, 249–258.

Robertson, I.M. (1970) Assessment of damage in potato tubers. Departmental Note SSN/60. Scottish Centre of Agricultural Engineering, Scottish Agricultural College, Edinburgh.

Robertson, J.F. (1993) Exposure to inspirable dust and crystalline quartz during potato handling processes. *Farm Building Progress* 114 (October), 7–11. Scottish Agricultural College, Aberdeen, UK.

Rodger-Brown, J., Rollett, A.C., Cunnington, A.C. and Ingram, G.H. (1999) A buffer feed system for use prior to application of sprays to potato tubers. In: *Proceedings of the 14th Triennial Conference of European Association for Potato Research.* Wageningen Pers, Wageningen, The Netherlands, pp. 647–648.

Rollett, A.C., Cunnington, A.C. and Rodger-Brown, J. (2001) A buffer feed system to provide an even flow of potato tubers for efficient spray treatment. Presented at *BCPC Conference on Seed Treatment – Challenges and Opportunities*, Wishaw, UK.

Rowe, W.P. (2005) *Feasibility Study and Project Design for Small/Medium Scale Potato Storage in Bamyan for CNFA Afghanistan Agriculture Development Project (AADP).* CNFA, Washington, DC; available at http://www.cnfa.org/ (accessed March 2008).

Rylski, L., Rappaport, L. and Pratt, H.K. (1974) Dual effects of ethylene on potato dormancy and sprout growth. *Plant Physiology* 53, 658–662.

SAC (1990) *Fertiliser Application for Potato Production.* Technical Note T207. Scottish Agricultural College, Edinburgh.

SAC (2005/6) *The Farm Management Handbook*, 26th edn. Scottish Agricultural College, Edinburgh.

SAM (2005) *Thermal Fog Generator Operating Instruction Manual.* Sands Agricultural Machinery Ltd, Norfolk, UK.

Schaper, L.A. and Yaeger, E.C. (1992) Coefficients of friction of Irish potatoes. *Transactions of the American Society of Agricultural Engineers* 35, 1647–1651.

Scheer, A. (1998) Storage systems for seed potatoes in the Netherlands. Presented at *EAPR Engineering Section Meeting*, Scottish Agricultural College, Aberdeen, UK, March 1998.

Searle (1975) *Searle Engineers Handbook (with acknowledgements to ASHRAE Guide).* International Cooling Group, Fareham, UK.

Stalham, M.A. (1992) Root growth of Estima, Record and Cara under contrasting irrigation regimes. In: *Cambridge University Potato Growers Research Association Annual Report 1991.* CUPGRA, Cambridge, pp. 12–16.

Stark, J.C. and Love, S.L. (eds) (2003) *Potato Production Systems.* University of Idaho Extension, Aberdeen, Idaho.

Statham, O. and Cunnington, A. (1993) Wet cooling systems for potato storage. In: *Proceedings of the 12th Triennial Conference of the European Association for Potato Research.* Wageningen Pers, Wageningen, The Netherlands, pp. 78–79.

Tao, Y., Morrow, C.T., Heineman, P. and Sommer, H.J. (1990) *Automated Machine Vision Inspection of Potatoes.* Proceedings of the ASAE, Paper No. 90-3531. American Society of Agricultural Engineers, St Joseph, Michigan.

Thompson, J.E., Mitchell, G.F. and Kasmire, R.F. (1992) Cooling horticultural commodities. In: Kader, A.A. (ed.) *Postharvest Technology of Horticultural Crops*, 2nd edn. Division of Agriculture and Natural Resources Publication No. 3311. University of California, Oakland, California, pp. 63–68.

Trenckmann, S. (1988) Versuche zur Verminderung der Schwarzfleckigkeit von Kartoffeln durch verschiedene Verfahren der Erwaruing. Dissertation No. 148, University of Göttingen, Göttingen, Germany.

Vanderplank, J.E. (1963) *Plant Diseases: Epidemics and Control.* Academic Press, New York.

Vollbracht, O. and Kuhnke, U. (1956) Mechanische Beschidigungen an Kartoffeln. *Kartoffelbau* 7, 74–77.

Waelti, H. (1989) Potato storage and ventilation in the Pacific Northwest. In: Cargill, B.F., Brook, R.C. and Forbush, T.D. (eds) *Potato Storage Technology and Practice – Proceedings of an International Symposium, June 27–29, 1985, Michigan State University, East Lansing, MI.* ASAE Publication No. 01-89. American Society of Agricultural Engineers, St Joseph, Michigan.

Wale, S.J. and Clayton, R.C. (1999) Population dynamics of *Fusarium coeruleum* in soils during potato cropping. In: *Abstracts of the 14th Triennial Conference of the European Association for Potato Research.* Wageningen Pers, Wageningen, The Netherlands, pp. 180–181.

Wiant, J.S., Findlen, H. and Kaufman, J. (1951) Effects of temperature on blackspot in Long Island and Red River Valley potatoes. *American Potato Journal* 28, 753–765.

Wilson, A.R. and Boyd, A.E.W. (1945) Potato wastage in clamps. *Agriculture, London* 51, 507–512.

Wiltshire, J., Turley, D., Milne, F. and Peters, J. (2003) *Improving the Understanding and Management of Skin Set, Bloom and Netting in Potatoes.* Report No. 2003/5. British Potato Council, Oxford, UK.

Wiltshire, J., Milne, F. and Peters, J. (2005) *Improving the Understanding and Management of Skin Set and Bloom in Potatoes.* Report No. 2006/1. British Potato Council, Oxford, UK.

Winfield, M., Lloyd, D., Griffiths, D.W., Bradshaw, J.E., Muir, D., Nevison, I.M. and Bryan, G.J. (2005) Assessing organoleptic attributes of *Solanum tuberosum* and *S. phureja* potatoes. *Aspects of Applied Biology* 76, 279–308.

Wright, R.C. (1942) *The Freezing Temperature of Some Fruits, Vegetables, and Florists' Stocks.* Circular No. 447. US Department of Agriculture, Washington, DC.

Wright, R.C. and Diehl, H.C. (1927) *Freezing Injury to Potatoes.* Technical Bulletin No. 27. US Department of Agriculture, Washington, DC, p. 23.

Wustman, R., Booth, R.H. and Rhoades, R.E. (1985) *Possibilities for the Application of Small Scale Potato Storage Techniques in Developing Countries.* Food and Agriculture Organization of the United Nations, Rome.

Xu, Y. and Burfoot, D. (1999) Predicting condensation in bulks of foodstuffs. *Journal of Food Engineering* 40, 121–127.

Xu, Y. and Burfoot, D. (2000) Modelling the application of chemicals in box potato stores. *Pest Management Science* 56, 111–119.

Xu, Y., Burfoot, D. and Huxtable, P. (2002) Improving the quality of stored potatoes using computer modelling. *Computers and Electronics in Agriculture* 34, 159–171.

Zapp, H.R., Ehler, S.H., Brown, G.K. and Armstrong, P.R. (1989) Advanced instrumented sphere (IS) for impact measurements. *Transactions of the American Society of Agricultural Engineers* 33, 955–960.

Zegers, D. and van den Berg, V. (1988) The effect of lighting quality and intensity on the efficiency of seed potato inspection. *The Agricultural Engineer* 43, 5–11.

Index